Calculations for Molecular Biology and Biotechnology

WITHDRAWN
FROM LIBRARY

BRITISH MEDICAL ASSOCIATION
0804035

To my parents Mary and Dude and to my wife Laurie and my beautiful daughter Myla.

Calculations for Molecular Biology and Biotechnology

A Guide to Mathematics in the Laboratory

Second Edition

Frank H. Stephenson

AMSTERDAM • BOSTON • HEIDELBERG • LONDON • NEW YORK • OXFORD • PARIS
SAN DIEGO • SAN FRANCISCO • SINGAPORE • SYDNEY • TOKYO
Academic Press is an imprint of Elsevier

Academic Press is an imprint of Elsevier
32 Jamestown Road, London NW1 7BY, UK
30 Corporate Drive, Suite 400, Burlington, MA 01803, USA
525 B Street, Suite 1800, San Diego, CA 92101-4495, USA

First edition 2003
Second edition 2010

Copyright © 2010 Elsevier Inc. All rights reserved

No part of this publication may be reproduced, stored in a retrieval system or transmitted in any form or by any means electronic, mechanical, photocopying, recording or otherwise without the prior written permission of the publisher

Permissions may be sought directly from Elsevier's Science & Technology Rights Department in Oxford, UK: phone (+44) (0) 1865 843830; fax (+44) (0) 1865 853333; email: permissions@elsevier.com. Alternatively, visit the Science and Technology Books website at www.elsevierdirect.com/rights for further information

Notice
No responsibility is assumed by the publisher for any injury and/or damage to persons or property as a matter of products liability, negligence or otherwise, or from any use or operation of any methods, products, instructions or ideas contained in the material herein.

Because of rapid advances in the medical sciences, in particular, independent verification of diagnoses and drug dosages should be made

British Library Cataloguing-in-Publication Data
A catalogue record for this book is available from the British Library

Library of Congress Cataloging-in-Publication Data
A catalog record for this book is available from the Library of Congress

ISBN : 978-0-12-375690-9

For information on all Academic Press publications
visit our website at www.elsevierdirect.com

Typeset by MPS Limited, a Macmillan Company, Chennai, India
www.macmillansolutions.com

Printed and bound in the United States of America

Transferred to Digital Printing in 2013

Working together to grow
libraries in developing countries

www.elsevier.com | www.bookaid.org | www.sabre.org

ELSEVIER BOOK AID International Sabre Foundation

Contents

Chapter 1

Scientific notation and metric prefixes

INTRODUCTION

There are some 3 000 000 000 base pairs (bp) making up human genomic DNA within a haploid cell. If that DNA is isolated from such a cell, it will weigh approximately 0.000 000 000 003 5 grams (g). To amplify a specific segment of that purified DNA using the polymerase chain reaction (PCR), 0.000 000 000 01 moles (*M*) of each of two primers can be added to a reaction that can produce, following some 30 cycles of the PCR, over 1 000 000 000 copies of the target gene.

On a day-to-day basis, molecular biologists work with extremes of numbers far outside the experience of conventional life. To allow them to more easily cope with calculations involving extraordinary values, two shorthand methods have been adopted that bring both enormous and infinitesimal quantities back into the realm of manageability. These methods use scientific notation and metric prefixes. They require the use of exponents and an understanding of significant digits.

1.1 SIGNIFICANT DIGITS

Certain techniques in molecular biology, as in other disciplines of science, rely on types of instrumentation capable of providing precise measurements. An indication of the level of precision is given by the number of digits expressed in the instrument's readout. The numerals of a measurement representing actual limits of precision are referred to as **significant digits**.

Although a zero can be as legitimate a value as the integers one through nine, significant digits are usually nonzero numerals. Without information on how a measurement was made or on the precision of the instrument used to make it, zeros to the left of the decimal point trailing one or more nonzero numerals are assumed not to be significant. For example, in stating that the human genome is 3 000 000 000 bp in length, the only significant digit in the number is the 3. The nine zeros are not significant. Likewise, zeros to the right of the decimal point preceding a set of nonzero numerals are assumed not to be significant. If we determine that the DNA within a

Calculations for Molecular Biology and Biotechnology. DOI: 10.1016/B978-0-12-375690-9.00001-2
© 2010 Elsevier Inc. All rights reserved.

sperm cell weighs 0.000 000 000 003 5 g, only the 3 and the 5 are significant digits. The 11 zeros preceding these numerals are not significant.

Problem 1.1 How many significant digits are there in each of the following measurements?

a) 3 001 000 000 bp
b) 0.003 04 g
c) 0.000 210 liters (L) (volume delivered with a calibrated micropipettor).

Solution 1.1

a) Number of significant digits: 4; they are: 3001
b) Number of significant digits: 3; they are: 304
c) Number of significant digits: 3; they are: 210

1.1.1 Rounding off significant digits in calculations

When two or more measurements are used in a calculation, the result can only be as accurate as the least precise value. To accommodate this necessity, the number obtained as solution to a computation should be rounded off to reflect the weakest level of precision. The guidelines in the following box will help determine the extent to which a numerical result should be rounded off.

Guidelines for rounding off significant digits

1. When adding or subtracting numbers, the result should be rounded off so that it has the same number of significant digits to the right of the decimal as the number used in the computation with the fewest significant digits to the right of the decimal.
2. When multiplying or dividing numbers, the result should be rounded off so that it contains only as many significant digits as the number in the calculation with the fewest significant digits.

Problem 1.2 Perform the following calculations, and express the answer using the guidelines for rounding off significant digits described in the preceding box

a) 0.2884 g + 28.3 g
b) 3.4 cm × 8.115 cm
c) 1.2 L × 0.155 L

Solution 1.2

a) $0.2884\,\text{g} + 28.3\,\text{g} = 28.5884\,\text{g}$

The sum is rounded off to show the same number of significant digits to the right of the decimal point as the number in the equation with the fewest significant digits to the right of the decimal point. (In this case, the value 28.3 has one significant digit to the right of the decimal point.)

28.5884 g is rounded off to 28.6 g

b) $3.4\,\text{cm} \times 8.115\,\text{cm} = 27.591\,\text{cm}^2$

The answer is rounded off to two significant digits since there are as few as two significant digits in one of the multiplied numbers (3.4 cm).

$27.591\,\text{cm}^2$ is rounded off to $28\,\text{cm}^2$

c) $1.2\,\text{L} \div 0.155\,\text{L} = 7.742\,\text{L}$

The quotient is rounded off to two significant digits since there are as few as two significant digits in one of the values (1.2 L) used in the equation.

7.742 L is rounded off to 7.7 L

1.2 EXPONENTS AND SCIENTIFIC NOTATION

An **exponent** is a number written above and to the right of (and smaller than) another number (called the **base**) to indicate the power to which the base is to be raised. Exponents of base 10 are used in scientific notation to express very large or very small numbers in a shorthand form. For example, for the value 10^3, 10 is the base and 3 is the exponent. This means that 10 is multiplied by itself three times ($10^3 = 10 \times 10 \times 10 = 1000$). For numbers less than 1.0, a negative exponent is used to express values as a reciprocal of base 10. For example,

$$10^{-3} = \frac{1}{10^3} = \frac{1}{10 \times 10 \times 10} = \frac{1}{1000} = 0.001$$

1.2.1 Expressing numbers in scientific notation

To express a number in scientific notation:

1. Move the decimal point to the right of the leftmost nonzero digit. Count the number of places the decimal has been moved from its original position.

2. Write the new number to include all numbers between the leftmost and rightmost significant (nonzero) figures. Drop all zeros lying outside these integers.
3. Place a multiplication sign and the number 10 to the right of the significant integers. Use an exponent to indicate the number of places the decimal point has been moved.
   a. For numbers greater than 10 (where the decimal was moved to the left), use a positive exponent.
   b. For numbers less than one (where the decimal was moved to the right), use a negative exponent.

Problem 1.3 Write the following numbers in scientific notation

a) 3 001 000 000

b) 78

c) 60.23×10^{22}

Solution 1.3

a) Move the decimal to the left nine places so that it is positioned to the right of the leftmost nonzero digit.

$$3.001\,000\,000$$

Write the new number to include all nonzero significant figures, and drop all zeros outside of these numerals. Multiply the new number by 10, and use a positive 9 as the exponent since the given number is greater than 10 and the decimal was moved to the left nine positions.

$$3\,001\,000\,000 = 3.001 \times 10^{9}$$

b) Move the decimal to the left one place so that it is positioned to the right of the leftmost nonzero digit. Multiply the new number by 10, and use a positive 1 as an exponent since the given number is greater than 10 and the decimal was moved to the left one position.

$$78 = 7.8 \times 10^{1}$$

c) 60.23×10^{22}

Move the decimal to the left one place so that it is positioned to the right of the leftmost nonzero digit. Since the decimal was moved one position to the left, add 1 to the exponent ($22 + 1 = 23$ = new exponent value).

$$60.23 \times 10^{22} = 6.023 \times 10^{23}$$

Problem 1.4 Write the following numbers in scientific notation

a) 0.000 000 000 015

b) 0.000 050 004 2

c) 437.28×10^{-7}

Solution 1.4

a) Move the decimal to the right 11 places so that it is positioned to the right of the leftmost nonzero digit. Write the new number to include all numbers between the leftmost and rightmost significant (nonzero) figures. Drop all zeros lying outside these numerals. Multiply the number by 10, and use a negative 11 as the exponent since the original number is less than 1 and the decimal was moved to the right by 11 places.

$$0.000000000015 = 1.5 \times 10^{-11}$$

b) Move the decimal to the right five positions so that it is positioned to the right of the leftmost nonzero digit. Drop all zeros lying outside the leftmost and rightmost nonzero digits. Multiply the number by 10 and use a negative 5 exponent since the original number is less than 1 and the decimal point was moved to the right five positions.

$$0.0000500042 = 5.00042 \times 10^{-5}$$

c) Move the decimal point two places to the left so that it is positioned to the right of the leftmost nonzero digit. Since the decimal is moved two places to the left, add a positive 2 to the exponent value ($-7 + 2 = -5$).

$$437.28 \times 10^{-7} = 4.3728 \times 10^{-5}$$

1.2.2 Converting numbers from scientific notation to decimal notation

To change a number expressed in scientific notation to decimal form:

1. If the exponent of 10 is positive, move the decimal point to the right the same number of positions as the value of the exponent. If necessary, add zeros to the right of the significant digits to hold positions from the decimal point.
2. If the exponent of 10 is negative, move the decimal point to the left the same number of positions as the value of the exponent. If necessary, add zeros to the left of the significant digits to hold positions from the decimal point.

Problem 1.5 Write the following numbers in decimal form

a) 4.37×10^5

b) 2×10^1

c) 23.4×10^7

d) 3.2×10^{-4}

Solution 1.5

a) Move the decimal point five places to the right, adding three zeros to hold the decimal's place from its former position.

$$4.37 \times 10^5 = 437\,000.0$$

b) Move the decimal point one position to the right, adding one zero to the right of the significant digit to hold the decimal point's new position.

$$2 \times 10^1 = 20.0$$

c) Move the decimal point seven places to the right, adding six zeros to hold the decimal point's position.

$$23.4 \times 10^7 = 234\,000\,000.0$$

d) The decimal point is moved four places to the left. Zeros are added to hold the decimal point's position.

$$3.2 \times 10^{-4} = 0.000\,32$$

1.2.3 Adding and subtracting numbers written in scientific notation

When adding or subtracting numbers expressed in scientific notation, it is simplest first to convert the numbers in the equation to the same power of 10 as that of the highest exponent. The exponent value then does not change when the computation is finally performed.

Problem 1.6 Perform the following computations

a) $(8 \times 10^4) + (7 \times 10^4)$

b) $(2 \times 10^3) + (3 \times 10^1)$

c) $(6 \times 10^{-2}) + (8 \times 10^{-3})$

d) $(3.9 \times 10^{-4}) - (3.7 \times 10^{-4})$

e) $(2.4 \times 10^{-3}) - (1.1 \times 10^{-4})$

Solution 1.6

a) $(8 \times 10^4) + (7 \times 10^4)$

$= 15 \times 10^4$	Numbers added.
$= 1.5 \times 10^5$	Number rewritten in standard scientific notation form.
$= 2 \times 10^5$	Number rounded off to one significant digit.

b) $(2 \times 10^3) + (3 \times 10^1)$

$= (2 \times 10^3) + (0.03 \times 10^3)$	Number with lowest exponent value expressed in terms of that of the largest exponent value.
$= 2.03 \times 10^3$	Numbers are added.
$= 2 \times 10^3$	Number rounded off to one significant digit.

c) $(6 \times 10^{-2}) + (8 \times 10^{-3})$

$= (6 \times 10^{-2}) + (0.8 \times 10^{-2})$	Exponents converted to the same values.
$= 6.8 \times 10^{-2}$	Numbers are added.
$= 7 \times 10^{-2}$	Number rounded off to one significant digit.

d) $(3.9 \times 10^{-4}) - (3.7 \times 10^{-4})$

$= 0.2 \times 10^{-4}$	Numbers are subtracted.
$= 2 \times 10^{-5}$	Numbers rewritten in standard scientific notation.

e) $(2.4 \times 10^{-3}) - (1.1 \times 10^{-4})$

$= (2.4 \times 10^{-3}) - (0.11 \times 10^{-3})$	Exponents converted to the same values.
$= 2.29 \times 10^{-3}$	Numbers are subtracted.
$= 2.3 \times 10^{-3}$	Number rounded off to show only one significant digit to the right of the decimal point.

■

1.2.4 Multiplying and dividing numbers written in scientific notation

Exponent laws used in multiplication and division for numbers written in scientific notation include:

The Product Rule: When multiplying using scientific notation, the exponents are added.

The Quotient Rule: When dividing using scientific notation, the exponent of the denominator is subtracted from the exponent of the numerator.

When working with the next set of problems, the following laws of mathematics will be helpful:

The Commutative Law for Multiplication: The result of a multiplication is not dependent on the order in which the numbers are multiplied. For example,

$$3 \times 2 = 2 \times 3$$

The Associative Law for Multiplication: The result of a multiplication is not dependent on how the numbers are grouped. For example,

$$3 \times (2 \times 4) = (3 \times 2) \times 4$$

Problem 1.7 Calculate the product

a) $(3 \times 10^4) \times (5 \times 10^2)$
b) $(2 \times 10^3) \times (6 \times 10^{-5})$
c) $(4 \times 10^{-2}) \times (2 \times 10^{-3})$

Solution 1.7

a) $(3 \times 10^4) \times (5 \times 10^2)$

$= (3 \times 5) \times (10^4 \times 10^2)$	Use Commutative and Associative laws to group like terms.
$= 15 \times 10^6$	Exponents are added.
$= 1.5 \times 10^7$	Number written in standard scientific notation.
$= 2 \times 10^7$	Number rounded off to one significant digit.

b) $(2 \times 10^3) \times (6 \times 10^{-5})$

$= (2 \times 6) \times (10^3 \times 10^{-5})$	Use Commutative and Associative laws to group like terms.
$= 12 \times 10^{-2}$	Exponents are added.
$= 1.2 \times 10^{-3}$	Number written in standard scientific notation.
$= 1 \times 10^{-3}$	Number rounded off to one significant digit.

c) $(4 \times 10^{-2}) \times (2 \times 10^{-3})$

$= (4 \times 2) \times (10^{-2} \times 10^{-3})$ Use Commutative and Associative laws to group like terms.

$= 8 \times 10^{-2+(-3)}$

$= 8 \times 10^{-5}$ Exponents are added.

Problem 1.8 Find the quotient

a) $\dfrac{8 \times 10^4}{2 \times 10^2}$

b) $\dfrac{5 \times 10^8}{3 \times 10^{-4}}$

c) $\dfrac{8.2 \times 10^{-6}}{3.6 \times 10^4}$

d) $\dfrac{9 \times 10^{-5}}{2.5 \times 10^{-3}}$

Solution 1.8

a) $\dfrac{8 \times 10^4}{2 \times 10^2}$

$= \dfrac{8}{2} \times 10^{4-2}$ The exponent of the denominator is subtracted from the exponent of the numerator.

$= 4 \times 10^2$

b) $\dfrac{5 \times 10^8}{3 \times 10^{-4}}$

$= \dfrac{5}{3} \times 10^{8-(-4)}$ The exponent of the denominator is subtracted from the exponent of the numerator.

$= 1.67 \times 10^{8-(-4)}$ Exponents: $8 - (-4) = 8 + 4 = 12$.

$= 2 \times 10^{12}$ Number rounded off to one significant digit.

c) $\dfrac{8.2 \times 10^{-6}}{3.6 \times 10^{4}}$

$= \dfrac{8.2}{3.6} \times 10^{-6-(+4)}$ — The exponent of the denominator is subtracted from the exponent of the numerator.

$= 2.3 \times 10^{-10}$ — Number rounded off to two significant digits. Exponent: $-6 - (+4) = -6 + (-4) = -10$.

d) $\dfrac{9 \times 10^{-5}}{2.5 \times 10^{-3}}$

$= \dfrac{9}{2.5} \times 10^{-5-(-3)}$ — The exponent of the denominator is subtracted from the exponent of the numerator.

$= 3.6 \times 10^{-2}$

$= 4 \times 10^{-2}$ — Number rounded off to one significant digit.

1.3 METRIC PREFIXES

A **metric prefix** is a shorthand notation used to denote very large or vary small values of a basic unit as an alternative to expressing them as powers of 10. Basic units frequently used in the biological sciences include meters, grams, moles, and liters. Because of their simplicity, metric prefixes have found wide application in molecular biology. The following table lists the most frequently used prefixes and the values they represent.

As shown in Table 1.1, one nanogram (ng) is equivalent to 1×10^{-9} g. There are, therefore, 1×10^{9} ngs per g (the reciprocal of 1×10^{-9}; $1/1 \times 10^{9} = 1 \times 10^{-9}$). Likewise, since one microliter (μL) is equivalent to 1×10^{-6} L, there are 1×10^{6} mL per liter.

When expressing quantities with metric prefixes, the prefix is usually chosen so that the value can be written as a number greater than 1.0 but less than 1000. For example, it is conventional to express 0.000 000 05 g as 50 ng rather than 0.05 mg or 50 000 pg.

1.3.1 Conversion factors and canceling terms

Translating a measurement expressed with one metric prefix into an equivalent value expressed using a different metric prefix is called a **conversion**.

Table 1.1 Metric prefixes, their abbreviations, and their equivalent values as exponents of 10.

Metric prefix	Abbreviation	Power of 10
giga-	G	10^9
mega-	M	10^6
kilo-	k	10^3
milli-	m	10^{-3}
micro-	μ	10^{-6}
nano-	n	10^{-9}
pico-	p	10^{-12}
femto-	f	10^{-15}
atto-	a	10^{-18}

These are performed mathematically by using a conversion factor relating the two different terms. A conversion factor is a numerical ratio equal to 1. For example,

$$\frac{1 \times 10^6\ \mu\text{g}}{\text{g}} \text{ and } \frac{1\ \text{g}}{1 \times 10^6\ \mu\text{g}}$$

are conversion factors, both equal to 1. They can be used to convert grams to micrograms or micrograms to grams, respectively. The final metric prefix expression desired should appear in the equation as a numerator value in the conversion factor. Since multiplication or division by the number 1 does not change the value of the original quantity, any quantity can be either multiplied or divided by a conversion factor and the result will still be equal to the original quantity; only the metric prefix will be changed.

When performing conversions between values expressed with different metric prefixes, the calculations can be simplified when factors of 1 or identical units are canceled. A factor of 1 is any expression in which a term is divided by itself. For example, $1 \times 10^6/1 \times 10^6$ is a factor of 1. Likewise, 1 L/1 L is a factor of 1. If, in a conversion, identical terms appear anywhere in the equation on one side of the equals sign as both a numerator and a denominator, they can be canceled. For example, if converting 5×10^{-4} L to microliters, an equation can be set up so that identical terms (in this case, liters) can be canceled to leave mL as a numerator value.

$5 \times 10^{-4}\text{L} = n\,\mu\text{L}$ Solve for n.

$$5 \times 10^{-4}\ \text{L} \times \frac{1 \times 10^{6}\ \mu\text{L}}{\text{L}} = n\,\mu\text{L}$$

Use the conversion factor relating liters and microliters with microliters as a numerator value. Identical terms in a numerator and a denominator are canceled. (Remember, 5×10^{-4} L is the same as 5×10^{-4}L/1. 5×10^{-4} L, therefore, is a numerator.)

$(5 \times 1)(10^{-4} \times 10^{6})\mu\text{L} = n\,\mu\text{L}$ Group like terms.

$5 \times 10^{-4+6}\mu\text{L} = n\,\mu\text{L}$ Numerator values are multiplied.

$5 \times 10^{2}\mu\text{L} = n\,\mu\text{L}$ Therefore, 5×10^{-4}L is equivalent to $5 \times 10^{2}\,\mu$L.

Problem 1.9

a) There are approximately 6×10^{9} bp per human diploid genome. What is this number expressed as kilobase pairs (kb)?

b) Convert 0.03 mg into ng.

c) Convert 0.0025 mL into mL.

Solution 1.9

a) $6 \times 10^{9}\,\text{bp} = n\,\text{kb}$ Solve for n.

Multiply by a conversion factor relating kb to bp with kb as a numerator:

$$6 \times 10^{9}\ \text{bp} \times \frac{1\ \text{kb}}{1 \times 10^{3}\ \text{bp}} = n\,\text{kb}$$

Cancel identical terms (bp) appearing as numerator and denominator, leaving kb as a numerator value.

$$\frac{(6 \times 10^{9})(1\ \text{kb})}{1 \times 10^{3}} = n\,\text{kb}$$

The exponent of the denominator is subtracted from the exponent of the numerator.

$$\frac{6}{1} \times 10^{9-3}\ \text{kb} = 6 \times 10^{6}\ \text{kb} = n\,\text{kb}$$

Therefore, 6×10^{9} bp is equivalent to 6×10^{6} kb.

b) $0.03\,\mu g = n$ ng. Solve for n.
Multiply by conversion factors relating g to mg and ng to g with ng as a numerator. Convert 0.03 mg to its equivalent in scientific notation (3×10^{-2} mg):

$$3 \times 10^{-2}\ \mu g \times \frac{1\,g}{1 \times 10^{6}\ \mu g} \times \frac{1 \times 10^{9}\ ng}{g} = n\,ng$$

Cancel identical terms appearing as numerator and denominator, leaving ng as a numerator value; multiply numerator and denominator values; and then group like terms.

$$\frac{(3 \times 1 \times 1)(10^{-2} \times 10^{9})\,ng}{(1 \times 1)(10^{6})} = n\,ng$$

Numerator exponents are added.

$$\frac{3 \times 10^{-2+9}\ ng}{1 \times 10^{6}} = \frac{3 \times 10^{7}\ ng}{1 \times 10^{6}} = n\,ng$$

The denominator exponent is subtracted from the numerator exponent.

$$\frac{3}{1} \times 10^{7-6}\ ng = 3 \times 10^{1}\ ng = n\,ng$$

Therefore, 0.03 mg is equivalent to 30 (3×10^{1}) ng.

c) $0.0025\,mL = n\,\mu L$. Solve for n.
Convert 0.0025 mL into scientific notation. Multiply by conversion factors relating L to mL and mL to L with mL as a numerator.

$$2.5 \times 10^{-3}\ mL \times \frac{1L}{1 \times 10^{3}\ mL} \times \frac{1 \times 10^{6}\ \mu L}{1L} = n\,\mu L$$

Cancel identical terms appearing as numerator and denominator, leaving mL as a numerator value. Multiply numerator values and denominator values. Group like terms.

$$\frac{(2.5 \times 1 \times 1)(10^{-3} \times 10^{6})\,\mu L}{(1 \times 1)(10^{3})} = n\,\mu L$$

Numerator exponents are added.

$$\frac{2.5 \times 10^{-3+6}\ \mu L}{1 \times 10^{3}} = \frac{2.5 \times 10^{3}\ \mu L}{1 \times 10^{3}} = n\,\mu L$$

The denominator exponent is subtracted from the numerator exponent.

$$\frac{2.5}{1} \times 10^{3-3}\,\mu\text{L} = 2.5 \times 10^{0}\,\mu\text{L} = 2.5\,\mu\text{L} = n\,\mu\text{L}$$

Therefore, 0.0025 mL is equivalent to 2.5 mL.

CHAPTER SUMMARY

Significant digits are numerals representing actual limits of precision. They are usually nonzero digits. Zeros to the left of the decimal point trailing a nonzero numeral are assumed not to be significant. Zeros to the right of the decimal point preceding a nonzero numeral are also assumed not to be significant.

When rounding off the sum or difference of two numbers, the calculated value should have the same number of significant digits to the right of the decimal as the number in the computation with the fewest significant digits to the right of the decimal. A product or quotient should have only as many significant digits as the number in the calculation with the fewest significant digits.

When expressing numbers in **scientific notation**, move the decimal point to the right of the leftmost nonzero digit, drop all zeros lying outside the string of significant figures, and express the new number as being multiplied by 10 having an exponent equal to the number of places the decimal point was moved from its original position (using a negative exponent if the decimal point was moved to the right).

When adding or subtracting numbers expressed in scientific notation, rewrite the numbers such that they all have the same exponent value as that having the highest exponent, then perform the calculation. When multiplying numbers expressed in scientific notation, add the exponents. When dividing numbers expressed in scientific notation, subtract the exponent of the denominator from the exponent of the numerator to obtain the new exponent value.

Numbers written in scientific notation can also be written using **metric prefixes** that will bring the value down to its lowest number of significant digits.

Chapter 2

Solutions, mixtures, and media

INTRODUCTION

Whether it is an organism or an enzyme, most biological activities function optimally only within a narrow range of environmental conditions. From growing cells in culture to sequencing of a cloned DNA fragment or assaying an enzyme's activity, the success or failure of an experiment can hinge on paying careful attention to a reaction's components. This section outlines the mathematics involved in making solutions.

2.1 CALCULATING DILUTIONS – A GENERAL APPROACH

Concentration is defined as an amount of some substance per a set volume:

$$\text{concentration} = \frac{\text{amount}}{\text{volume}}$$

Most laboratories have found it convenient to prepare concentrated stock solutions of commonly used reagents, those found as components in a large variety of buffers or reaction mixes. Such stock solutions may include 1 *M* (mole)Tris, pH 8.0, 500 m*M* ethylenediaminetetraacetic acid (EDTA), 20% sodium dodecylsulfate (SDS), 1 *M* $MgCl_2$, and any number of others. A specific volume of a stock solution at a particular concentration can be added to a buffer or reagent mixture so that it contains that component at some concentration less than that in the stock. For example, a stock solution of 95% ethanol can be used to prepare a solution of 70% ethanol. Since a higher percent solution (more concentrated) is being used to prepare a lower percent (less concentrated) solution, a **dilution** of the stock solution is being performed.

There are several methods that can be used to calculate the concentration of a diluted reagent. No one approach is necessarily more valid than another. Typically, the method chosen by an individual has more to do with how his or her brain approaches mathematical problems than with the legitimacy of the procedure. One approach is to use the equation $C_1V_1 = C_2V_2$, where

Calculations for Molecular Biology and Biotechnology. DOI: 10.1016/B978-0-12-375690-9.00002-4
© 2010 Elsevier Inc. All rights reserved.

C_1 is the initial concentration of the stock solution, V_1 is the amount of stock solution taken to perform the dilution, C_2 is the concentration of the diluted sample, and V_2 is the final, total volume of the diluted sample.

For example, if you were asked how many μL of 20% sugar should be used to make 2 mL of 5% sucrose, the $C_1V_1 = C_2V_2$ equation could be used. However, to use this approach, all units must be the same. Therefore, you first need to convert 2 mL into a microliter amount. This can be done as follows:

$$2\ \text{mL} \times \frac{1000\ \mu\text{L}}{1\ \text{mL}} = 2000\ \mu\text{L}$$

C_1, then, is equal to 20%, V_1 is the volume you wish to calculate, C_2 is 5%, and V_2 is 2000 μL. The calculation is then performed as follows:

$$\begin{aligned} C_1V_1 &= C_2V_2 \\ (20\%)V_1 &= (5\%)(2000\ \mu\text{L}) \end{aligned}$$

Solving for V_1 gives the following result:

$$V_1 = \frac{(5\%)(2000\ \mu\text{L})}{20\%} = 500\ \mu\text{L}$$

The percent units cancel since they are in both the numerator and the denominator of the equation, leaving mL as the remaining unit. Therefore, you would need 500 μL of 20% sucrose plus 1500 μL (2000 μL − 500 μL = 1500 μL) of water to make a 5% sucrose solution from a 20% sucrose solution.

Dimensional analysis is another general approach to solving problems of concentration. In this method, an equation is set up such that the known concentration of the stock and all volume relationships appear on the left side of the equation and the final desired concentration is placed on the right side. Conversion factors are actually part of the equation. Terms are set up as numerator or denominator values such that all terms cancel except for that describing concentration. A dimensional analysis equation is set up in the following manner.

$$\text{starting concentration} \times \text{conversion factor} \times \frac{\text{unknown volume}}{\text{final volume}} = \text{desired concentration}$$

Using the dimensional analysis approach, the problem of discovering how many microliters of 20% sucrose are needed to make 2 mL of 5% sucrose is written as follows:

$$20\% \times \frac{1\ \text{mL}}{1000\ \mu\text{L}} \times \frac{x\ \mu\text{L}}{2\ \text{mL}} = 5\%$$

Notice that all terms on the left side of the equation will cancel except for the percent units. Solving for x μL gives the following result:

$$\frac{(20\%)x}{2000} = 5\%$$

$$x = \frac{(5\%)(2000)}{20\%} = 500$$

Since x is a μL amount, you need 500 μL of 20% sucrose in a final volume of 2 mL to make 5% sucrose. Notice how similar the last step of the solution to this equation is to the last step of the equation using the $C_1V_1 = C_2V_2$ approach.

Making a conversion factor part of the equation obviates the need for performing two separate calculations, as is required when using the $C_1V_1 = C_2V_2$ approach. For this reason, dimensional analysis is the method used for solving problems of concentration throughout this book.

2.2 CONCENTRATIONS BY A FACTOR OF X

The concentration of a solution can be expressed as a multiple of its standard working concentration. For example, many buffers used for agarose or acrylamide gel electrophoresis are prepared as solutions 10-fold (10X) more concentrated than their standard running concentration (1X). In a 10X buffer, each component of that buffer is 10-fold more concentrated than in the 1X solution. To prepare a 1X working buffer, a dilution of the more concentrated 10X stock is performed in water to achieve the desired volume. To prepare 1000 mL (1 L) of 1X Tris-borate-EDTA (TBE) gel running buffer from a 10X TBE concentrate, for example, add 100 mL of 10X solution to 900 mL of distilled water. This can be calculated as follows:

$$10\text{X Buffer} \times \frac{n\ \text{mL}}{1000\ \text{mL}} = 1\text{X Buffer}$$

n mL of 10X buffer is diluted into a total volume of 1000 mL to give a final concentration of 1X. Solve for n.

$$\frac{10Xn}{1000} = 1X$$

Multiply numerator values.

$$(1000) \times \frac{10Xn}{1000} = 1X\,(1000)$$

Use the Multiplication Property of Equality (see the following box) to multiply each side of the equation by 1000. This cancels out the 1000 in the denominator on the left side of the equals sign.

$$10Xn = 1000X$$

$$\frac{10Xn}{10X} = \frac{1000X}{10X}$$

Divide each side of the equation by 10X. (Again, this uses the Multiplication Property of Equality.)

$$n = 100$$

The X terms cancel since they appear in both the numerator and the denominator. This leaves n equal to 100.

Therefore, to make 1000 mL of 1X buffer, add 100 mL of 10X buffer stock to 900 mL of distilled water (1000 mL − 100 mL contributed by the 10X buffer stock = 900 mL).

Multiplication property of equality

Both sides of an equation may be multiplied by the same nonzero quantity to produce equivalent equations. This property also applies to division: both sides of an equation can be divided by the same nonzero quantity to produce equivalent equations.

Problem 2.1 How is 640 mL of 0.5X buffer prepared from an 8X stock?

Solution 2.1

We start with a stock of 8X buffer. We want to know how many mL of the 8X buffer should be in a final volume of 640 mL to give us a buffer having a concentration of 0.5X. This relationship can be expressed mathematically as follows:

$$8X \text{ buffer} \times \frac{n \text{ mL}}{640 \text{ mL}} = 0.5X \text{ buffer}$$

Solve for n.

$\frac{8Xn}{640} = 0.5X$ — Multiply numerator values on the left side of the equation. Since the mL terms appear in both the numerator and the denominator, they cancel out.

$8Xn = 320X$ — Multiply each side of the equation by 640.

$n = \frac{320X}{8X} = 40$ — Divide each side of the equation by 8X. The X terms, since they appear in both the numerator and the denominator, cancel.

Therefore, add 40 mL of 8X stock to 600 mL of distilled water to prepare a total of 640 mL of 0.5X buffer (640 mL final volume − 40 mL 8X stock = 600 mL volume to be taken by water).

■

2.3 PREPARING PERCENT SOLUTIONS

Many reagents are prepared as a percent of solute (such as salt, cesium chloride, or sodium hydroxide) dissolved in solution. Percent, by definition, means 'per 100.' 12%, therefore, means 12 per 100, or 12 out of every 100. 12% may also be written as the decimal 0.12 (derived from the fraction 12/100 = 0.12).

Depending on the solute's initial physical state, its concentration can be expressed as a weight per volume percent (% w/v) or a volume per volume percent (% v/v). A percentage in weight per volume refers to the weight of solute (in grams) in a total of 100 mL of solution. A percentage in volume per volume refers to the amount of liquid solute (in mL) in a final volume of 100 mL of solution.

Most microbiology laboratories will stock a solution of 20% (w/v) glucose for use as a carbon source in bacterial growth media. To prepare 100 mL of 20% (w/v) glucose, 20 grams (g) of glucose are dissolved in enough distilled water that the final volume of the solution, with the glucose completely dissolved, is 100 mL.

Problem 2.2 How can the following solutions be prepared?

a) 100 mL of 40% (w/v) polyethylene glycol (PEG) 8000

b) 47 mL of a 7% (w/v) solution of sodium chloride (NaCl)

c) 200 mL of a 95% (v/v) solution of ethanol

Solution 2.2

a) Weigh out 40 g of PEG 8000 and dissolve in distilled water so that the final volume of the solution, with the PEG 8000 completely dissolved, is 100 mL. This is most conveniently done by initially dissolving the PEG 8000 in approximately 60 mL of distilled water. When the granules are dissolved, pour the solution into a 100 mL graduated cylinder and bring the volume up to the 100 mL mark with distilled water.

b) First, 7% of 47 must be calculated. This is done by multiplying 47 by 0.07 (the decimal form of 7%; 7/100 = 0.07):

$$0.07 \times 47 = 3.29$$

Therefore, to prepare 47 mL of 7% NaCl, weigh out 3.29 g of NaCl and dissolve the crystals in some volume of distilled water less than 47 mL, a volume measured so that, when the 3.29 g of NaCl are added, it does not exceed 47 mL. When the NaCl is completely dissolved, dispense the solution into a 50 mL graduated cylinder and bring the final volume up to 47 mL with distilled water.

c) 95% of 200 mL is calculated by multiplying 0.95 (the decimal form of 95%) by 200:

$$0.95 \times 200 = 190$$

Therefore, to prepare 200 mL of 95% ethanol, measure 190 mL of 100% (200 proof) ethanol and add 10 mL of distilled water to bring the final volume to 200 mL.

2.4 DILUTING PERCENT SOLUTIONS

When approaching a dilution problem involving percentages, express the percent solutions as fractions of 100. The problem can be written as an equation in which the concentration of the stock solution ('what you have') is positioned on the left side of the equation and the desired final concentration ('what you want') is on the right side of the equation. The unknown volume (x) of the stock solution to add to the volume of the final mixture should also be expressed as a fraction (with x as a numerator and the final desired volume as a denominator). This part of the equation should also be positioned on the left side of the equals sign. For example, if 30 mL of 70% ethanol is to be prepared from a 95% ethanol stock solution, the following equation can be written:

$$\frac{95}{100} \times \frac{x \text{ mL}}{30 \text{ mL}} = \frac{70}{100}$$

You then solve for x.

$$\frac{95x}{3000} = \frac{70}{100}$$

Multiply numerators together and multiply denominators together. The mL terms, since they are present in both the numerator and the denominator, cancel.

$$\frac{3000}{1} \times \frac{95x}{3000} = \frac{70}{100} \times \frac{3000}{1}$$

Multiply both sides of the equation by 3000.

$$95x = \frac{210\,000}{100}$$

Simplify the equation.

$$95x = 2100$$

$$\frac{95x}{95} = \frac{2100}{95}$$

Divide each side of the equation by 95.

$$x = 22$$

Round off to two significant figures.

Therefore, to prepare 30 mL of 70% ethanol using a 95% ethanol stock solution, combine 22 mL of 95% ethanol stock with 8 mL of distilled water.

Problem 2.3 If 25 g of NaCl are dissolved into a final volume of 500 mL, what is the percent (w/v) concentration of NaCl in the solution?

Solution 2.3

The concentration of NaCl is 25 g/500 mL (w/v). To determine the percent (w/v) of the solution, we need to know how many grams of NaCl are in 100 mL. We can set up an equation of two ratios in which x represents the unknown number of grams. This relationship is read 'x g is to 100 mL as 25 g is to 500 mL:'

$$\frac{x\ \text{g}}{100\ \text{mL}} = \frac{25\ \text{g}}{500\ \text{mL}}$$

Solving for x gives the following result:

$$x\ \text{g} = \frac{(25\ \text{g})(100\ \text{mL})}{500\ \text{mL}} = \frac{2500\ \text{g}}{500} = 5\ \text{g}$$

Therefore, there are 5 g of NaCl in 100 mL (5 g/100 mL), which is equivalent to a 5% solution.

Problem 2.4 If 8 mL of distilled water is added to 2 mL of 95% ethanol, what is the concentration of the diluted ethanol solution?

Solution 2.4

The total volume of the solution is 8 mL + 2 mL = 10 mL. This volume should appear as a denominator on the left side of the equation. This dilution is the same as if 2 mL of 95% ethanol were added to 8 mL of water. Either way, it is a quantity of the 95% ethanol stock that is used to make the dilution. The '2 mL,' therefore, should appear as the numerator in the volume expression on the left side of the equation:

$$\frac{95}{100} \times \frac{2\ \text{mL}}{10\ \text{mL}} = \frac{x}{100}$$

$$\frac{190}{1000} = \frac{x}{100}$$

The mL terms cancel. Multiply numerator values and denominator values on the left side of the equation.

$$0.19 = \frac{x}{100}$$

Simplify the equation.

$$19 = x$$

Multiply both sides of the equation by 100.

If x in the original equation is replaced by 19, it is seen that the new concentration of ethanol in this diluted sample is 19/100, or 19%.

Problem 2.5 How many microliters of 20% SDS are required to bring 1.5 mL of solution to 0.5%?

Solution 2.5

In previous examples, there was control over how much water we could add in preparing the dilution to bring the sample to the desired concentration. In this example, however, a fixed volume (1.5 mL) is used as a starting sample and must be brought to the desired concentration. Solving this problem will require the use of the **Addition Property of Equality** (see the following box).

The addition property of equality

You may add (or subtract) the same quantity to (from) both sides of an equation to produce equivalent equations. For any real numbers a, b, and c, if $a = b$, then $a + c = b + c$, and $a - c = b - c$.

Since concentration, by definition, is the amount of a particular component in a specified volume, by adding a quantity of a stock solution to a fixed volume, the final volume is changed by that amount and the concentration is changed accordingly. The amount of stock solution (x mL) added in the process of the dilution must also be figured into the final volume, as follows.

$$\frac{20}{100} \times \frac{x\,\text{mL}}{1.5\,\text{mL} + x\,\text{mL}} = \frac{0.5}{100}$$

$$\frac{20x}{150 + 100x} = \frac{0.5}{100}$$

Multiply numerators and denominators. The mL terms cancel out.

$$\frac{150 + 100x}{1} \times \frac{20x}{150 + 100x} = \frac{0.5}{100} \times \frac{150 + 100x}{1}$$

Multiply both sides of the equation by $150 + 100x$.

$$20x = \frac{75 + 50x}{100}$$

Simplify the equation.

$$\frac{100}{1} \times 20x = \frac{75 + 50x}{100} \times \frac{100}{1}$$

Multiply both sides of the equation by 100.

$$2000x = 75 + 50x$$

Simplify.

$$1950x = 75$$

Subtract $50x$ from both sides of the equation (Addition Property of Equality).

$$x = \frac{75}{1950} = 0.03846\,\text{mL}$$

Divide both sides of the equation by 1950.

$$0.03846\,\text{mL} \times \frac{1000\,\mu\text{L}}{\text{mL}} = 38.5\,\mu\text{L}$$

Convert mL to μL and round off to one significant figure to the right of the decimal point.

Therefore, if 38.5 μL of 20% SDS are added to 1.5 mL, the SDS concentration of that sample will be 0.5% in a final volume of 1.5385 mL. If there were some other component in that initial 1.5 mL, the concentration of that component would change by the addition of the SDS. For example, if NaCl were present at a concentration of 0.2%, its concentration would be altered by the addition of more liquid. The initial solution of 1.5 mL would contain the following amount of NaCl:

$$\frac{0.2\,\text{g}}{100\,\text{mL}} \times 1.5\,\text{mL} = 0.003\,\text{g}$$

Therefore, 1.5 mL of 0.2% NaCl contains 0.003 g of NaCl.

In a volume of 1.5385 mL (the volume after the SDS solution has been added), 0.003 g of NaCl is equivalent to a 0.195% NaCl solution, as shown here:

$$\frac{0.003}{1.5385} \times 100 = 0.195\%$$

2.5 MOLES AND MOLECULAR WEIGHT – DEFINITIONS

A **mole** is equivalent to 6.023×10^{23} molecules. That molecule may be a pure elemental atom or a molecule consisting of a bound collection of atoms. For example, a mole of hydrogen is equivalent to 6.023×10^{23} molecules of hydrogen. A mole of glucose ($C_6H_{12}O_6$) is equivalent to 6.023×10^{23} molecules of glucose. The value 6.023×10^{23} is also known as **Avogadro's number**.

The **molecular weight** (**MW**, or **gram molecular weight**) of a substance is equivalent to the sum of its atomic weights. For example, the gram molecular weight of NaCl is 58.44: the atomic weight of Na (22.99 g) plus the atomic weight of chlorine (35.45 g). Atomic weights can be found in the periodic table of the elements. The molecular weight of a compound, as obtained commercially, is usually provided by the manufacturer and is printed on the container's label. On many reagent labels, a **formula weight** (**FW**) is given. For almost all applications in molecular biology, this value is used interchangeably with molecular weight.

Problem 2.6 What is the molecular weight of sodium hydroxide (NaOH)?

Solution 2.6

Atomic weight of Na	22.99
Atomic weight of O	16.00
Atomic weight of H	+1.01
Molecular weight of NaOH	40.00

Problem 2.7 What is the molecular weight of glucose ($C_6H_{12}O_6$)?

Solution 2.7

The atomic weight of each element in this compound must be multiplied by the number of times it is represented in the molecule:

Atomic weight of C = 12.01	
$12.01 \times 6 =$	72.06
Atomic weight of H = 1.01	
$1.01 \times 12 =$	12.12
Atomic weight of O = 16.00	
$16 \times 6 =$	+ 96.00
Molecular weight of $C_6H_{12}O_6$	180.18

Therefore, the molecular weight of glucose is 180.18 g/*m*.

2.5.1 Molarity

A 1 molar (1 *M*) solution contains the molecular weight of a substance (in grams) in 1 L of solution. For example, the molecular weight of NaCl is 58.44. A 1 *M* solution of NaCl, therefore, contains 58.44 g of NaCl dissolved in a final volume of 1000 mL (1 L) water. A 2 *M* solution of NaCl contains twice that amount (116.88 g) of NaCl dissolved in a final volume of 1000 mL water.

Note: Many protocols instruct to 'q.s. with water.' This means to bring it up to the desired volume with water (usually in a graduated cylinder or other volumetric piece of labware). Q.s. stands for *quantum sufficit*.

Problem 2.8 How is 200 mL of 0.3 *M* NaCl prepared?

Solution 2.8

The molecular weight of NaCl is 58.44. The first step in solving this problem is to calculate how many grams are needed for 1 L of a 0.3 *M* solution. This can be done by setting up a ratio stating '58.44 g is to 1 *M* as *x* g is to 0.3 *M*.' This relationship, expressed mathematically, can be written as follows. We then solve for *x*.

$$\frac{58.44\text{ g}}{1M} = \frac{x\text{ g}}{0.3M}$$

Because units on both sides of the equation are equivalent (if we were to multiply one side by the other, all terms would cancel), we will disregard them. Multiplying both sides of the equation by 0.3 gives

$$\frac{0.3}{1} \times \frac{58.44}{1} = \frac{x}{0.3} \times \frac{0.3}{1} \qquad 17.53 = x$$

Therefore, to prepare 1 L of 0.3 *M* NaCl, 17.53 g of NaCl are required.

Another ratio can now be written to calculate how many grams of NaCl are needed if 200 mL of a 0.3 *M* NaCl solution are being prepared. It can be expressed verbally as '17.53 g is to 1000 mL as *x* g is to 200 mL,' or written in mathematical terms:

$$\frac{17.53\ \text{g}}{1000\ \text{mL}} = \frac{x\ \text{g}}{200\ \text{mL}}$$

$$\frac{17.53 \times 200}{1000} = x$$

Multiply both sides of the equation by 200 to cancel out the denominator on the right side of the equation and to isolate *x*.

$$\frac{3506}{1000} = x = 3.51$$

Simplify the equation.

Therefore, to prepare 200 mL of 0.3 *M* NaCl solution, 3.51 g of NaCl are dissolved in distilled water to a final volume of 200 mL.

Problem 2.9 How is 50 mL of 20 millimolar (m*M*) sodium hydroxide (NaOH) prepared?

Solution 2.9

First, because it is somewhat more convenient to deal with terms expressed as molarity (*M*), convert the 20 m*M* value to an *M* value:

$$20\ \text{m}M \times \frac{1\,M}{1000\ \text{m}M} = 0.02\ M$$

Next, set up a ratio to calculate the amount of NaOH (40.0 g molecular weight) needed to prepare 1 L of 0.02 *M* NaOH. Use the expression '40.0 g is to 1 *M* as *x* g is to 0.02 *M*.' Solve for *x*.

$$\frac{40.0\text{ g}}{1\,M} = \frac{x\text{ g}}{0.02\,M}$$

$$\frac{(40.0)(0.02)}{1} = x$$ Multiply both sides of the equation by 0.02.

$$0.8 = x$$

Therefore, if 1 L of 0.02 *M* NaOH is to be prepared, add 0.8 g of NaOH to water to a final volume of 1000 mL.

Now, set up a ratio to determine how much is required to prepare 50 mL and solve for *x*. (The relationship of ratios should read as follows: 0.8 g is to 1000 mL as *x* g is to 50 mL.)

$$\frac{0.8\text{ g}}{1000\text{ mL}} = \frac{x\text{ g}}{50\text{ mL}}$$

$$\frac{(0.8)(50)}{1000} = \frac{(x)(50)}{50}$$ Multiply both sides of the equation by 50. This cancels the 50 in the denominator on the right side of the equation.

$$\frac{40.0}{1000} = x = 0.04$$ Simplify the equation.

Therefore, to prepare 50 mL of 20 m*M* NaOH, 0.04 g of NaOH is dissolved in a final volume of 50 mL of distilled water.

Problem 2.10 How many moles of NaCl are present in 50 mL of a 0.15 *M* solution?

Solution 2.10

A 0.15 *M* solution of NaCl contains 0.15 moles of NaCl per liter. Since we want to know how many moles are in 50 mL, we need to use a conversion factor to convert liters to milliliters. The equation can be written as follows:

$$x\text{ mol} = 50\text{ mL} \times \frac{1\text{L}}{1000\text{ mL}} \times \frac{0.15\text{ mol}}{\text{L}}$$

Notice that terms on the right side of the equation cancel except for moles. Multiplying numerator and denominator values gives

$$x\text{ mol} = \frac{(50) \times (1) \times (0.15)\text{ mol}}{1000} = \frac{7.5\text{ mol}}{1000} = 0.0075\text{ mol}$$

Therefore, 50 mL of 0.15 *M* NaCl contains 0.0075 moles of NaCl.

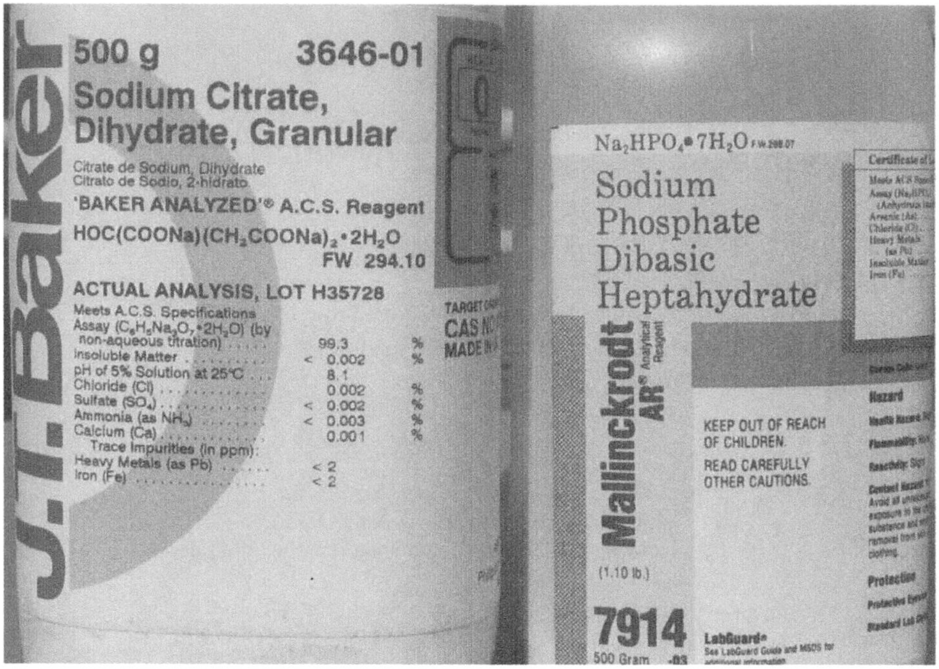

■ **FIGURE 2.1** Sodium citrate and sodium phosphate dibasic can come in hydrated forms.

2.5.2 Preparing molar solutions in water with hydrated compounds

Many compounds used in the laboratory, though dry and in crystalline or powder form, come 'hydrated.' That is, the compounds have water molecules attached to them. Examples are sodium citrate dihydrate (with two water molecules) and sodium phosphate dibasic heptahydrate (with seven water molecules) (Figure 2.1). You will find the compound's molecular weight listed on the container's label as the FW. In the case of hydrated compounds, the FW includes the weight contributed by the molecular weight of water (18.015 g/mole). These water molecules, therefore, contribute to the water you are using to dissolve the compound. They must be taken into consideration so that the final solution is not diluted to a concentration less than is desired. The following problem will demonstrate the calculation.

Problem 2.11 You want to prepare a 500 mL solution of 250 m*M* sodium phosphate dibasic heptahydrate ($Na_2HPO_4 \cdot H_2O$; FW 268.07). How much compound and how much water should be mixed?

Solution 2.11

We will first calculate how much of the Na_2HPO_4 compound should be added to make a 500 mL (final volume) solution at 250 m*M*.

250 m*M* is equivalent to 0.25 *M*, as shown below:

$$250\ \text{m}M \times \frac{1\,M}{1000\ \text{m}M} = 0.25\,M$$

We now calculate how much of the compound is equivalent to 0.25 *M*:

$$\frac{0.25\,M}{\text{liter}} \times \frac{268.07\ \text{g}}{\text{mole}} = \frac{67.02\ \text{g}}{\text{liter}}$$

Therefore, if we were making one liter (1000 mL) of solution, we would need 67.02 g of Na_2HPO_4. However, we're going to make 500 mL. We set up a relationship of ratios:

$$\frac{67.02\ \text{g}}{1000\ \text{mL}} = \frac{x\ \text{g}}{500\ \text{mL}}$$

Solving for x gives

$$x = \frac{(67.02\ \text{g})(500\ \text{mL})}{1000\ \text{mL}} = 33.51\ \text{g}$$

Therefore, we need 33.51 g of Na_2HPO_4 in a final volume of 500 mL to make a 0.25 *M* solution. We now determine how many moles 33.51 g represents. Using the FW of 268.07 g/mol, we have

$$33.51\ \text{g} \times \frac{1\ \text{mole}}{268.07\ \text{g}} = 0.125\ \text{moles}$$

Therefore, we will be placing 0.125 moles of sodium phosphate dibasic heptahydrate into a final volume of 500 mL. However, since water is part of the compound (and part of its FW), we will also be adding 0.875 moles of water in the process:

$$0.125\ \text{moles compound} \times \frac{7\ \text{moles of water}}{1\ \text{mole of compound}} = 0.875\ \text{moles of water}$$

Since the molecular weight of water is 18.015 g/mole and since one gram of water is equivalent to one mL of water, when we add the 33.51 g of sodium phosphate dibasic heptahydrate, we are also adding the equivalent of 15.76 mL of water:

$$0.875\ \text{moles} \times \frac{18.015\ \text{mL}}{\text{mole}} = 15.76\ \text{mL}$$

Therefore, to make 500 mL of 250 m*M* sodium phosphate dibasic heptahydrate, you would combine 33.51 g of compound and 484.24 mL (500 mL − 15.76 mL) of water.

As an inclusive relationship to calculate the amount of water contributed by the hydrated compound in this example, we have

$$0.5 \text{ liters} \times \frac{0.25 \text{ moles}}{\text{liter}} \times \frac{7 \text{ moles water}}{1 \text{ mole compound}} \times \frac{18.015 \text{ mL water}}{\text{mole}} = 15.76 \text{ mL}$$

The relationship above can be used as a general way to calculate the amount of water contributed by a hydrated compound, as follows:

$$V \times M \times \text{molecules of water} \times 18.015 \text{ mL water} = \text{mL water from hydrated compound}$$

where V is the final volume in liters, M is the desired molarity (in moles/liter), and 'molecules of water' is the number of water molecules attached to the compound.

Note: This calculation is not necessary as long as you weigh out the desired amount of compound you need to produce the correct molarity and then bring the volume up (q.s.) to the desired amount in a graduated cylinder.

2.5.3 Diluting molar solutions

Diluting stock solutions prepared in molar concentration into volumes of lesser molarity is performed as described in Section 2.1.

Problem 2.12 From a 1 *M* Tris solution, how is 400 mL of 0.2 *M* Tris prepared?

Solution 2.12

The following equation is used to solve for *x*, the amount of 1 *M* Tris added to 400 mL to yield a 0.2 *M* solution.

$$1\,M \times \frac{x\text{ mL}}{400\text{ mL}} = 0.2\,M$$

$$\frac{1x\,M}{400} = 0.2\,M$$

The mL units cancel since they appear in both the numerator and the denominator on the left side of the equation. Multiply numerator values.

$$x = (0.2)(400)$$

Multiply both sides of the equation by 400.

$$x = 80$$

Therefore, to 320 mL (400 mL − 80 mL = 320 mL) of distilled water, add 80 mL of 1 *M* Tris, pH 8.0, to bring the solution to 0.2 *M* and a final volume of 400 mL.

Problem 2.13 How is 4 mL of 50 m*M* NaCl solution prepared from a 2 *M* NaCl stock?

Solution 2.13

In this example, a conversion factor must by included in the equation so that molarity (*M*) can be converted to millimolarity.

$$2\,M \times \frac{1000\text{ m}M}{M} \times \frac{x\text{ mL}}{4\text{ mL}} = 50\text{ m}M$$

$$\frac{2000x\text{ m}M}{4} = 50\text{ m}M$$

Multiply numerators. (On the left side of the equation, the *M* and mL units both cancel since these terms appear in both the numerator and the denominator.)

$$2000x\text{ mM} = 200\text{ m}M$$

Multiply both sides of the equation by 4.

$$x = \frac{200\text{ m}M}{2000\text{ m}M} = 0.1$$

Divide each side of the equation by 2000 m*M*.

Therefore, to 3.9 mL of distilled water, add 0.1 mL of 2 *M* NaCl stock solution to produce 4 mL final volume of 50 m*M* NaCl.

2.5.4 Converting molarity to percent

Since molarity is a concentration of grams per 1000 mL, it is a simple matter to convert it to a percent value, an expression of a gram amount in 100 mL. The method is demonstrated in the following problem.

Problem 2.14 Express 2.5 *M* NaCl as a percent solution.

Solution 2.14

The gram molecular weight of NaCl is 58.44. The first step in solving this problem is to determine how many grams of NaCl are in a 2.5 *M* NaCl solution. This can be accomplished by using an equation of ratios: '58.44 g is to 1 *M* as *x* g is to 2.5 *M*.' This relationship is expressed mathematically as follows:

$$\frac{58.44\text{ g}}{1\,M} = \frac{x\text{ g}}{2.5\,M}$$

Solve for *x*.

$$\frac{(58.44)(2.5)}{1} = x$$ Multiply both sides of the equation by 2.5.

$$146.1 = x$$ Simplify the equation.

Therefore, to prepare a 2.5 *M* solution of NaCl, 146.1 g of NaCl are dissolved in a total volume of 1 L.

Percent is an expression of concentration in parts per 100. To determine the relationship between the number of grams of NaCl present in a 2.5 *M* NaCl solution and the equivalent percent concentration, ratios can be set up that state '146.1 g is to 1000 mL as *x* g is to 100 mL.'

$$\frac{146.1\text{g}}{1000\text{ mL}} = \frac{x\text{ g}}{100\text{ mL}}$$

Solve for *x*.

$$\frac{(146.1)(100)}{1000} = x$$ Multiply both sides of the equation by 100 (the denominator on the right side of the equation).

$$\frac{14610}{1000} = x = 14.6$$ Simplify the equation.

Therefore, a 2.5 *M* NaCl solution contains 14.6 g of NaCl in 100 mL, which is equivalent to a 14.6% NaCl solution.

2.5.5 Converting percent to molarity

Converting a solution expressed as percent to one expressed as a molar concentration is a matter of changing an amount per 100 mL to an equivalent amount per liter (1000 mL), as demonstrated in the following problem.

Problem 2.15 What is the molar concentration of a 10% NaCl solution?

Solution 2.15

A 10% solution of NaCl, by definition, contains 10 g of NaCl in 100 mL of solution. The first step to solving this problem is to calculate the amount of NaCl in 1000 mL of a 10% solution. This is accomplished by setting up a relationship of ratios as follows:

$$\frac{10\text{ g}}{100\text{ mL}} = \frac{x\text{ g}}{1000\text{ mL}}$$

Solve for *x*.

$$\frac{(10)(1000)}{100} = x$$

Multiply both sides of the equation by 1000 (the denominator on the right side of the equation).

$$\frac{10\ 000}{100} = x = 100$$

Simplify the equation.

Therefore, a 1000 mL solution of 10% NaCl contains 100 g of NaCl.

Using the gram molecular weight of NaCl (58.44), an equation of ratios can be written to determine molarity. In the following equation, we determine the molarity (*M*) equivalent to 100 g:

$$\frac{x\,M}{100\text{ g}} = \frac{1\,M}{58.44\text{ g}}$$

Solve for *x*.

$$x = \frac{100}{58.44} = 1.71$$

Multiply both sides of the equation by 100 and divide by 58.44.

Therefore, a 10% NaCl solution is equivalent to 1.71 *M* NaCl.

2.6 NORMALITY

A 1 normal (1 *N*) solution is equivalent to the gram molecular weight of a compound divided by the number of hydrogen ions present in solution (i.e., dissolved in one liter of water). For example, the gram molecular weight of hydrochloric acid (HCl) is 36.46. Since, in a solution of HCl, one H^+ ion can combine with Cl^- to form HCl, a 1 *N* HCl solution contains 36.46/1 = 36.46 g HCl in 1 L. A 1 *N* HCl solution, therefore, is equivalent to a 1 *M* HCl solution. As another example, the gram molecular weight of sulfuric acid (H_2SO_4) is 98.0. Since, in a H_2SO_4 solution, two H^+ ions can combine with SO_4^{2-} to form H_2SO_4, a 1 *N* H_2SO_4 solution contains 98.0/2 = 49.0 g of H_2SO_4 in 1 L. Since half the gram molecular weight of H_2SO_4 is used to prepare a 1 *N* H_2SO_4 solution, a 1 *N* H_2SO_4 solution is equivalent to a 0.5 *M* H_2SO_4 solution.

Normality and molarity are related by the equation

$$N = nM$$

where n is equal to the number of replaceable H^+ (or Na^+) or OH^- ions per molecule.

Problem 2.16 What is the molarity of a 1 *N* sodium carbonate (Na_2CO_3) solution?

Solution 2.16

Sodium carbonate has two replaceable Na^+ ions. The relationship between normality and molarity is

$$N = nM$$

Solving for *M* gives the following result:

$$\frac{N}{n} = M$$

Divide both sides of the equation by *n* (the number of replaceable Na^+ ions).

$$\frac{1}{2} = M$$

In this problem, the number of replaceable + ions, *n*, is 2. The normality, *N*, is 1.

$$0.5 = M$$

Therefore, a 1 *N* sodium carbonate solution is equivalent to a 0.5 *M* sodium carbonate solution.

2.7 pH

The first chemical formula most of us learn, usually during childhood, is that for water, H_2O. A water molecule is composed of two atoms of hydrogen and one atom of oxygen. The atoms of water, however, are only transiently associated in this form. At any particular moment, a certain number of water molecules will be dissociated into hydrogen (H^+) and hydroxyl (OH^-) ions. These ions will reassociate within a very short time to form the H_2O water molecule again. The dissociation and reassociation of the atoms of water can be depicted by the following relationship:

$$H_2O \rightleftharpoons H^+ + OH^-$$

In actuality, the hydrogen ion is donated to another molecule of water to form a hydronium ion (H_3O^+). A more accurate representation of the dissociation of water, therefore, is as follows:

$$H_2O + H_2O \rightleftharpoons H_3O^+ + OH^-$$

For most calculations in chemistry, however, it is simpler and more convenient to think of the H^+ hydrogen ion (rather than the H_3O^+ hydronium ion) as being the dissociation product of H_2O.

A measure of the hydrogen ion (H^+) concentration in a solution is given by its **pH** value. A solution's pH is defined as the negative logarithm to the base 10 of its hydrogen ion concentration:

$$\mathrm{pH} = -\log[H^+]$$

In this nomenclature, the brackets signify concentration. A **logarithm (log)** is an exponent, a number written above, smaller, and to the right of another number, called the **base**, to which the base should be raised. For example, for 10^2, the 2 is the exponent and the 10 is the base. In 10^2, the base, 10, should be raised to the second power, which means that 10 should be multiplied by itself a total of two times. This will give a value of 100, as shown here:

$$10^2 = 10 \times 10 = 100$$

The log of 100 is 2 because that is the exponent of 10 that yields 100. The log of 1000 is 3 since $10^3 = 1000$:

$$10^3 = 10 \times 10 \times 10 = 1000$$

The log of a number can be found on most calculators by entering the number and then pressing the **log** key.

In pure water, the H^+ concentration ($[H^+]$) is equal to $10^{-7} M$. In other words, in 1 L of water, 0.000 000 1 M of hydrogen ion will be present. The pH of water, therefore, is calculated as follows:

$$pH = -\log(1 \times 10^{-7}) = -(-7) = 7$$

Pure water, therefore, has a pH of 7.0.

Likewise, the concentration of the hydroxyl (OH^-) ion in water is equivalent to that of the hydrogen (H^+) concentration. The OH^- concentration is also $1 \times 10^{-7} M$.

pH values range from 0 to 14. Solutions having pH values less than 7 are acidic. Solutions having pH values greater than 7 are alkaline, or basic. Water, with a pH of 7, is considered a neutral solution.

In more specific terms, an **acid** is defined as a substance that donates a proton (or hydrogen ion). When an acid is added to pure water, the hydrogen ion concentration increases above $1 \times 10^{-7} M$. The molecule to which the proton is donated is called the **conjugate base**. When an acid loses a proton, it becomes ionized (charged). The loss of a proton leaves the molecule with a negative charge. A substance is referred to as a 'strong' acid if it becomes almost completely ionized in water – its H^+ ion almost completely dissociates. For example, the strong acid, HCl dissociates as follows:

$$HCl \rightleftharpoons H^+ + Cl^-$$

The hydrogen ion of a 'weak' acid dissociates but a little.

A **base** is a substance that accepts a proton. A 'strong' base is one that almost completely ionizes in water to give OH^- ions. Sodium hydroxide, for example, is a strong base and dissociates in water as follows:

$$NaOH \rightleftharpoons Na^+ + OH^-$$

When a base is added to pure water, the OH^- ion is dissociated from the base. This hydroxyl ion can associate with the H^+ ions already in the water to form H_2O molecules, reducing the solution's hydrogen ion concentration and increasing the solution's pH.

Water is a unique molecule. Since it can dissociate into H^+ *and* OH^- ions, it acts as both an acid and as a base. We can define the dissociation of the two types of ions in water as a constant called $\mathbf{K}_w$. Since both the H^+ and OH^- ions in water have a concentration of $1 \times 10^{-7} M$, the dissociation constant of water is equal to 1×10^{-14}.

$$[H^+][OH^-] = [1 \times 10^{-7}][1 \times 10^{-7}] = 1 \times 10^{-14}$$

We have, therefore

$$K_w = [H^+][OH^-]$$

We can rearrange this equation such that

$$[H^+] = \frac{K_w}{[OH^-]}$$

or

$$[OH^-] = \frac{K_w}{[H^+]}$$

Just as water has a pH, so does it have a **pOH**, which is defined as the negative logarithm of the OH^- concentration.

$$pOH = -\log[OH]$$

and

$$pH + pOH = 14$$

or, rearranging

$$pOH = 14 - pH$$

Therefore, by knowing either the pH or the pOH, the counterpart value can be calculated.

Problem 2.17 A solution of a strong acid has a hydrogen ion concentration of 1×10^{-5} *M*. What is the solution's pH?

Solution 2.17

pH is the negative logarithm of 1×10^{-5}.

$$pH = -\log[H^+] = -\log(1 \times 10^{-5}) = -(-5) = 5$$

Therefore, the pH of the solution is 5. It is acidic.

Problem 2.18 The hydrogen ion concentration in a solution of a strong acid is $2.5 \times 10^{-4} M$. What is the solution's pH?

Solution 2.18

The **Product Rule for Logarithms** states that, for any positive numbers *M*, *N*, and *a* (where *a* is not equal to 1), the logarithm of a product is the sum of the logarithms of the factors:

$$\log_a MN = \log_a M + \log_a N$$

Since we are working in base 10, *a* is equal to 10.

The Product Rule of Logarithms will be used to solve this problem.

$$\begin{aligned} pH &= -\log(2.5 \times 10^{-8}) \\ &= -(\log 2.5 + \log 10^{-5}) \\ &= -[0.40 + (-5)] \\ &= -(0.40 - 5) \\ &= -(-4.6) \\ &= 4.6 \end{aligned}$$

Therefore, the solution has a pH of 4.6. Although the math shown here demonstrates a fundamental concept of logarithms as revealed by the Product Rule and therefore has worth in any instruction in mathematics, the answer can be determined straight away by taking the log value of 2.5×10^{-8} on a calculator.

Problem 2.19 The pH of a solution of strong acid is 3.75. What is the hydrogen ion concentration in the solution?

Solution 2.19

To calculate the hydrogen ion concentration in this problem will require that we determine the **antilog** of the pH. An antilog is found by doing the reverse process of that used to find a logarithm. The log of 100 is 2. The antilog of 2 is 100. The log of 1000 is 3. The antilog of 3 is 1000. For those calculators that do not have an antilog key, this can usually be obtained by entering the value, pressing the **10^x** key, and then pressing the = sign. (Depending on the type of calculator you are using, you may need to press the **SHIFT** key to gain access to the **10^x** function.)

$pH = -\log[H^+]$	Equation for calculating pH.
$-\log[H^+] = 3.75$	The pH is equal to 3.75.
$\log[H^+] = -3.75$	Multiply each side of the equation by −1.
$[H^+] = 1.8 \times 10^{-4}$	Take the antilog of each side of the equation. **Note:** Taking the antilog of the log of a number, since they are opposite and canceling operations, is equivalent to doing nothing to that number. For example, the antilog of the log of 100 is 100.

Therefore, the hydrogen ion concentration is $1.8 \times 10^{-4} M$.

Problem 2.20 A solution has a pH of 4.5. What is the solution's pOH?

Solution 2.20

The pOH is obtained by subtracting the pH from 14.

$$pOH = 14 - pH$$
$$pOH = 14 - 4.5 = 9.5$$

Therefore, the pOH of the solution is 9.5.

Problem 2.21 What are the OH^- concentration and the pH of a 0.01 *M* solution of HCl?

Solution 2.21

Since HCl is a strong acid, all hydrogen atoms will dissociate into H^+ ions. The H^+ ion concentration, therefore, is equivalent to the molar concentration; that is, $1 \times 10^{-2} M$. We can then use the following relationship to determine the OH^- ion concentration.

$$[OH^-] = \frac{K_w}{[H^+]}$$

$$[OH^-] = \frac{1 \times 10^{-14}}{1 \times 10^{-2}}$$

$$= 1 \times 10^{-14-(-2)}$$

$$= 1 \times 10^{-12}$$

Therefore, a 0.01 *M* solution of HCl has a OH^- concentration of 1×10^{-12} *M*. Its pH is calculated as

$$pH = -\log[H^+] = -\log(1 \times 10^{-2}) = -(-2) = 2$$

Or, calculated another way

$$pOH = -\log[OH^-] = -\log[1 \times 10^{12}] = 12$$

and, since pH = 14 − pOH, we have

$$pH = 14 - 12 = 2$$

The solution has a pH of 2.0.

Problem 2.22 What is the pH of a 0.02 *M* solution of sodium hydroxide (NaOH)?

Solution 2.22

Sodium hydroxide is a strong base and, as such, is essentially ionized completely to Na^+ and OH^- in dilute solution. The OH^- concentration, therefore, is 0.02 *M*, the same as the concentration of NaOH. For a strong base, the H^+ ion contribution from water is negligible and so will be ignored. The first step to solving this problem is to determine the pOH. The pOH value will then be subtracted from 14 to obtain the pH.

$$\begin{aligned} pOH &= -\log(0.02) \\ &= -(-1.7) = 1.7 \\ pH &= 14 - 1.7 = 12.3 \end{aligned}$$

Therefore, the pH of the 0.02 *M* NaOH solution is 12.3.

2.8 pK_a AND THE HENDERSON–HASSELBALCH EQUATION

In the Bronsted concept of acids and bases, an **acid** is defined as a substance that donates a proton (a hydrogen ion). A **base** is a substance that accepts a proton. When a Bronsted acid loses a hydrogen ion, it becomes a

Bronsted base. The original acid is called a **conjugate acid**. The base created from the acid by loss of a hydrogen ion is called a **conjugate base**.

Dissociation of an acid in water follows the general formula

$$HA + H_2O \rightleftharpoons H_3O^+ + A^-$$

where HA is a conjugate acid, H_2O is a conjugate base, H_3O^+ is a conjugate acid, and A^- is a conjugate base.

The acid's ionization can be written as a simple dissociation, as follows:

$$HA \rightleftharpoons H^+ + A^-$$

The dissociation of the HA acid will occur at a certain rate characteristic of the particular acid. Notice, however, that the arrows go in both directions. The acid dissociates into its component ions, but the ions come back together again to form the original acid. When the rate of dissociation into ions is equal to the rate of ion reassociation, the system is said to be in **equilibrium**. A strong acid will reach equilibrium at the point where it is completely dissociated. A weak acid will have a lower percentage of molecules in a dissociated state and will reach equilibrium at a point less than 100% ionization. The concentration of acid at which equilibrium occurs is called the **acid dissociation constant**, designated by the symbol $\boldsymbol{K_a}$. It is represented by the following equation:

$$K_a = \frac{[H^+][A^-]}{[HA]}$$

A measure of K_a for a weak acid is given by its pK_a, which is equivalent to the negative logarithm of K_a:

$$pK_a = -\log K_a$$

pH is related to pK_a by the **Henderson–Hasselbalch equation**:

$$\begin{aligned} pH &= pK_a + \log\frac{[\text{conjugate base}]}{[\text{acid}]} \\ &= pK_a + \log\frac{[A^-]}{[HA]} \end{aligned}$$

The Henderson–Hasselbalch equation can be used to calculate the amount of acid and conjugate base to be combined for the preparation of a buffer solution having a particular pH, as demonstrated in the following problem.

Problem 2.23 You wish to prepare 2 L of 1 *M* sodium phosphate buffer, pH 8.0. You have stocks of 1 *M* monobasic sodium phosphate (NaH_2PO_4) and 1 *M* dibasic sodium phosphate (Na_2HPO_4). How much of each stock solution should be combined to make the desired buffer?

Solution 2.23

Monobasic sodium phosphate (NaH_2PO_4) in water exists as Na^+ and $H_2PO_4^-$ ions. $H_2PO_4^-$ (phosphoric acid, the conjugate acid) dissociates further to HPO_4^{2-} (the conjugate base) + H^+ and has a pK_a of 6.82 at 25°C. (pK_a values can be found in the Sigma chemical catalogue (Sigma, St. Louis, MO) or in *The CRC Handbook of Chemistry and Physics*, David P. Lide, Editor-in-Chief, 87th Edition, 2006, CRC Press, Boca Raton, Florida, USA. The pH and pK_a values will be inserted into the Henderson–Hasselbalch equation to derive a ratio of the conjugate base and acid to combine to give a pH of 8.0. Note that the stock solutions are both at a concentration of 1 *M*. No matter in what ratio the two solutions are combined, there will always be one mole of phosphate molecules per liter.

$$\text{pH} = \text{p}K_a + \log\frac{[\text{A}^-]}{[\text{HA}]}$$

$$8.0 = 6.82 + \log\frac{[\text{HPO}_4^{2-}]}{[\text{H}_2\text{PO}_4^-]}$$

Insert pH and pK_a values into Henderson–Hasselbalch equation.

$$1.18 = \log\frac{[\text{HPO}_4^{2-}]}{[\text{H}_2\text{PO}_4^-]}$$

Subtract 6.82 from both sides of the equation.

$$\text{antilog}\,1.18 = \frac{[\text{HPO}_4^{2-}]}{[\text{H}_2\text{PO}_4^-]} = 15.14$$

Take the antilog of each side of the equation.

Therefore, the ratio of HPO_4^{2-} to $H_2PO_4^-$ is equal to 15.14. To make 1 *M* sodium phosphate buffer, 15.14 parts Na_2HPO_4 should be combined with one part NaH_2PO_4; 15.14 parts Na_2HPO_4 plus one part NaH_2PO_4 is equal to a total of 16.14 parts. The amount of each stock to combine to make 2 L of the desired buffer is then calculated as follows:

For Na_2HPO_4, the amount is equal to

$$\frac{15.14}{16.14} \times 2\text{ L} = 1.876\text{ L}$$

For NaH_2PO_4, the amount is equal to

$$\frac{1}{16.14} \times 2\text{ L} = 0.124\text{ L}$$

When these two volumes are combined, you will have 1 *M* sodium phosphate buffer having a pH of 8.0.

CHAPTER SUMMARY

Concentration is defined as an amount of some substance per a set volume:

$$\text{concentration} = \frac{\text{amount}}{\text{volume}}$$

Dilutions can be calculated using the $C_1V_1 = C_2V_2$ equation or by the **dimensional analysis** method. The molecular weight of a compound is calculated as the sum of the atomic weights of its component molecules. A **1 molar** (1 *M*) solution of a chemical contains the molecular weight of the substance dissolved in a total volume of 1 L.

A general way to calculate the amount of water contributed by a hydrated compound when preparing molar solutions from hydrated compounds is given by the relationship

$$V \times M \times \text{molecules of water} \times 18.015\text{mL water} = \text{mL water from hydrated compound}$$

where V is the final volume in liters, M is the desired molarity (in moles/liter), and 'molecules of water' is the number of water molecules attached to the compound.

A **1 normal** (1 *N*) solution of a chemical is equivalent to its gram molecular weight divided by the number of hydrogen ions present in the solution. pH is a measure of the hydrogen ion concentration in a solution and is calculated as the negative logarithm of the solution's hydrogen ion concentration.

$$\text{pH} = -\log[\text{H}^+]$$

The pK_a of an acid is equivalent to the negative logarithm of its dissociation constant (its K_a), calculated as

$$K_a = \frac{[\text{H}^+][\text{A}^-]}{[\text{HA}]}$$

Chapter 3

Cell growth

3.1 THE BACTERIAL GROWTH CURVE

Where bacteria or other unicellular organisms are concerned, **cell growth** refers to cell division and the increase in cell quantity rather than to the size of an individual cell. It's about numbers. The rate at which bacterial cells divide in culture is influenced by several factors, including the level of available nutrients, the temperature at which the cells are incubated, and the degree of aeration. Depending on their genotypes, different strains of bacteria may also have different growth rates in any particular defined medium.

An understanding of the characteristics of cell growth is important for several applications in molecular biology and biotechnology:

- For the most efficient and reproducible transformations of *Escherichia coli* by recombinant plasmid, it is important that the cells be harvested and prepared at a particular point in their growth, at a particular cell concentration.
- *E. coli* is most receptive to bacteriophage infection during a certain period of growth.
- The highest yield of a cellular or recombinant protein can be achieved during a particular window of cell growth.

A common procedure for determining the rate of cell growth is to start with a large volume (50 to 100 mL) of a defined medium containing a small inoculum (1 mL) of cells from a liquid culture that has been grown over the previous night. The point at which the larger culture is inoculated is considered to be time zero. The culture is then incubated at the proper temperature (37°C with aeration for most laboratory strains). At various times, a sample of the culture is withdrawn and its optical density (OD) is determined. That sample is then diluted to a concentration designed to give easily countable, well-isolated colonies when spread onto Petri plates containing a solid agar-based medium. The plates are incubated overnight and the number of colonies is counted. If colonies on the plate are well separated, it can be assumed that each colony arises from a single viable

Calculations for Molecular Biology and Biotechnology. DOI: 10.1016/B978-0-12-375690-9.00003-6
© 2010 Elsevier Inc. All rights reserved.

cell. The OD at the various sampling times can then be correlated with cell number and a growth rate can be derived. For subsequent experiments, if all conditions are carefully reproduced, the number of cells present at any particular time during incubation can be extrapolated by measuring the culture's OD.

As an example of the type of math involved in this procedure, imagine that a growth experiment has been set up as just described in which a 100 mL volume of tryptone broth is inoculated with 0.5 mL of overnight culture of *E. coli*. At the time of inoculation, a 0.75 mL sample is withdrawn and its OD at 550 nm (OD_{550}) is determined. (Beyond a certain OD_{550} reading, an increase in cell density no longer results in a linear increase in absorbance. For many spectrophotometers, this value is an OD_{550} of 1.5. Extremely dense cultures, therefore, should be diluted to keep the absorbance at 550 nm below 1.5.)

Following the reading of a sample's absorbance, a **serial dilution** is then performed by transferring 0.1 mL from the 0.75 mL sample into 9.9 mL of tryptone broth. That diluted sample is vortexed, and 0.2 mL is withdrawn and diluted into a second tube, containing 9.8 mL of tryptone broth. The second diluted sample is vortexed to ensure proper mixing and 0.1 mL is withdrawn and spread onto an agar plate. Following overnight incubation of that plate, 420 colonies are counted.

Problem 3.1 In the example just given, what is the dilution of cells in the second tube?

Solution 3.1

The dilutions can be represented as fractions. A 0.1 mL aliquot diluted into 9.9 mL can be written as the fraction $^{0.1}/_{10}$, where the numerator is the volume of sample being transferred (0.1 mL) and the denominator is equal to the final volume of the dilution (0.1 mL + 9.9 mL = 10.0 mL). Multiple dilutions are multiplied together to give the dilution in the final sample. This approach can be simplified if the fractions are converted to scientific notation (see Chapter 1 for a discussion of how to convert fractions to scientific notation):

$$\frac{0.1\,\text{mL}}{10\,\text{mL}} \times \frac{0.2\,\text{mL}}{10\,\text{mL}} = (1 \times 10^{-2})(2 \times 10^{-2})$$

$$= 2 \times 10^{-4}$$

Therefore, the sample in the second tube has been diluted to 2×10^{-4} (see Figure 3.1).

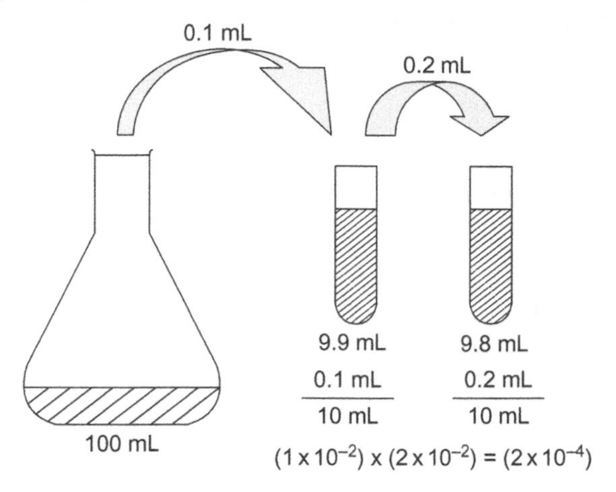

■ **FIGURE 3.1** The dilution series described in Problem 3.1 in which 0.1 mL from a 100 mL culture is diluted into a final volume of 10 mL and 0.2 mL from that dilution is dispensed into a final volume of 10 mL. These dilutions, when considered as scientific notation, show that an overall dilution of 2×10^{-4} was made.

Problem 3.2 What is the dilution of cells spread onto the agar plate?

Solution 3.2

One-tenth mL of the dilution is spread on the plate. This gives

$$\frac{0.1\,\text{mL}}{10\,\text{mL}} \times \frac{0.2\,\text{mL}}{10\,\text{mL}} \times 0.1\,\text{mL} = (1 \times 10^{-2})(2 \times 10^{-2})(1 \times 10^{-1}\,\text{mL})$$
$$= 2 \times 10^{-5}\,\text{mL}$$

Therefore, a dilution of 2×10^{-5} mL is spread on the agar plate (Figure 3.2). Note that this amount is equivalent to withdrawing 0.000 02 mL (20 nL)

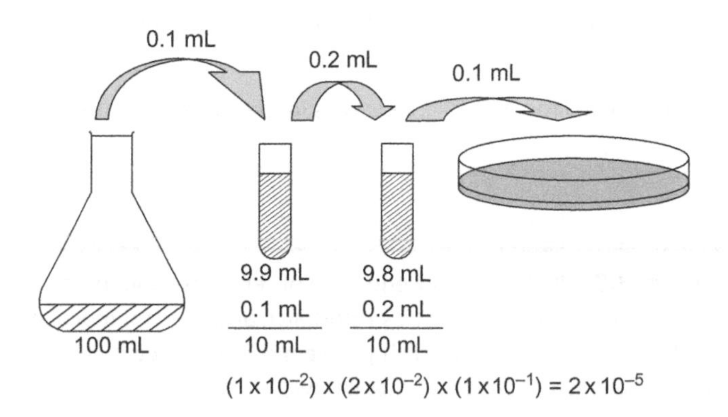

■ **FIGURE 3.2** The dilution and plating protocol for Problem 3.2.

directly from the 100 mL culture for plating. Performing dilutions, therefore, allows the experimenter to withdraw, in manageable volumes, what is essentially a very small amount of a solution.

Problem 3.3 What is the concentration of viable cells in the 100 mL culture immediately following inoculation?

Solution 3.3

Each colony on the spread plate can be assumed to have arisen from a viable cell. To determine the number of viable cells in the 100 mL culture, the number of colonies counted on the spread plate following overnight incubation is divided by the dilution:

$$\frac{420 \text{ cells}}{2 \times 10^{-5} \text{ mL}} = \frac{4.2 \times 10^{2} \text{ cells}}{2 \times 10^{-5} \text{ mL}} = 2.1 \times 10^{7} \text{ cells/mL}$$

Therefore, there are 2.1×10^7 cells/mL in the 100 mL culture after inoculation.

Problem 3.4 In total, how many viable cells are present in the 100 mL culture?

Solution 3.4

Multiply the cell concentration by the culture volume:

$$\frac{2.1 \times 10^{7} \text{ cells}}{\text{mL}} \times 100 \text{ mL} = 2.1 \times 10^{9} \text{ total cells}$$

Therefore, there are 2.1×10^9 viable cells in the 100 mL culture.

Problem 3.5 What is the concentration of cells in the overnight culture from which the 0.5 mL sample was withdrawn and used for inoculating the 100 mL of growth media? (This problem can be solved by two methods, shown as 3.5(a) and 3.5(b) below.)

Solution 3.5(a)

In Problem 3.4, it was determined that the 100 mL culture, at time zero, contained 2.1×10^9 cells. Those cells came from 0.5 mL of the overnight culture, or

$$\frac{2.1 \times 10^9 \text{ cells}}{0.5 \text{ mL}} = \frac{2.1 \times 10^9 \text{ cells}}{5 \times 10^{-1} \text{ mL}} = 4.2 \times 10^9 \text{ cells/mL}$$

Solution 3.5(b)

An equation can be written that describes the dilution of the overnight culture (0.5 mL of cells of unknown concentration (x cells/mL) transferred into 100 mL of fresh tryptone broth to give a concentration of 2.1×10^7 cells/mL):

$$\frac{x \text{ cells}}{\text{mL}} \times \frac{0.5 \text{ mL}}{100 \text{ mL}} = \frac{2.1 \times 10^7 \text{ cells}}{\text{mL}}$$

Now we solve for x. In the 0.5 mL/100 mL fraction, the mL terms cancel since they are present in both the numerator and the denominator. Multiplying numerators on the left side of the equation and multiplying both sides of the equation by 100 (written in scientific notation as 1×10^2) gives

$$\frac{0.5x \text{ cells}}{\text{mL}} = \frac{(2.1 \times 10^7 \text{ cells})(1 \times 10^2)}{\text{mL}} = \frac{2.1 \times 10^9 \text{ cells}}{\text{mL}}$$

Dividing each side of the equation by 0.5 (written in scientific notation as 5×10^{-1}) yields the following equation:

$$\frac{x \text{ cells}}{\text{mL}} = \frac{2.1 \times 10^9 \text{ cells}}{5 \times 10^{-1} \text{ mL}} = \frac{0.42 \times 10^{9-(-1)} \text{ cells}}{\text{mL}} = \frac{4.2 \times 10^9 \text{ cells}}{\text{mL}}$$

Therefore, the overnight culture has a concentration of 4.2×10^9 cells/mL.

■

3.1.1 Sample data

For the example just described, at 1 hr intervals, 0.75 mL samples are withdrawn, their optical densities are determined, and each sample is diluted and plated to determine viable cell count. The dataset in Table 3.1 is obtained.

Table 3.1 Sample data for Problems 3.6 through 3.15.

Hours after inoculation	OD_{550}	Cells/mL
0	0.008	2.1×10^7
1	0.020	4.5×10^7
2	0.052	1.0×10^8
3	0.135	2.6×10^8
4	0.200	4.0×10^8
5	0.282	5.8×10^8
6	0.447	9.6×10^8
7	0.661	1.5×10^9
8	1.122	2.0×10^9

3.2 MANIPULATING CELL CONCENTRATION

The following problems demonstrate how cell cultures can be manipulated by dilutions to provide desired cell concentrations.

Problem 3.6 At $t = 3$ hours (hr), the cell density is determined to be 2.6×10^8 cells/mL. If 3.5 mL of culture is withdrawn at that time, the aliquot is centrifuged to pellet the cells, and the pellet is then resuspended in 7 mL of tryptone broth, what is the new cell concentration?

Solution 3.6

It is first necessary to determine the number of cells in the 3.5 mL aliquot. This is done by multiplying the cell concentration (2.6×10^8 cells/mL) by the amount withdrawn (3.5 mL). The following equation is used to solve for x, the number of cells in 3.5 mL:

$$\frac{2.6 \times 10^8 \text{ cells}}{\text{mL}} \times 3.5\,\text{mL} = x \text{ cells}$$

Solving for x yields

$$\frac{(2.6 \times 10^8 \text{ cells}) \times (3.5\,\text{mL})}{\text{mL}} = x \text{ cells}$$

$$9.1 \times 10^8 \text{ cells} = x \text{ cells}$$

Therefore, there are 9.1×10^8 cells in the 3.5 mL aliquot. When these 9.1×10^8 cells are resuspended in 7 mL, the new concentration can be calculated as follows:

$$\frac{x \text{ cells}}{\text{mL}} = \frac{9.1 \times 10^8 \text{ cells}}{7 \text{ mL}} = \frac{1.3 \times 10^8 \text{ cells}}{\text{mL}}$$

Therefore, the cells resuspended in 7 mL of tryptone broth have a concentration of 1.3×10^8 cells/mL.

Problem 3.7 In Problem 3.6, 3.5 mL of culture at 2.6×10^8 cells/mL is pelleted by centrifugation. Into what volume should the pellet be resuspended to obtain a cell concentration of 4.0×10^9 cells/mL?

Solution 3.7

In Problem 3.6, it was determined that there are a total of 9.1×10^8 cells in a 3.5 mL sample of a culture having a concentration of 2.6×10^8 cells/mL. The following equation can be used to determine the volume required to resuspend the centrifuged pellet to obtain a concentration of 4.0×10^9 cells/mL:

$$\frac{9.1 \times 10^8 \text{ cells}}{x \text{ mL}} = \frac{4.0 \times 10^9 \text{ cells}}{\text{mL}}$$

Now solve for *x*. Multiply both sides of the equation by *x*:

$$9.1 \times 10^8 \text{ cells} = \frac{(4.0 \times 10^9 \text{ cells}) \times (x \text{ mL})}{\text{mL}}$$

Divide each side of the equation by 4.0×10^9 cells:

$$\frac{9.1 \times 10^8 \text{ cells}}{4.0 \times 10^9 \text{ cells}} = x = 0.23$$

Therefore, to obtain a concentration of 4.0×10^9 cells/mL, the pelleted cells should be resuspended in 0.23 mL of tryptone broth.

Problem 3.8 At six hours following inoculation, the cell concentration is 9.6×10^8 cells/mL. If a sample is withdrawn at this time, how should it be diluted to give 300 colonies when spread on an agar plate?

Solution 3.8

Since each colony on a plate is assumed to arise from an individual cell, the first step to solving this problem is to determine how many milliliters of culture contains 300 cells. The following equation will provide this answer. The equation is written so that the mL terms cancel.

$$\frac{9.6 \times 10^8 \text{ cells}}{\text{mL}} \times x \text{ mL} = 300 \text{ cells}$$

Solving for x gives

$$x = \frac{3 \times 10^2 \text{ cells}}{9.6 \times 10^8 \text{ cells}} = 3.1 \times 10^{-7}$$

Therefore, 300 cells are contained in 3.1×10^{-7} mL of culture at $t = 6$ hr. Assume that 0.1 mL of the dilution series to be made will be plated. To perform an equivalent dilution to 2.1×10^{-7} mL, the following serial dilution can be performed.

$$(1 \times 10^{-2})(1 \times 10^{-2})(1 \times 10^{-1})(3.1 \times 10^{-1})(1 \times 10^{-1}) = 3.1 \times 10^{-7}$$

This is equivalent to performing the following dilution series and spreading 0.1 mL on an agar plate (see Figure 3.3):

$$\frac{0.1 \text{ mL}}{10 \text{ mL}} \times \frac{0.1 \text{ mL}}{10 \text{ mL}} \times \frac{1.0 \text{ mL}}{10 \text{ mL}} \times \frac{3.1 \text{ mL}}{10 \text{ mL}} \times 0.1 \text{ mL}$$

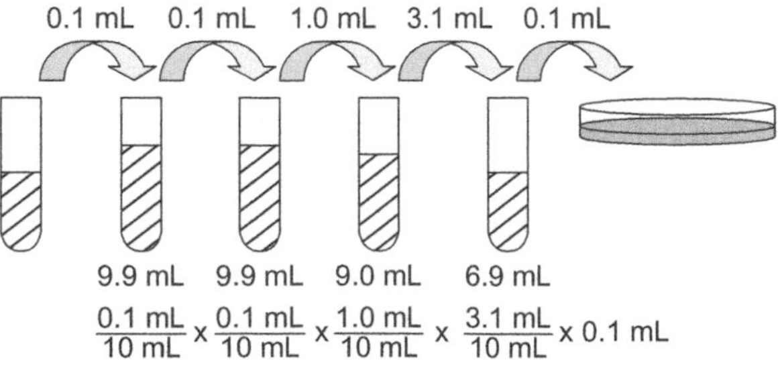

■ **FIGURE 3.3** The dilution series described in Problem 3.8.

Note: There is no one correct way to determine a dilution series that will achieve the dilution you want. The volumes you use for the dilutions are dictated by several parameters, including the amount of dilution buffer available, the accuracy of the pipettes, and the size of the available dilution tubes.

3.3 PLOTTING OD_{550} VS. TIME ON A LINEAR GRAPH

A linear plot of the Table 3.1 sample data, in which OD_{550} is placed on the vertical (y) axis and time on the horizontal (x) axis, will produce the curve shown in Figure 3.4.

When cells from an overnight culture are inoculated into fresh medium, as described in this chapter, they experience a period during which they adjust to the new environment. Very little cell division occurs during this time. This period, called the **lag phase**, may last from minutes to hours, depending on the strain and the richness of the medium. The culture then enters a period during which rapid cell division occurs. During this period, each cell in the population gives rise to two daughter cells. These two daughter cells give rise to four cells, those four to eight, etc. Because of the continuous doubling in cell number, this period of growth is called the **exponential** or **logarithmic (log) phase**. As the supply of nutrients is consumed and as inhibitory waste products accumulate, the rate of cell division slows. At this point, the culture enters the **stationary phase**, during

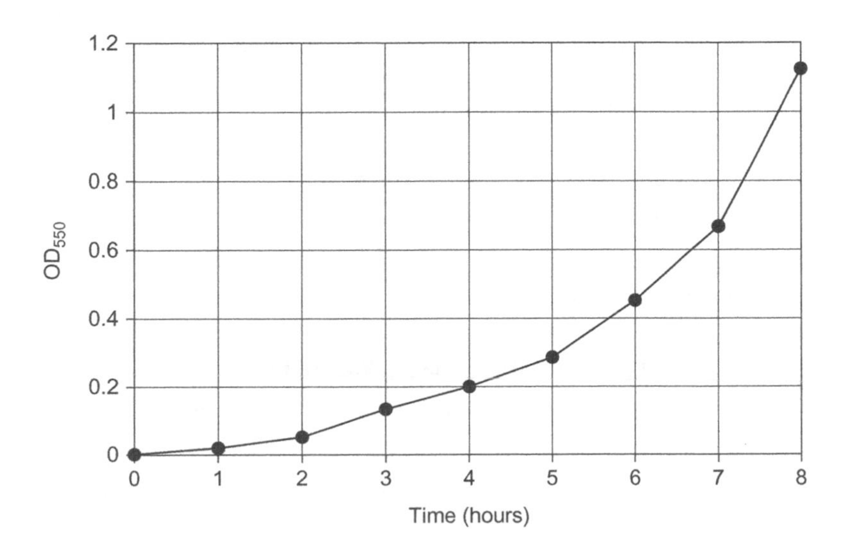

FIGURE 3.4 Plot of OD_{550} vs. time for sample data from Table 3.1.

which time cell death and cell growth occur at equivalent rates. The overall viable cell number during this stage remains constant. (For the data plotted in the experiment outlined earlier, the graph is seen to be entering this period.) Further depletion of nutrients leads to increased cell death. As more cells die than divide, the culture enters the **death phase**. Most strains have entered the cell death phase by 24 hr.

The phases of cell growth are not easily distinguishable in the plot in Figure 3.4. They become more discernible in a logarithmic plot (see Section 3.5).

3.4 PLOTTING THE LOGARITHM OF OD_{550} VS. TIME ON A LINEAR GRAPH

3.4.1 Logarithms

An **exponent** is a number written above and to the right of (and smaller than) another number, called the **base**, to indicate the power to which the base is to be raised. A **logarithm** (**log**) is an exponent. Most mathematics in molecular biology and in most basic sciences are performed using base 10, the decimal system for naming numbers (from 0 to 9) at positions around the decimal point. Since this is the 'commonly used' system, logarithms of base 10 numbers are called **common logarithms**, and, unless specified otherwise, the base for a logarithm is assumed to be 10. The log, base 10, of a number *n* is the exponent to which 10 must be raised in order to get *n*. For example, $10^2 = 100$. Here, the exponent is 2 and the base is 10. The log of 100, therefore, is equal to the exponent 2. The log of 1000 is 3, since $10^3 = 1000$. The log of 10 is 1, since $10^1 = 10$. The log of 20 is 1.3, since $10^{1.3} = 20$. Log values can be found in a log table or, on a calculator, by entering the *n* value and then pressing the **log** key. Plotting the logarithm of the OD_{550} readings of the bacterial culture vs. time should result in a linear plot for those readings taken during the logarithmic phase of cell growth.

3.4.2 Sample OD_{550} data converted to logarithm values

From the sample data (Table 3.1), the log values obtained for the OD_{550} readings are as given in Table 3.2.

3.4.3 Plotting logarithm OD_{550} vs. time

When the log of the OD_{550} readings vs. time are plotted, the curve in Figure 3.5 is generated.

In Figure 3.5, the plot is linear from zero to three hours after inoculation. This is the period of exponential growth. An inflection in the line occurs

Table 3.2 OD_{550} readings from Table 3.1 converted to log values.

Hours after inoculation	OD_{550}	Log of OD_{550}
0	0.008	−2.10
1	0.020	−1.70
2	0.052	−1.28
3	0.135	−0.87
4	0.200	−0.70
5	0.282	−0.55
6	0.447	−0.35
7	0.661	−0.18
8	1.122	0.05

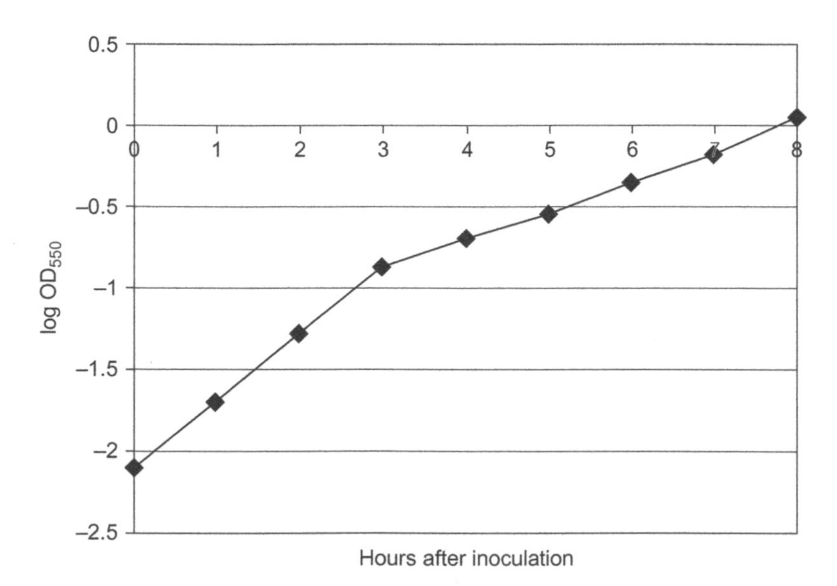

■ **FIGURE 3.5** Plot of log OD_{550} vs. incubation time for sample data.

between three and four hours. This represents the end of exponential growth and the very beginning of stationary phase. This type of plot can give a better visual representation of the length of the exponential phase of growth and the point at which growth enters the stationary phase than that provided by the linear plot of OD_{550} vs. time shown in Figure 3.1.

3.5 PLOTTING THE LOGARITHM OF CELL CONCENTRATION VS. TIME

Plotting the log of cell concentration vs. time should give a plot similar to that obtained when plotting the log of OD_{550} vs. time.

3.5.1 Determining logarithm values

Log values can be obtained on a calculator by entering a number and then pressing the ***log*** key. The log values of the cell concentrations at various times after inoculation are shown in Table 3.3.

When the log of cell concentration vs. hours after inoculation is plotted, the graph shown in Figure 3.6 is generated.

Table 3.3 Log of cell concentration for Table 3.1 sample data.

Hours after inoculation	Cells/mL	Log of cell concentration
0	2.1×10^7	7.32
1	4.5×10^7	7.65
2	1.0×10^8	8.00
3	2.6×10^8	8.41
4	4.0×10^8	8.60
5	5.8×10^8	8.76
6	9.6×10^8	8.98
7	1.5×10^9	9.18
8	2.0×10^9	9.30

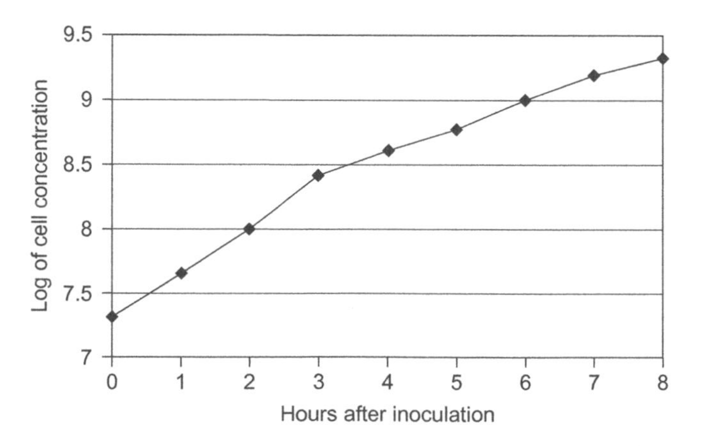

■ **FIGURE 3.6** Data from Table 3.3 graphed on a linear plot.

3.6 CALCULATING GENERATION TIME

3.6.1 Slope and the growth constant

The **generation time** of cells in culture is defined as the time required for the cell number to double. When the log of cell concentration vs. time is plotted on linear graph paper, a line is generated rising from left to right. The vertical rise on the y axis divided by the horizontal length on the x axis over which the vertical rise occurs is the **slope** of the line. Slope, stated in a more popular way, is 'the rise over the run.' The slope of the line during exponential growth is designated as $\boldsymbol{K}$, the growth constant.

The increase in the number of cells (N) per unit time (t) is proportional to the number of cells present in the culture at the beginning of the time interval (N_0). This relationship is expressed by the formula

$$\log(N) = \log(N_0) + Kt$$

Subtracting $\log(N_0)$ from both sides of the equation and dividing both sides of the equation by t yields the slope of the line, K:

$$\frac{\log(N) - \log(N_0)}{t} = K$$

Problem 3.9 What is the slope (the K value) for the sample data in Table 3.1?

Solution 3.9

The preceding equation will be used. In the numerator, the log of an earlier cell concentration is subtracted from the log of a later cell concentration. An interval during the exponential phase of growth should be used. For example, since exponential growth is occurring during the time from zero to three hours following inoculation, values obtained for the time interval from $t = 0$ to $t = 1$ or from $t = 1$ to $t = 2$ or from $t = 2$ to $t = 3$ can be used. A number closest to the actual K value can be obtained by using the values calculated from the $t = 0$ and $t = 3$ time points (a total of 180 minutes (min)):

$$\frac{\log(N) - \log(N_0)}{t} = K$$

The cell concentration at $t = 3$ hr (180 min) is 2.6×10^8 cells/mL. The cell concentration at $t = 0$ is 2.1×10^7 cells/mL. These values are entered into the equation for calculating K.

$$\frac{\log(2.6 \times 10^8) - \log(2.1 \times 10^7)}{180\,\text{min}} = K$$

Substitute the log values for N and N_0 into the equation.

$$\frac{8.41 - 7.32}{180\,\text{min}} = K$$

Simplify the equation.

$$\frac{1.09}{180\,\text{min}} = K = \frac{0.0061}{\text{min}}$$

Therefore, the K value for this dataset if 0.0061/min.

$$\frac{8.41 - 7.32}{180\,\text{min}} = K$$

Substitute the log values for N and N_0 into the equation.

$$\frac{1.09}{180\,\text{min}} = K = \frac{0.0061}{\text{min}}$$

Simplify the equation.

3.6.2 Generation time

As shown earlier, the equation describing cell growth is given by the formula

$$\log(N) = \log(N_0) + Kt$$

The generation time of a particular bacterial strain under a defined set of conditions is the time required for the cell number to double. In this case, N would equal 2 and N_0 would equal 1 (or $N = 4$, $N_0 = 2$, or any other values representing a doubling in number). The formula then becomes

$$\log(2) = \log(1) + Kt$$

Converting to log values:

$$0.301 = 0 + Kt$$

Dividing each side of the equation by K gives the generation time t:

$$t = \frac{0.301}{K}$$

Problem 3.10 What is the generation time for the example culture?

Solution 3.10

From Problem 3.9, the growth constant, K, was calculated to be 0.0061/min. Placing this value into the expression for determining generation time gives

$$\text{generation time} = t = \frac{0.301}{K} = \frac{0.301}{0.0061/\text{min}} = 49.3\,\text{min}$$

Therefore, the generation time for the cells of this dataset is 49.3 min.

Problem 3.11 What is the cell concentration 150 min after inoculation?

Solution 3.11

In Problem 3.9, a growth constant, K, of 0.0061/min was calculated. The initial cell concentration, 2.1×10^7 cells/mL, is found in Table 3.3. These values can be entered into the equation describing cell growth to determine the cell concentration at 150 min.

$$\log(N) = \log(N_0) + Kt$$

The values for K, initial cell concentration, and t are placed into the equation:

$$\log(N) = \log(2.1 \times 10^7) + \left(\frac{0.0061}{\text{min}}\right)(150\,\text{min})$$

The values for K, initial cell concentration, and t are placed into the equation.

$$\log(N) = 7.32 + 0.92 = 8.24$$

Simplify the equation.

$$\log(N) = \log(2.1 \times 10^7) + \left(\frac{0.0061}{\text{min}}\right)(150\,\text{min})$$

Simplifying the equation yields

$$\log(N) = 7.32 + 0.92 = 8.24$$

To determine cell number, it must first be determined what number, when converted to a logarithm, is equal to 8.24. This number is the inverse of the log function and is called the **antilogarithm** (**antilog**) or **inverse logarithm**. The antilog of a number x (antilog x) is equal to 10^x.

Most calculators do not have a key marked 'antilog.' To find an antilog, you must use the **10^x** key. If this key does not exist, then you need to raise 10 to the x power using the **x^y** key. On some calculators, the **log** key serves as the **10^x** key after a **shift** or **inverse** key is pressed.

To find the antilog of 8.24 on a calculator, enter **8**, **.**, **2**, **4**, **shift**, and **10^x**. Or enter **1**, **0**, **x^y**, **8**, **.**, **2**, **4**, and =. Either path should yield an answer, when rounded off, of 1.7×10^8 cells/mL.

Therefore, at 150 min following inoculation, the culture will have a concentration of 1.7×10^8 cells/mL.

3.7 PLOTTING CELL GROWTH DATA ON A SEMILOG GRAPH

Semilog graph paper uses a log scale on the y (vertical) axis. The y axis is configured in 'cycles.' Each cycle is labeled 1 through 9 and corresponds to a power of 10. The scale between 1 and 2 is expanded compared with that from 2 to 3. The scale between 2 and 3 is expanded compared to that from 3 to 4. The divisions up the y axis become progressively smaller through 9, until 1 is reached and a new cycle begins. Semilog graph paper can be obtained commercially with as many as six cycles. Using semilog graph paper to plot growth data allows a quick visual determination of cell concentration at any point in time and an estimation of doubling time without having to perform log and antilog calculations. In the graphing methods outlined here, the sample data described in Table 3.1 will be used.

3.7.1 Plotting OD_{550} vs. time on a semilog graph

Plotting OD_{550}, on the y axis, vs. the hours following inoculation, on the x axis, is one way of representing the data. Since the OD_{550} values cover four places around the decimal (four decimal points), four-cycle semilog graph paper should be used (Figure 3.7).

Depending on how long following inoculation samples are taken, the curve may begin to level off (plateau) at later times as the OD_{550} value approaches 1.5. This can result from the culture reaching its maximal absorbance level. Even though the cells may still be growing, the spectrophotometer is incapable of discerning a change.

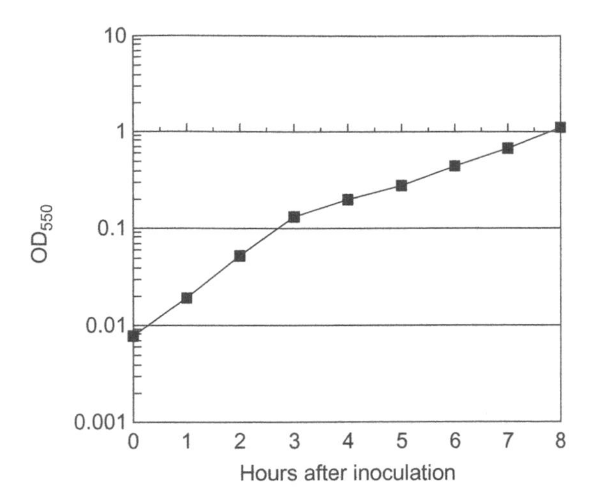

■ **FIGURE 3.7** Plot of OD_{550} vs. time on a semilog graph.

3.7.2 Estimating generation time from a semilog plot of OD_{550} vs. time

Generation time is also referred to as the **doubling time**. As shown in Problem 3.12, the generation time can be derived directly from a semilog plot by determining the time required for the OD_{550} value on the *y* axis to double.

Problem 3.12 Using a plot of OD_{550} vs. time, determine what the generation time is for those cells represented by the sample data.

Solution 3.12

Choose a point on the *y* axis (OD_{550}) in the region corresponding to the exponential growth phase. For example, an OD_{550} value of 0.03 can be chosen. Twice this value is 0.06. Draw horizontal lines from these two positions on the *y* axis to the plotted curve. From those intersection points on the curve, draw vertical lines to the *x* axis (Figure 3.8).

The difference between the points where these lines meet the *x* axis is approximately 0.8 hr. Since 0.8 hr is equivalent to 48 min (0.8 hr × 60 min/hr = 48 min), the generation time is approximately 48 min.

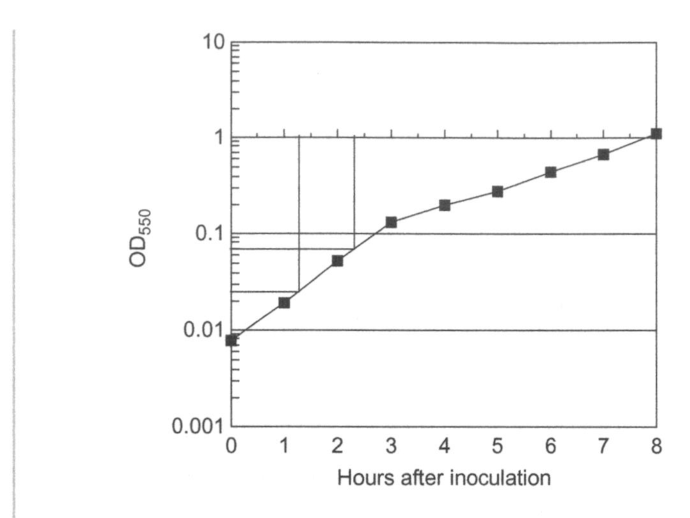

■ **FIGURE 3.8** Determination of generation time for Problem 3.12.

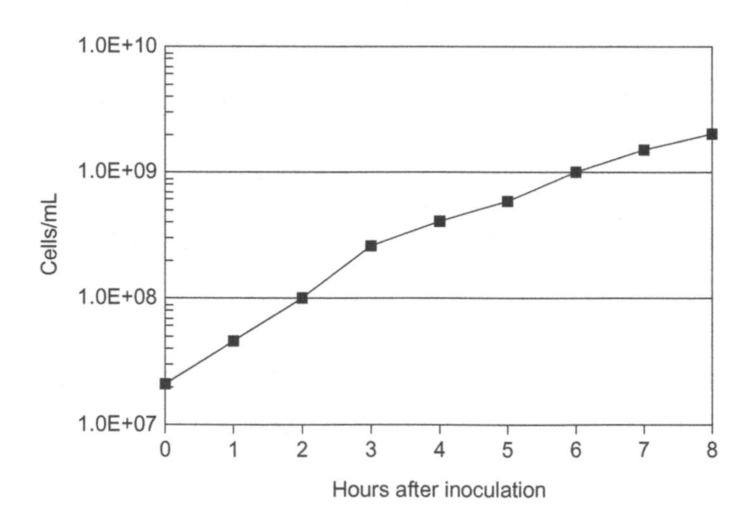

■ **FIGURE 3.9** Semilog plot of cell concentration vs. incubation time for sample data. The notation 1.0E+07 on the *y* axis is shorthand for 1.0×10^7. The E in this notation stands for 'exponent.'

3.8 PLOTTING CELL CONCENTRATION VS. TIME ON A SEMILOG GRAPH

When cell concentration is plotted against time (the number of hours following inoculation) on a semilog graph, the plot in Figure 3.9 is generated. (Since cell concentration ranges with exponent values between seven and nine (covering three exponent values), three-cycle semilog graph paper is to be used.)

Problem 3.13 Assuming that a culture of cells is grown again under identical conditions to those used to generate the sample data in Table 3.1, at what point will the culture have a concentration of 2×10^8 cells/mL?

Solution 3.13

Drawing a horizontal line from the 2×10^8 cells/mL point on the *y* axis to the plotted curve and then vertically down from that position to the *x* axis leads to a value of 2.7 hrs (see Figure 3.10). Since 0.7 hr is equivalent to 42 min (0.7 hr × 60 min/hr = 42 min), the culture will reach a concentration of 2×10^8 cells/mL in 2 hr 42 min.

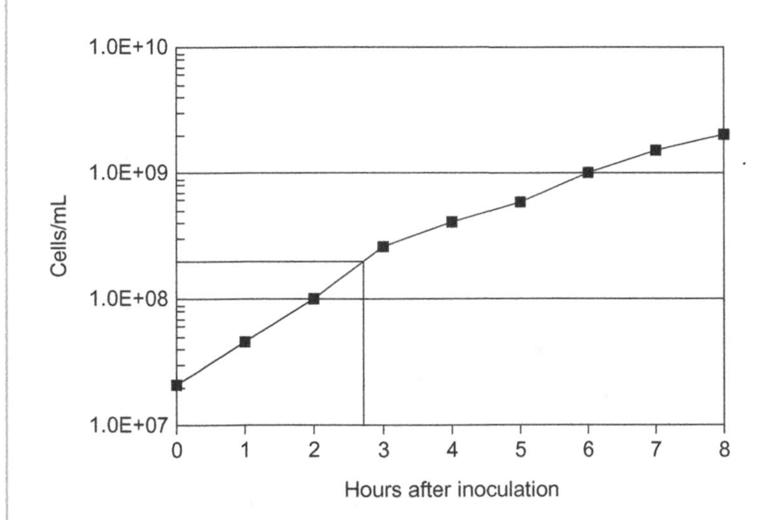

■ **FIGURE 3.10** Plot for Problem 3.13.

3.9 DETERMINING GENERATION TIME DIRECTLY FROM A SEMILOG PLOT OF CELL CONCENTRATION VS. TIME

The generation time for a particular bacterial strain in culture can be read directly from the semilog plot of cell concentration vs. time in a similar manner to that described in Problem 3.13.

Problem 3.14 Using a plot of OD_{550} vs. time, determine the generation time for those cells represented by the sample data in Table 3.1.

Solution 3.14

Choose a point on the *y* axis (cells/mL) in the region corresponding to the exponential growth phase. For example, a cell concentration value of 1×10^8 cells/mL can be chosen. Twice this value is 2×10^8 cells/mL. Draw horizontal lines from these two positions on the *y* axis to the plotted curve (see Figure 3.11). From those intersection points on the curve, draw vertical lines to the *x* axis. The difference between the positions where these lines meet the *x* axis is approximately 0.7 hr. Since 0.7 hr is equivalent to 42 min (0.7 hr $\times$ 60 min/hr $=$ 42 min), the generation time is approximately 42 min.

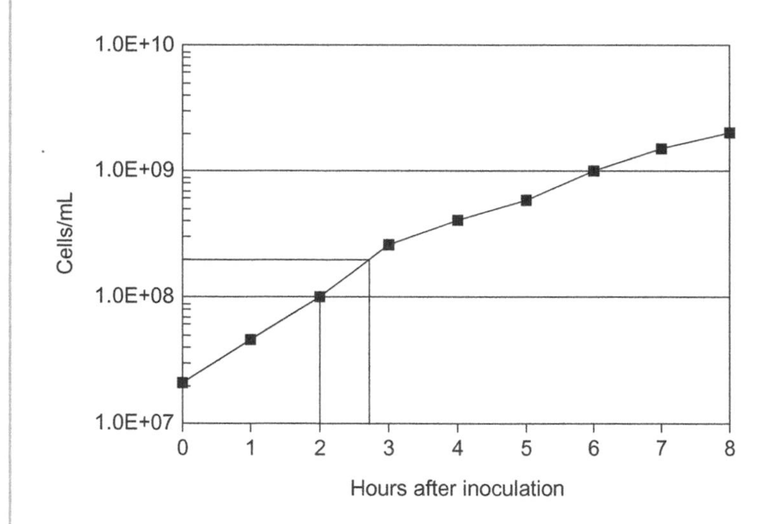

■ **FIGURE 3.11** Plot for Problem 3.14.

3.10 PLOTTING CELL DENSITY VS. OD_{550} ON A SEMILOG GRAPH

Plotting cell density vs. OD_{550} on a semilog graph allows the experimenter to quickly estimate cell concentration for any OD_{550} value. For the sample data in Table 3.1, this plot gives the curve shown in Figure 3.12.

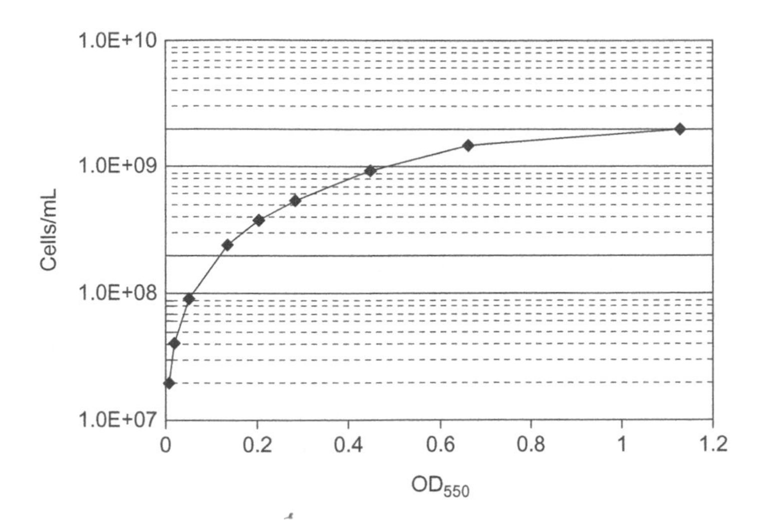

■ **FIGURE 3.12** Semilog plot of cell concentration vs. OD_{550} for sample data.

Problem 3.15 A culture is grown under identical conditions to those used to generate the sample data. If a sample is withdrawn and its OD_{550} is determined to be 0.31, what is the approximate cell concentration?

Solution 3.15

To determine the cell concentration corresponding to an OD_{550} value of 0.31, draw a vertical line from the 0.31 position on the *x* axis up to the plotted curve (Figure 3.13). Draw a horizontal line from this position on the curve to the *y* axis.

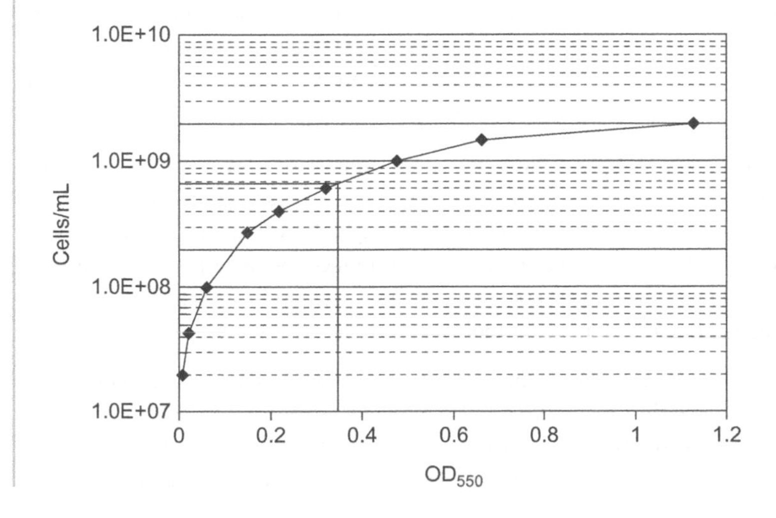

■ **FIGURE 3.13** Plot for Problem 3.15.

This point on the y axis corresponds to 6×10^8 cells/mL. Therefore, a culture having an OD_{550} of 0.31 will have a cell concentration of 6×10^8 cells/mL.

3.11 THE FLUCTUATION TEST

Bacteriophages (or **phages**) are viruses that infect bacteria. A number of different types of bacteriophage infect the bacterium *E. coli*. When infecting their host, many of the best-characterized bacteriophages undertake a lytic course of infection, in which the phage particle attaches to the host's cell wall and subsequently injects its genome into the cell's cytoplasm, where it replicates and expresses the genes necessary for the construction of progeny phage. During the infection process, progeny phage accumulate inside the cell. Ultimately, the cell bursts open (lyses), releasing the new phage particles into the environment, where they can infect other cells.

If bacteriophage are grown in liquid culture, infection and phage multiplication lead to a decrease in the culture's turbidity as more and more cells succumb to infection and lyse. If bacteriophages are plated with cells on a Petri plate, infection of a single cell by a single phage eventually leads to a circular area of lysis (called a **plaque**) within the bacterial lawn.

The bacteriophage T1 follows the infection pathway just described. T1 initiates infection by first binding to a protein complex, called the T1-phage receptor site, on the host's cell wall. Once attachment is accomplished, infection leads to the eventual destruction of the attacked cell. Not all cells in an *E. coli* population, however, are susceptible to infection. Within a population of cells, there is a chance that some members of that population carry a mutation that alters the T1 receptor. Such cells are resistant to T1 infection. A T1-resistant cell passes this characteristic on to daughter cells in a stable manner.

In the early 1940s, as more and more mutations were being characterized in both prokaryotic and eukaryotic systems, an essential question remained unanswered: What is the origin of such genetic variants? Two possibilities were considered: Either mutations are induced as a response to the environment, or they arise spontaneously; pre-existing in a population, the environment merely selects for their survival.

In 1943, Max Delbrück and Salvador Luria designed an experiment called the **fluctuation test** that used T1-resistant mutants of *E. coli* to determine which hypothesis was correct. The experiment entailed looking at the number of T1-resistant cells in a parallel series of identical cultures. By the first hypothesis, that of induced mutation, T1-resistant variants arise only upon exposure to T1 bacteriophage. All cells have an equal (but small) probability of acquiring T1 resistance following exposure to T1. Once acquired, the adaptation is passed on to their descendants. This hypothesis predicts that, no matter the stage of cell growth, the fraction of T1-resistant cells should fluctuate neither between a series of cultures nor between a series of samples taken from the same bulk culture.

By the second hypothesis, that of spontaneous mutation, the acquisition of T1 resistance may occur at any time during growth of a culture prior to T1 exposure. The number of T1-resistant cells in any particular population will depend upon whether the mutation occurred early or late in the growth of that culture. This theory predicts larger fluctuations in the number of T1-resistant cells between cultures in a parallel series than from a series of samples taken from the same bulk culture.

3.11.1 Fluctuation test example

A series of 10 tubes containing 0.2 mL of tryptone broth and a single 125 mL flask containing 10 mL of tryptone broth are each inoculated with *E. coli* to a concentration of 500 cells/mL. Cells are incubated at 37°C until the cultures reach the stationary phase of growth. One-tenth mL samples are taken from each of the small cultures and ten 0.1 mL aliquots are taken from the 10 mL bulk culture. These samples are then spread on agar plates saturated with T1 phage. The plates are incubated overnight and the number of colonies on each plate is counted. The results shown in Table 3.4 are obtained.

Table 3.4 Sample data for fluctuation test problems.

Small culture T1-resistant colonies	Bulk culture T1-resistant colonies
1	24
0	20
0	16
5	25
124	18
0	21
0	18
0	17
71	22
0	19

Interpretation of results

By visual inspection of the data, there is an obvious difference between the distribution of T1-resistant colonies derived from the different cultures. As predicted by the theory of spontaneous mutation, there are large differences in the numbers of T1-resistant colonies derived from the parallel series of small cultures, while the number of T1-resistant colonies per plate arising from samples of the bulk culture shows little variation. To examine these results mathematically, however, it is first necessary to determine the mean value for each set of cultures. A **mean** (also called an **average)** represents a value within the range of an entire set of numbers. It falls between the two extreme values of a number set and serves as a representative value for any set of measurements. It is calculated as the sum of the measures in a distribution divided by the number of measures. The mean value is represented by the symbol x with a line over it:

$$mean = \overline{x}$$

Problem 3.16 What are the mean values for each dataset?

Solution 3.16

For the parallel series of small cultures, the sum of T1-resistant colonies is

$$1 + 0 + 0 + 5 + 124 + 0 + 0 + 0 + 71 + 0 = 201$$

The mean value is then calculated as the sum divided by the number of measures taken:

$$\frac{201 \text{ T1-resistant colonies}}{10 \text{ samples}} = 20.1 \text{ T1-resistant colonies/sample}$$

For the samples taken from the 10 mL bulk culture, the sum of T1-resistant colonies is

$$24 + 20 + 16 + 25 + 18 + 21 + 18 + 17 + 22 + 19 = 200$$

The mean value is then calculated as this sum divided by the number of samples taken:

$$\frac{200 \text{ T1-resistant colonies}}{10 \text{ samples}} = 20.0 \text{ T1-resistant colonies/sample}$$

3.11.2 Variance

The spontaneous mutation hypothesis predicts a large fluctuation around the mean value for the samples taken from the individual cultures. Fluctuation within a dataset around the set's average value is referred to as the **variance**. Although the mean values of T1-resistant colonies between the two sets of data should be quite similar, as might be predicted by either hypothesis, to satisfy the spontaneous mutation hypothesis, the variance between the two sets of samples should be large.

Variance is calculated by the formula

$$\text{variance} = \frac{\sum(x - \bar{x})^2}{n - 1}$$

By this formula, the term $(x - \bar{x})^2$ requires that the mean value for each dataset be subtracted from each individual value in that dataset. The remainder of each subtraction is then squared. (**Squaring** a number means to multiply that number or quantity by itself. It is denoted by the exponent 2. It can be obtained on a calculator by entering the number to be squared and then pressing the x^2 key.) The symbol Σ (the Greek letter **sigma**) demands that all terms in a series be added. In this case, the series is the $(x - \bar{x})^2$ values obtained for each data point. The number n is equal to the number of measurements.

Problem 3.17 What is the variance for each dataset?

Solution 3.17

Calculation of the sum of squares for the small-cultures dataset is demonstrated in Table 3.5.

The sum of the values in the $(x - \bar{x})^2$ column $= \Sigma(x - \bar{x})^2 = 16403$.

Since there are 10 samples in this dataset, $n = 10$. Placing these values into the formula for calculating variance yields

$$\text{Variance} = \frac{\sum(x - \bar{x})^2}{n - 1} = \frac{16403}{10 - 1} = \frac{16403}{9} = 1823$$

Calculating the sum of squares for the bulk-culture dataset is shown in Table 3.6.

The sum of the values in the $(x - \bar{x})^2$ column $= \Sigma(x - \bar{x})^2 = 80$.

Table 3.5 Calculating the sum of squares of the mean subtracted from each value in the small-cultures data for Problem 3.17.

T1-resistant Colonies	$(x - \bar{x})$	$(x - \bar{x})^2$
1	$1 - 20 = -19$	361
0	$0 - 20 = -20$	400
0	$0 - 20 = -20$	400
5	$5 - 20 = -15$	225
124	$124 - 20 = 104$	10816
0	$0 - 20 = -20$	400
0	$0 - 20 = -20$	400
0	$0 - 20 = -20$	400
71	$71 - 20 = 51$	2601
0	$0 - 20 = -20$	400
	sum	**16403**

Table 3.6 Calculating the sum of squares of the mean subtracted from each value in the bulk-culture data for Problem 3.17.

T1-resistant Colonies	$(x - \bar{x})$	$(x - \bar{x})^2$
24	$24 - 20 = 4$	16
20	$20 - 20 = 0$	0
16	$16 - 20 = -4$	16
25	$25 - 20 = 5$	25
18	$18 - 20 = -2$	4
21	$21 - 20 = 1$	1
18	$18 - 20 = -2$	4
17	$17 - 20 = -3$	9
22	$22 - 20 = 2$	4
19	$19 - 20 = -1$	1
	sum	**80**

Since there are 10 samples in this dataset, $n = 10$.

Placing these values into the formula for calculating variance yields

$$\text{variance} = \frac{\sum(x - \bar{x})^2}{n - 1} = \frac{80}{10 - 1} = \frac{80}{9} = 8.9$$

Unlike the mean values for the two sets of samples, the variance values are drastically different and support the spontaneous mutation hypothesis.

3.12 MEASURING MUTATION RATE

Luria and Delbrück demonstrated that data from a fluctuation test can be used, by two different methods, to calculate a spontaneous mutation rate. One of these methods utilizes the law of probabilities as described by a Poisson distribution. The other utilizes a graphical approach, taking into account the frequency distribution of mutant cells in the parallel series of small cultures.

3.12.1 The Poisson distribution

The **Poisson distribution** is used to describe the distribution of rare events in a large population. For example, at any particular time, there is a certain probability that a particular cell within a large population of cells will acquire a mutation. Mutation acquisition is a rare event. If the large population of cells is divided into smaller cultures, as is done in the fluctuation test, the Poisson distribution can be used to determine the probability that any particular small culture will contain a mutated cell.

Calculating a Poisson distribution probability requires the use of the number ***e***, described in the following box.

The number *e*

In molecular biology, statistics, physics, and engineering, most calculations employing the use of logarithms are in one of two bases, either base 10 or base *e*. The number *e* is the base of the **natural logarithms**, designated as **ln**. For example, ln 2 is equivalent to $\log_e 2$. The value of *e* is roughly equal to 2.7182818. *e* is called an irrational number because its decimal representation neither terminates nor repeats. In that regard, it is like the number pi (p) (the ratio of the circumference of a circle to its diameter). In fact, pi and ***e*** are related by the expression $e^{ip} = 1$, where *i* is equal to the square root of −1.

Many calculators have an **ln** key for finding natural logarithms. Many calculators also have an $\mathbf{e^x}$ key, used to find the antilogarithm base *e*.

The Poisson distribution is written mathematically as

$$P = \frac{e^{-m}m^r}{r!}$$

where P is the fraction of samples that will contain r objects each, if an average of m objects per sample is distributed at random over the collection of samples. (The m component is sometimes referred to as the **expectation**.)

The component e is the base of the **natural** system of logarithms (see previous box). The exclamation mark, **!**, is the symbol for factorial. The **factorial** of a number is the product of the specified number and each positive integer less than itself down to and including 1. For example, 5! (read '5 factorial') is equal to $5 \times 4 \times 3 \times 2 \times 1 = 120$. **0!** is equal to **1**.

3.12.2 Calculating mutation rate using the Poisson distribution

To use the fluctuation test to determine mutation rate via the Poisson distribution, the small cultures must be prepared with a sufficiently small inocula such that, following incubation, some of them will have no mutants at all. In addition, it is important to assay the total number of cells per culture at the time the culture is spread on selective plates.

Mutation rate is calculated using the fraction of small parallel cultures that contain no mutants. To solve the Poisson distribution for the zero case, the equation is

$$P_0 = \frac{e^{-m}m^0}{0!}$$

For our example, six of the ten small cultures contained no T1-resistant bacteria. P_0, the fraction of cultures containing no mutants, is therefore $6/10 = 0.6$. The Poisson distribution relationship can now be written as

$$\frac{6}{10} = 0.6 = \frac{e^{-m}m^0}{0!}$$

The component m^0 is equivalent to 1. (Any number raised to the exponent 0 is equal to 1.) Likewise, 0! is equal to 1. The equation then reduces to

$$0.6 = e^{-m}$$

We can solve for m using the following relationship:

$$x = e^y \text{ is equivalent to } y = \ln x$$

and

$$x = e^{-y} \text{ is equivalent to } y = -\ln x$$

$$m = -\ln 0.6$$

To determine the natural log of 0.6 on the calculator, press **.**, **6**, and **ln**. This gives a value, when rounded, of -0.51.

$$m = -(-0.51)$$

The negative of a negative number is a positive value. Therefore,

$$m = 0.51$$

That is, m, the average number of mutant bacteria in a series of cultures representing an entire population, is equivalent to 0.51 mutants per small culture.

To determine a mutation rate as mutations/cell/cell division, the number of mutants per small culture (in our example, 0.51) must be divided by the number of bacteria per cell division. The following relationship can be used to calculate that value.

> **Calculating the increase in bacteria from generation to generation**
> The average number of bacteria in the next subsequent generation is calculated by dividing the number of bacteria at the beginning of the current generation by the natural logarithm of 2.

Therefore, if there were 2.6×10^8 bacteria per small culture (as assayed at the time the cultures were spread on the selective plates containing T1 bacteriophage), then the number of bacteria per cell division is equal to

$$\begin{aligned} 2.6 \times 10^8 / \ln 2 &= 2.6 \times 10^8/0.69 \\ &= 3.8 \times 10^8 \end{aligned}$$

The mutation rate is then calculated as

$$0.51/(3.8 \times 10^8) = 1.3 \times 10^{-9} \text{ mutations/cell/cell division}$$

3.12.3 Using a graphical approach to calculate mutation rate from fluctuation test data

Using the Poisson distribution method to calculate mutation rate does not make use of all the information available from a fluctuation test; it makes no use of the frequency distribution of cultures that do contain mutant (T1-resistant) cells. A second approach developed by Luria and Delbrück relies on the assumption that, once a population of cells reaches a certain size, at least one cell will acquire a mutation. In the subsequent generation, as the number of cells doubles (and assuming a constant mutation frequency), two cells should newly acquire mutations. In the next generation, following another doubling, four cells should newly acquire mutations.

The fraction of cells acquiring mutations is assumed to stay constant over time. As a simplified example, if a single cell in a population of four cells acquires a mutation, then ¼ of the cells have acquired a mutation. In the next generation, after the cells have doubled, two out of the eight cells (¼) would be expected to acquire mutations. During the next doubling, four out of the sixteen daughter cells (¼) would be expected to acquire mutations. Although the mutation rate should remain constant, the overall proportion of mutant cells in the population will increase as those cells acquiring mutations pass that new trait along to daughter cells in subsequent generations.

The equation developed by Luria and Delbrück to account for this scenario and to determine the time at which there is likely to be a single mutant somewhere in the series of small cultures is

$$r = aN_t \ln(aN_tC)$$

where r is the average number of mutants per small culture, C is the number of small cultures used in the fluctuation test, a is the mutation rate, and N_t is the total number of cells in each small culture.

In the example provided in this text, $r = 20$, $C = 10$, and $N_t = 2.6 \times 10^8$.

The value of aN_t in the equation can be solved by interpolation through the systematic substitution of arbitrary values of a and values of N_t. A graph of r vs. aN_t when C is equal to 10 is then used to solve for the actual mutation rate, a. Mutation rates (a) typically vary from 1×10^{-8} to 1×10^{-12}. The number of bacteria (N_t) in the small cultures, depending on the strain and the media used, may vary from 5×10^7 to 5×10^{10}. To prepare a graph of r vs. aN_t, various values of a are multiplied by various values of N_t, shown in Tables 3.7–3.11. Once aN_t is determined, r can be solved via the mutation rate equation. For example, when $a = 5 \times 10^{-7}$ and $N_t = 5 \times 10^7$, then $aN_t = 25$. Using the mutation rate equation,

$$r = aN_t \ln(aN_tC)$$

Table 3.7 r for various values of a when $N_t = 5 \times 10^7$ and $C = 10$.

a	N_t	aN_t	r
5×10^{-7}	5×10^7	25.0	138.0
5×10^{-8}	5×10^7	2.5	8.0
2×10^{-8}	5×10^7	1.0	2.3
1×10^{-8}	5×10^7	0.5	0.8
5×10^{-9}	5×10^7	0.25	0.23

Table 3.8 r for various values of a when $N_t = 1 \times 10^8$ and $C = 10$.

a	N_t	aN_t	r
5×10^{-7}	1×10^8	50.0	310.0
5×10^{-8}	1×10^8	5.0	19.6
2×10^{-8}	1×10^8	2.0	6.0
1×10^{-8}	1×10^8	1.0	2.3
5×10^{-9}	1×10^8	0.5	0.8
1×10^{-9}	1×10^8	0.1	0.0

Table 3.9 r for various values of a when $N_t = 2 \times 10^8$ and $C = 10$.

a	N_t	aN_t	r
5×10^{-7}	2×10^8	100.0	691.0
5×10^{-8}	2×10^8	10.0	46.1
2×10^{-8}	2×10^8	4.0	14.8
1×10^{-8}	2×10^8	2.0	6.0
5×10^{-9}	2×10^8	1.0	2.3
1×10^{-9}	2×10^8	0.2	0.14
5×10^{-10}	2×10^8	0.1	0.0

Table 3.10 r for various values of a when $N_t = 5 \times 10^8$ and $C = 10$.

a	N_t	aN_t	r
5×10^{-7}	5×10^8	250.0	1956.0
5×10^{-8}	5×10^8	25.0	138.0
2×10^{-8}	5×10^8	10.0	46.1
1×10^{-8}	5×10^8	5.0	19.6
5×10^{-9}	5×10^8	2.5	8.0
1×10^{-9}	5×10^8	0.5	0.8
5×10^{-10}	5×10^8	0.25	0.23

Table 3.11 r for various values of a when N_t is 1×10^9 and $C = 10$.

a	N_t	aN_t	r
5×10^{-7}	1×10^9	500.0	4259.0
5×10^{-8}	1×10^9	50.0	311.0
2×10^{-8}	1×10^9	20.0	106.1
1×10^{-8}	1×10^9	10.0	46.1
5×10^{-9}	1×10^9	5.0	19.6
1×10^{-9}	1×10^9	1.0	2.3
5×10^{-10}	1×10^9	0.5	0.8
1×10^{-10}	1×10^9	0.1	0.0

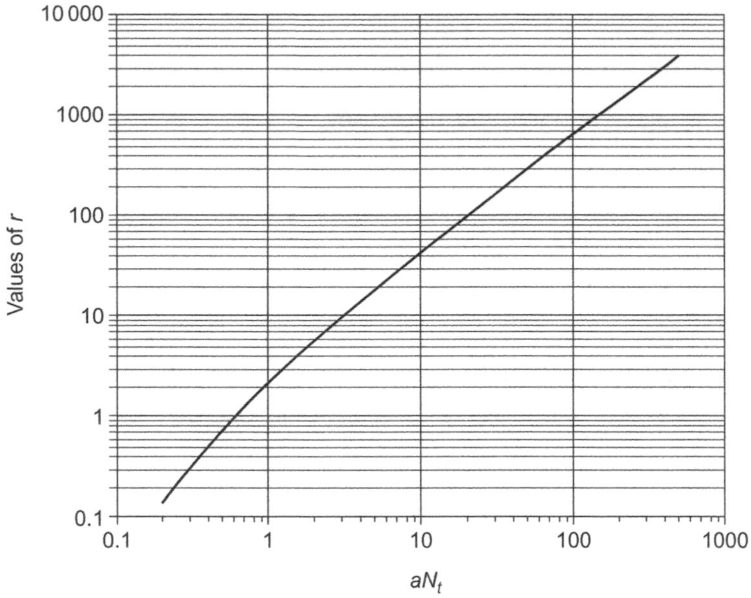

FIGURE 3.14 Graph of r plotted against aN_t for $C = 10$.

Substituting our known values yields

$$r = 25\ \ln(25 \times 10) = 25\ \ln\ 250 = (25)(5.5) = 138$$

Tables 3.7–3.11 show values of r obtained when various values of a and N_t are used.

Plotting r vs. aN_t for $C = 10$ gives the graph presented in Figure 3.14.

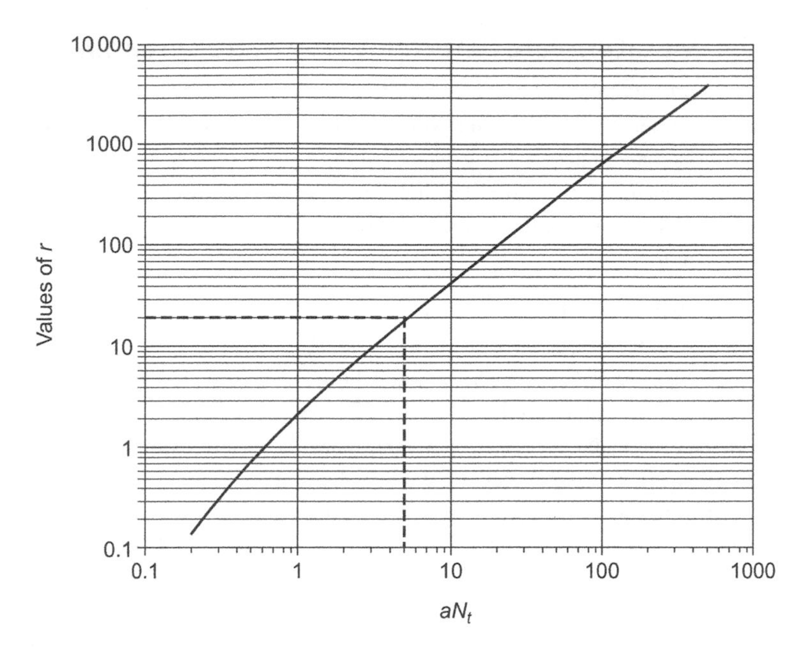

■ **FIGURE 3.15** Determining aN_t when $r = 20$.

From Figure 3.15, it can be seen that when $r = 20$, $aN_t = 5$.

Therefore, using the equation $r = aN_t \ln(aN_tC)$, where $r = 20$, $N_t = 2.6 \times 10^8$, $C = 10$, and aN_t is equivalent to 5, we have

$$20 = a(2.6 \times 10^8)\ln(5 \times 10)$$
$$20 = a(2.6 \times 10^8)\ln(50)$$
$$20 = a(2.6 \times 10^8)(3.9)$$

Solving for a, the mutation rate, gives the following solution:

$$a = \frac{20}{(2.6 \times 10^8)(3.9)} = \frac{20}{1.014 \times 10^9} = 2 \times 10^{-8}$$

This mutation rate (2×10^{-8}) is slightly higher than that calculated by the Poisson distribution method. This is because this method is sensitive to the appearance of 'jackpot' tubes, in which a mutation occurs very early following inoculation such that a large number of mutant cells appear in the final population. Jackpot tubes result in excessively large values of r and an overestimation of the actual number of mutant cells arising during a defined period of time.

3.12.4 Mutation rate determined by plate spreading

T1-resistant mutants can be detected by spreading cells on an agar plate, spraying an aerosol of T1 bacteriophage onto the plate, and then allowing time for the appearance of T1-resistant colonies. This approach to mutation detection can be used, with slight modifications, to determine a mutation frequency. For example, the following protocol could be followed to arrive at such a number.

1. Begin with a culture of *E. coli* cells at a concentration of 2×10^5 cells/mL.
2. With a sterile plate spreader, spread 0.1 mL (2×10^4 cells) onto each of 24 tryptone plates and onto each of 20 plates previously spread with T1 bacteriophage. This latter series of plates is used to assay for the number of T1-resistant cells in the initial culture.
3. Allow the spread plates to incubate for 6 hr. (Assuming a generation time of 40 min, this allows for nine doublings (360 min ÷ 40 min = 9).)
4. At 6 hr, spray an aerosol of T1 bacteriophage onto 20 of the 24 plates not previously treated with T1, and incubate these sprayed plates for 12 hr to allow the growth of resistant colonies. From the remaining four untreated plates, collect and assay the number of cells.

This experiment might give the following results. Initial number of T1-resistant cells (as determined by the number of colonies appearing on the 20 plates previously spread with T1 phage) = 0.

Initial number of bacteria =

$$\frac{2 \times 10^5 \text{ cells}}{\text{mL}} \times \frac{0.1\,\text{mL}}{\text{plate}} \times 20 \text{ plates} = 4 \times 10^5 \text{ cells}$$

After six hours of growth, the four untreated plates are assayed and found to have 1×10^7 cells/plate. Therefore, on the 20 plates sprayed with a T1 aerosol, there are a total of 2×10^8 cells:

$$20 \text{ plates} \times \frac{1 \times 10^7 \text{ cells}}{\text{plate}} = 2 \times 10^8 \text{ total cells}$$

Following incubation of the 20 sprayed plates, five colonies arise.

The mutation rate can be calculated using the following formula:

$$\text{mutation rate} = \frac{\dfrac{\text{change in number of resistant colonies}}{\text{change in total number of cells}}}{\ln 2}$$

In the example provided, the change in the number of T1-resistant colonies is

$$5 - 0 = 5$$

The change in the number of total cells is

$$2 \times 10^8 \text{ cells} - 4 \times 10^5 \text{ cells} = 2 \times 10^8 \text{ cells}$$

(Subtracting 4×10^5 from 2×10^8 yields 2×10^8 when significant figures are considered.)

Using the formula given above for mutation rate yields

$$\text{mutation rate} = \frac{5/2 \times 10^8}{\ln 2} = \frac{2.5 \times 10^{-8}}{0.693} = 3.6 \times 10^{-8}$$

Therefore, the mutation rate is 3.6×10^{-8} mutations/bacterium/cell division.

3.13 MEASURING CELL CONCENTRATION ON A HEMOCYTOMETER

A hemocytometer is a microscope slide engraved with a $1\,\text{mm} \times 1\,\text{mm}$ grid. A coverslip rests on supports 0.1 mm above the grid such that $0.1\,\text{mm}^3$ chambers are formed by the $1\,\text{mm} \times 1\,\text{mm}$ squares of the grid and the 0.1 mm space between the slide and the coverslip. A drop of cell suspension is drawn into the chamber below the coverslip and the slide is examined under a microscope. The observer then counts the number of cells in several of the $1\,\text{mm} \times 1\,\text{mm}$ squares. (For best accuracy, a total of over 200 cells should be counted.) Cell count is then converted to a concentration, as demonstrated in the following problem.

Problem 3.18 A suspension of hybridoma cells is diluted 1 mL/10 mL and an aliquot of the dilution is counted on a hemocytometer. In a total of 10 grid squares, 320 cells are counted. What is the cell concentration of the cell suspension?

Solution 3.18

To determine cell concentration, the conversion of area (mm^3) to volume must be used. One milliliter of liquid is equal to $1\,\text{cm}^3$ (one cubic centimeter).

One cm^3 = 1000 mm^3. There are ten 0.1 mm^3 per cubic millimeter. These conversion factors and the dilution factor will be used to determine cell concentration, as shown in the following equation:

$$\frac{320 \text{ cells}}{10\ 0.1 - \text{mm}^3 \text{ squares}} \times \frac{10\ 0.1 - \text{mm}^3 \text{ squares}}{\text{mm}^3} \times \frac{1000 \text{ mm}^3}{\text{cm}^3} \times \frac{1 \text{ cm}^3}{\text{mL}} \times \frac{10 \text{ mL}}{1 \text{ mL}} = \frac{3.2 \times 10^6 \text{ cells}}{\text{mL}}$$

Therefore, the cell suspension has a concentration of 3.2×10^6 cells/mL.

CHAPTER SUMMARY

Cell concentration is determined by making a serial dilution of a cell culture, spreading aliquots of the dilutions onto Petri dishes for the growth of individual colonies, and calculating the number of cells per volume in the undiluted culture. The increase in cell number (N) per unit time (t) is proportional to the number of cells at the beginning of the time interval (N_0) and is expressed by the relationship

$$\log(N) = \log(N_0) + Kt$$

where K is the growth constant calculated as the slope of the line plotting the log of the cell concentration vs. time.

The **generation time** of cells in culture is the time required for the cell number to double. It is described by the equation

$$t = \frac{0.301}{K}$$

Generation time can also be estimated from a plot of either cell culture OD vs. time after inoculation or cell concentration vs. time after inoculation.

The **fluctuation test**, an experiment that measures the variability in the appearance of mutant bacteria between a bulk culture and a group of small, individual cultures, can be used to discern whether or not mutants pre-exist in a population or arise spontaneously as a result of exposure to a particular environment. Variability about a mean value is measured by variance, given by the expression

$$\text{variance} = \frac{\sum (x - \bar{x})^2}{n - 1}$$

If there is a large variance between the numbers of mutants within a set of small, individual cultures, then those mutants are most likely pre-existing in the population. The fluctuation test can also be used to estimate mutation rate by using Poisson distribution for the zero case – evaluating the number of small individual cultures having no mutants within them and solving for m, the average number of mutant bacteria per small culture, using the expression

$$P_0 = \frac{e^{-m} m^0}{0!}$$

The mutation rate is calculated as

$$\text{mutation rate} = \frac{m}{\text{number of cells per small culture}/\ln\ 2}$$

Mutation rate can also be measured by plate spreading using the equation

$$\text{mutation rate} = \frac{\dfrac{\text{change in number of resistant colonies}}{\text{change in total number of cells}}}{\ln 2}$$

The concentration of eukaryotic cells in suspension can be determined using a hemocytometer.

REFERENCES

Luria, S.E., and M. Delbrück (1943). Mutations of bacteria from virus sensitivity to virus resistance. *Genetics* 28:491–511.

Chapter 4

Working with bacteriophages

INTRODUCTION

A **bacteriophage** (**phage**) is a virus that infects bacteria. It is little more than nucleic acid surrounded by a protein coat. To infect a cell, a bacteriophage attaches to a receptor site on the bacteria's cell wall. Upon attachment, the phage injects its DNA into the cell's cytoplasm, where it is replicated. Phage genes are expressed for the production and assembly of coat proteins that encapsulate the replicated phage DNA. When a critical number of virus particles has been assembled, the host cell lyses and the newly made phage are released into the environment, where they can infect new host cells. Their simple requirements for propagation, their short generation time, and their relatively simple genetic structure have made bacteriophages ideal subjects of study for elucidating the basic mechanisms of transcription, DNA replication, and gene expression. A number of bacteriophages have been extensively characterized. Several, such as the bacteriophages λ and M13, have been genetically engineered to serve as vectors for the cloning of exogenous genetic material.

4.1 MULTIPLICITY OF INFECTION (moi)

An experiment with bacteriophage typically begins with an initial period during which the virus is allowed to adsorb to the host cells. It is important to know the ratio of the number of bacteriophages to the number of cells at this stage of the infection process. Too many bacteriophages attaching to an individual cell can cause cell lysis, even before the infection process can yield progeny virus particles ('lysis from without'). If too few bacteriophages are used for the infection, it may be difficult to detect or measure the response being tested. The bacteriophage to cell ratio is called the **multiplicity of infection** (**moi**).

Problem 4.1 A 0.1 mL aliquot of a bacteriophage stock having a concentration of 4×10^9 phage/mL is added to 0.5 mL of *E. coli* cells having a concentration of 2×10^8 cells/mL. What is the moi?

Calculations for Molecular Biology and Biotechnology. DOI: 10.1016/B978-0-12-375690-9.00004-8
© 2010 Elsevier Inc. All rights reserved.

Solution 4.1

First, calculate the total number of bacteriophage and the total number of bacteria.

Total number of bacteriophage:

$$0.1\,\text{mL} \times \frac{4 \times 10^9\ \text{phage}}{\text{mL}} = 4 \times 10^8\ \text{bacteriophage}$$

Total number of cells:

$$0.5\,\text{mL} \times \frac{2 \times 10^8\ \text{cells}}{\text{mL}} = 1 \times 10^8\ \text{cells}$$

The moi is then calculated as bacteriophage per cell:

$$\text{moi} = \frac{4 \times 10^8\ \text{phage}}{1 \times 10^8\ \text{cells}} = 4\ \text{phage/cell}$$

Therefore, the moi is 4 phage/cell.

Problem 4.2 A 0.25 mL aliquot of an *E. coli* culture having a concentration of 8×10^8 cells/mL is placed into a tube. What volume of a bacteriophage stock having a concentration of 2×10^9 phage/mL should be added to the cell sample to give an moi of 0.5?

Solution 4.2

First, calculate the number of cells in the 0.25 mL aliquot:

$$0.25\,\text{mL} \times \frac{8 \times 10^8\ \text{cells}}{\text{mL}} = 2 \times 10^8\ \text{cells}$$

Next, calculate how many bacteriophage particles are required for an moi of 0.5 when 2×10^8 cells are used.

$$\frac{x\ \text{phage}}{2 \times 10^8\ \text{cells}} = \frac{0.5\ \text{phage}}{\text{cell}}$$

Solve for *x*:

$$x = \left(\frac{0.5\ \text{phage}}{\text{cell}}\right)(2 \times 10^8\ \text{cells}) = 1 \times 10^8\ \text{phage}$$

Finally, calculate the volume of the bacteriophage stock that will contain 1×10^8 phage:

$$\frac{2 \times 10^9 \text{ phage}}{\text{mL}} \times x \text{ mL} = 1 \times 10^8 \text{ phage}$$

Solving for x yields

$$x = \frac{1 \times 10^8 \text{ phage}}{2 \times 10^9 \text{ phage/mL}} = 0.05 \text{ mL}$$

Therefore, 0.05 mL of the phage stock added to the aliquot of cells will give an moi of 0.5.

4.2 PROBABILITIES AND MULTIPLICITY OF INFECTION (moi)

In the previous chapter, probability was used to estimate the number of mutant cells in a culture of bacteria. Probability can also be used to examine infection at the level of the individual cell and to estimate the distribution of infected cells in culture. These methods are demonstrated in the following problems.

Problem 4.3 A culture of bacteria is infected with bacteriophage at an moi of 0.2. What is the probability that any one cell will be infected by two phage?

Solution 4.3

When the moi is less than 1, the math used to calculate the chance that any particular cell will be infected is similar to that used to predict a coin toss. The probability that any one cell will be infected by a single virus is equal to the moi (in this case 0.2). This also means that 20% of the cells will be infected ($100 \times 0.2 = 20\%$) or that each cell has a 20% chance of being infected. This value can further be expressed as a '1 in' number by taking its reciprocal:

$$\frac{1}{0.2} = 5$$

Therefore, one in every five cells will be infected.

If it is assumed that attachment of one phage does not influence the attachment of other phage, then the attachment of a second phage will have the same probability as the attachment of the first. The probabilities of both events can be multiplied.

The probability that a cell will be infected by two phage, therefore, is the product of the probabilities of each independent event:

$$0.2 \times 0.2 = 0.04$$

This can be expressed in several ways:

a) 4% of the cells will have two phage particles attached ($0.04 \times 100 = 4\%$),
b) a cell has a 4% chance of being infected by two phage particles, or
c) 1 in 25 cells will be infected by two phage particles (the reciprocal of $0.04 = 1/0.04 = 25$).

Problem 4.4 A culture of cells is infected by bacteriophage λ at an moi of 5. What is the probability that a particular cell will not be infected during the phage adsorption period?

Solution 4.4

In the previous chapter, the Poisson distribution was used to determine the number of mutants that might be expected in a population of cells. This distribution is represented by the equation

$$P_r = \frac{e^{-m}m^r}{r!}$$

where *P* is the probability, *r* is the number of successes (in this example, a cell with zero attached phage is a 'success'), *m* is the average number of phage/cell (the moi; 5 in this example), and *e* is the base of the natural logarithms.

For the zero case, the equation becomes

$$P_0 = \frac{e^{-5}5^0}{0!}$$

In solving for the zero case, the following two properties are encountered.

- Any number raised to the zero power is equal to 1.
- The exclamation symbol (!) designates the factorial of a number, which is the product of all integers from that number down to 1. For example, 4! (read '4 factorial' or 'factorial 4') is equal to $4 \times 3 \times 2 \times 1 = 24$. 0! is equal to 1.

The Poisson distribution for the zero case becomes

$$P_0 = \frac{e^{-5}(1)}{1} = e^{-5}$$

e^{-5} is equal to 0.0067 (on the calculator: **5**, **+/−**, then $\mathbf{e^x}$).

Therefore, at an moi of 5, the probability that a cell will not be infected is 0.0067. This is equivalent to saying that 0.67% of the culture will be uninfected (100 × 0.0067 = 0.67%) or that 1 in 149 cells will be uninfected (1/0.0067 = 149).

Problem 4.5 A culture of bacteria is infected with bacteriophage at an moi of 0.2. Twenty-cell aliquots of the infected culture are withdrawn. How many phage-infected cells should be expected in each aliquot?

Solution 4.5

The number of phage-infected cells should equal the moi multiplied by the number of cells in the sample. The product represents an average number of infected cells per aliquot:

$$\frac{0.2 \text{ phage infections}}{\text{cell}} \times 20 \text{ cells} = 4 \text{ phage infections}$$

Therefore, in each 20-cell aliquot, there should be an average of four infected cells.

Problem 4.6 In a 20-cell aliquot from a culture infected at an moi of 0.2, what is the probability that no cells in that aliquot will be infected?

Solution 4.6

As shown in Problem 4.5, there should be, on average, four infected cells in a 20-cell aliquot. Using the Poisson distribution for the zero case, where *r* is equal to the number of successes (in this case, 0 infected cells is a success) and *m* is equal to the average number of infected cells per 20-cell aliquot (four), the Poisson distribution becomes

$$P_0 = \frac{e^{-4}4^0}{0!} = \frac{e^{-4}(1)}{(1)} = e^{-4}$$

To find e^{-4} on the calculator, enter **4**, **+/−** , then $\mathbf{e^x}$. This gives a value of 0.018.

Therefore, the probability of finding no infected cells in a 20-cell aliquot taken from a culture infected at an moi of 0.2 is 0.018. This probability can also be expressed as a '1 in' number by taking its reciprocal:

$$\frac{1}{0.018} = 55.6$$

Or, there is a 1-in-55.6 chance that a 20-cell aliquot will contain no infected cells.

Problem 4.7 In a 20-cell aliquot taken from a culture infected at an moi of 0.2, what is the probability of finding 12 infected cells?

Solution 4.7

The equation for the Poisson distribution can be used. In Problem 4.5, it was shown that the average number of infected cells in a 20-cell aliquot taken from such a culture is four and that this value represents the *m* factor in the Poisson distribution. The *r* factor for this problem, the number of successes, is 12. The Poisson distribution is then written

$$P_r = \frac{e^{-m}m^r}{r!}$$

$$P_{12} = \frac{e^{-4}4^{12}}{12!}$$

On the calculator, e^{-4} is found by entering **4**, **+/−** , then $\mathbf{e^x}$. This yields a value of 0.018. A value for 4^{12} is found on the calculator by entering **4**, $\mathbf{x^y}$, **1**, **2**, then **=**. This gives a value of 16 777 216. Twelve factorial (12!) is equal to the product of all integers from 1 to 12:

$$(12! = 12 \times 11 \times 10 \times 9 \times 8 \times 7 \times 6 \times 5 \times 4 \times 3 \times 2 \times 1)$$

On the calculator, this number is found by entering **1**, **2**, then ***x!***. This gives a value for 12! of 479 001 600. Placing these values into the equation for the Poisson distribution yields

$$P_{12} = \frac{e^{-4}4^{12}}{12!} = \frac{(0.018)(16\,777\,216)}{479\,001\,600} = \frac{301\,990}{479\,001\,600} = 6.3 \times 10^{-4}$$

Therefore, the probability that 12 infected cells will be found in an aliquot of 20 cells taken from a culture infected at an moi of 0.2 is 0.00063. This value can be expressed as a '1 in' number by taking its reciprocal:

$$\frac{1}{6.3 \times 10^{-4}} = 1587.3$$

There is a 1-in-1587.3 chance that 12 infected cells will be found in such a 20-cell aliquot.

Problem 4.8 What is the probability that a sample of 20 cells taken from a culture infected at an moi of 0.2 will have four or more infected cells?

Solution 4.8

As a first step, the probabilities for the events not included (the probabilities for the cases in which zero, one, two, or three infected cells are found per 20-cell sample) are calculated. The probability of all possible infections (i.e., the probability of having no infected cells, one infected cell, two infected cells, three infected cells, four infected cells, five infected cells, etc.) should equal 1. If the probabilities of obtaining zero, one, two, and three infected cells are subtracted from 1.0, then the remainder will be equivalent to the probability of obtaining four or more infected cells. As shown in Problem 4.5, *m*, the average number of infected cells in a 20-cell aliquot from a culture infected at an moi of 0.2, is 4. The probability for the case in which no infected cells are present in a 20-cell aliquot was calculated in Problem 4.6 and was found to be 0.018.

The probability that the 20-cell aliquot will contain one infected cell is

$$P_1 = \frac{e^{-4}4^1}{1!} = \frac{(0.018)(4)}{1} = 0.072$$

The probability that the 20-cell aliquot will contain two infected cells is

$$P_2 = \frac{e^{-4}4^2}{2!} = \frac{(0.018)(16)}{2 \times 1} = \frac{0.288}{2} = 0.144$$

The probability that the 20-cell aliquot will contain three infected cells is

$$P_3 = \frac{e^{-4}4^3}{3!} = \frac{(0.018)(64)}{3 \times 2 \times 1} = \frac{1.152}{6} = 0.192$$

The probability of having four or more infected cells in a 20-cell aliquot is then

$$1 - P_0 - P_1 - P_2 - P_3 = 1 - 0.018 - 0.072 - 0.144 - 0.192 = 0.574$$

Therefore, 57.4% ($0.574 \times 100 = 57.4\%$) of the time, a 20-cell aliquot infected at an moi of 0.2 will contain four or more infected cells. Expressed as a '1 in' number, 1 in every 1.7 20-cell aliquots (the reciprocal of $0.574 = 1.7$) will contain four or more infected cells.

Problem 4.9 What is the probability that, in a 20-cell aliquot from a culture infected at an moi of 0.2, zero or one phage-infected cell will be found?

Solution 4.9

As shown in Problem 4.8, under these experimental conditions, the probability (P) of finding zero infected cells (P_0) is equal to 0.018 and the probability of finding one infected cell (P_1) is equal to 0.072. The probability of finding either zero or one phage-infected cell in a 20-cell aliquot is the sum of the two probabilities:

$$0.018 + 0.072 = 0.090$$

Therefore, 9% of the time ($100 \times 0.090 = 9\%$), either zero or one phage-infected cell will be found. Or, expressed as a '1 in' number, 1 in 11 20-cell aliquots will contain either zero or one infected cell (the reciprocal of 0.090 is equal to 11).

Problem 4.10 Following a period to allow for phage adsorption, cell culture samples are plated to determine the number of infected cells. It is found that 60% of the cells from the infected culture did not produce a burst of phage; i.e., they were not infected. What is the moi (the average number of phage-infected cells) of the culture?

Solution 4.10

For this problem, the equation for the Poisson distribution must be solved for m. Since the problem describes the zero case, the Poisson distribution,

$$P_r = \frac{e^{-m} m^r}{r!}$$

becomes

$$0.6 = \frac{e^{-m} m^0}{0!}$$

Since m^0 and 0! are both equal to 1, the equation is

$$0.6 = \frac{e^{-m}(1)}{1} = e^{-m}$$

To solve for m, the following relationship can be used: $P = e^{-m}$, which is equivalent to $m = -\ln P$. Solving for m yields

$$m = -\ln 0.6$$

To determine the natural log of 0.6 on a calculator, enter ., **6**, then **ln**. This gives a value, when rounded, of −0.51. Therefore,

$$m = -(-0.51)$$

Since the negative of a negative number is a positive value, $m = 0.51$.

Therefore, the moi of the culture is 0.51.

4.3 MEASURING PHAGE TITER

The latter stage of bacteriophage development involves release of progeny phage particles from the cell. Different bacteriophages accomplish this step in different ways. Late during infection, the bacteriophage λ encodes a protein, endolysin, which digests the bacterial cell wall. When the cell wall is critically weakened, the cell bursts open, releasing the 50–200 virus particles produced during the infection. The infected cell is killed by this process. Following replication of the M13 phage genome inside an infected cell, it is packaged into a protein coat as it passes through the cytoplasmic membrane on its way out of the cell. Mature M13 phage particles pass out of the cell individually rather than in a burst. Although M13 phage do not lyse infected cells, those harboring M13 grow more slowly than uninfected cells.

When plated with a lawn of susceptible cells, a bacteriophage will form a circular area of reduced turbidity called a **plaque**. Depending on the bacteriophage, a plaque may be clear (such as is formed by phage T1 or the clear mutants of λ) or a plaque can be an area containing slow-growing cells (as is formed by M13 infection).

When a phage stock is diluted and plated with susceptible cells such that individual plaques can be clearly discerned on a bacterial lawn, it is possible to determine the concentration of virus particles within the stock. If the phage stock is diluted to a degree that individual and well-isolated plaques appear on the bacterial lawn, it is assumed that each plaque results from one phage infecting one cell. The process of determining phage concentration by dilution and plating with susceptible cells is called **titering** or the **plaque assay**. This method determines the number of viable phage particles in a stock suspension. A bacteriophage capable of productively infecting a cell is called a **plaque-forming unit** (**PFU**).

The following problems demonstrate how to calculate phage titer and to perform dilutions of phage stocks.

Problem 4.11 A bacteriophage stock is diluted in the following manner: 0.1 mL of the phage stock is diluted into 9.9 mL of dilution buffer (making a total volume in this first dilution tube of 10.0 mL). From this first dilution tube, 0.1 mL is withdrawn and diluted into a second tube containing 9.9 mL of dilution buffer. From the second dilution tube, 0.1 mL is taken and diluted into a third tube containing 9.9 mL of dilution buffer. From this third tube, 1.0 mL is withdrawn and added to 9.0 mL in a fourth tube. Finally, 0.1 mL is withdrawn from the fourth dilution tube and plated with 0.2 mL of susceptible cells in melted top agar. After incubating the plate overnight, 180 plaques are counted. What is the titer of the bacteriophage stock?

Solution 4.11

The series of dilutions described in this problem can be written so that each dilution step is represented as a fraction. To obtain the overall dilution factor, multiply all fractions.

$$\frac{0.1\text{ mL}}{10\text{ mL}} \times \frac{0.1\text{ mL}}{10\text{ mL}} \times \frac{0.1\text{ mL}}{10\text{ mL}} \times \frac{1.0\text{ mL}}{10\text{ mL}} \times 0.1\text{ mL}$$

Expressing this series of fractions in scientific notation yields the following overall dilution (and see Figure 4.1):

$$(1 \times 10^{-2}) \times (1 \times 10^{-2}) \times (1 \times 10^{-2}) \times (1 \times 10^{-1}) \times (1 \times 10^{-1}\text{ mL}) = 1 \times 10^{-8}\text{ mL}$$

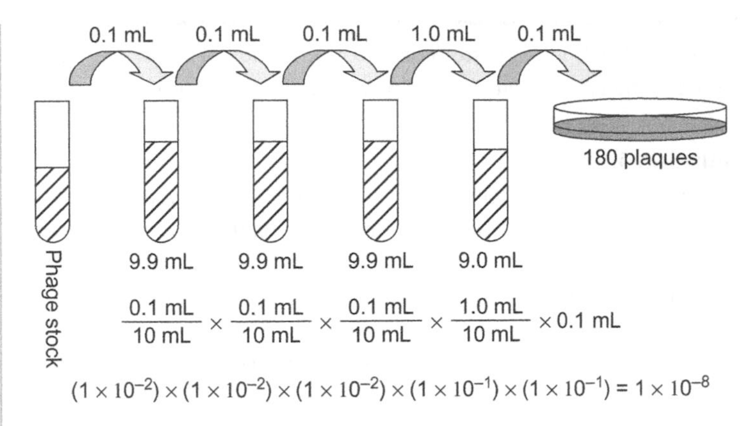

■ **FIGURE 4.1** The dilution series for Problem 4.11. A dilution of 1×10^{-8} yields 180 plaques.

To obtain the concentration of phage in the stock suspension, divide the number of plaques counted by the dilution factor.

$$\frac{180\text{ PFU}}{1 \times 10^{-8}\text{ mL}} = 1.8 \times 10^{10}\text{ PFU/mL}$$

Therefore, the bacteriophage stock has a titer of 1.8×10^{10} PFU/mL.

■

4.4 DILUTING BACTERIOPHAGE

It is often necessary to dilute a bacteriophage stock so that the proper amount (and a convenient volume) of virus can be added to a culture. The following considerations should be taken into account when planning a dilution scheme.

- A 0.1 mL aliquot taken from the last dilution tube is a convenient volume to plate with susceptible cells. This would comprise 1×10^{-1} of the 2×10^{-7} dilution factor.
- A 0.1 mL aliquot is a convenient volume to remove from the phage stock into the first dilution tube. If the first dilution tube contains 9.9 mL of buffer, then this would account for 1×10^{-2} of the 2×10^{-7} dilution factor.
- Since the significant digit of the dilution factor is 2 (the 2 in 2×10^{-7}), this number must be brought into the dilution series. This can be accomplished by making a 0.2 mL/10 mL dilution (equivalent to a dilution factor of 2×10^{-2}). Therefore, 0.2 mL taken from the first dilution tube can be added to a second dilution tube, containing 9.8 mL of buffer.

Problem 4.12 A phage stock has a concentration of 2.5×10^9 PFU/mL. How can the stock be diluted and plated to give 500 plaques on a plate?

Solution 4.12

This problem can be tackled by taking the reverse approach of that taken for Problem 4.11. If x represents the dilution factor required to form 500 plaques per plate from a phage stock having a concentration of 2.5×10^9 PFU/mL, then the following equation can be written:

$$\frac{500\ \text{PFU}}{x\ \text{mL}} = \frac{2.5 \times 10^9\ \text{PFU}}{\text{mL}}$$

Solving for x gives

$$500\ \text{PFU} = \frac{(2.5 \times 10^9\ \text{PFU})(x\ \text{mL})}{\text{mL}}$$

$$\frac{500\ \text{PFU}}{2.5 \times 10^9\ \text{PFU}} = x = 2 \times 10^{-7}$$

Therefore, the phage stock must be diluted by 2×10^{-7}.

The phage can be diluted as follows:

$$(1 \times 10^{-2}) \times (2 \times 10^{-2}) \times (1 \times 10^{-1}\ \text{mL}) = 2 \times 10^{-5}\ \text{mL}$$

The remaining amount of the dilution factor to be accounted for is

$$(2 \times 10^{-5}) \times (x) = 2 \times 10^{-7}$$

Solving for x yields

$$x = \frac{2 \times 10^{-7}}{2 \times 10^{-5}} = 1 \times 10^{-2}$$

The last of the dilutions of those remaining should be one in which 0.1 mL is taken into 9.9 mL of buffer. Therefore, if a phage stock at a concentration of 2.5×10^9 PFU/mL is diluted as follows:

$$\frac{0.1\ \text{mL}}{10\ \text{mL}} \times \frac{0.2\ \text{mL}}{10\ \text{mL}} \times \frac{0.1\ \text{mL}}{10\ \text{mL}} \times 0.1\ \text{mL plated}$$

then 500 plaques should appear on the plated lawn (see Figure 4.2).

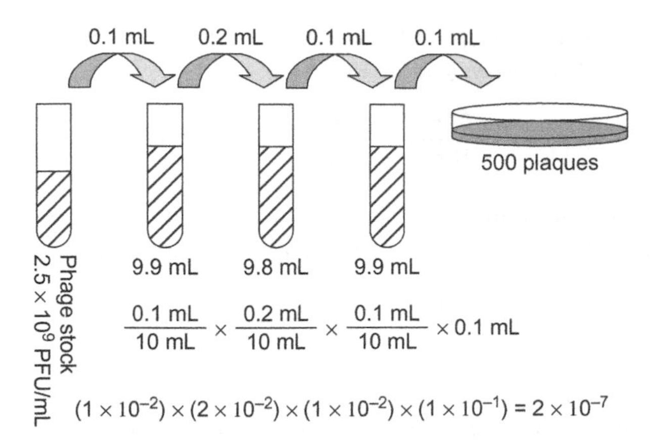

■ **FIGURE 4.2** The dilution series and phage plating of Problem 4.12 describes diluting the phage stock by 2×10^{-7} and the generation of 500 plaques from this dilution.

4.5 MEASURING BURST SIZE

Mutation of either the phage or the bacterial genome or changes in the conditions under which infection occurs can alter the phage/host interaction and the efficiency with which phage replication occurs. Measuring the number of progeny bacteriophage particles produced within and released from an infected cell is a simple way to gauge overall gene expression of the infecting phage. The number of mature virus particles released from an infected cell is called the **burst size**.

To perform an experiment measuring burst size, for example, phage are added to susceptible cells at an moi of 0.1 and allowed to adsorb for 10–30 min on ice. The cell/phage suspension is diluted into a large volume with cold growth media and centrifuged to pellet the cells. The supernatant is poured off to remove unadsorbed phage. The pelleted cells are resuspended in cold media and then diluted into a growth flask containing prewarmed media. A sample is taken immediately and assayed for PFUs. The plaques arising from this titration represent the number of infected cells (termed **infective centers (ICs)**); each infected cell gives rise to one plaque. After a period of time sufficient for cell lysis to occur, phage are titered again for PFUs. The burst size is calculated as PFUs released per IC (PFUs/IC).

Problem 4.13 In a burst size experiment, phage are added to 2×10^8 cells at an moi of 0.1 and allowed to adsorb to susceptible cells for 20 minutes. Dilution and centrifugation is performed to remove unadsorbed phage. The pelleted infected cells are resuspended in 10 mL of tryptone broth. At this point, ICs are assayed by diluting an aliquot of the infected cells as follows (and see Figure 4.3):

$$\frac{0.1\,\text{mL}}{10\,\text{mL}} \times \frac{0.1\,\text{mL}}{10\,\text{mL}} \times 0.1\,\text{mL plated}$$

The resuspended infected cells are then shaken at 37°C for 90 min. The number of PFUs is then assayed by diluting an aliquot as follows:

$$\frac{0.1\,\text{mL}}{10\,\text{mL}} \times \frac{0.1\,\text{mL}}{10\,\text{mL}} \times \frac{1\,\text{mL}}{10\,\text{mL}} \times \frac{0.5\,\text{mL}}{10\,\text{mL}} \times 0.1\,\text{mL plated}$$

From the assay of ICs, 200 plaques are counted on the bacterial lawn. From the assay of PFUs following the 90 min incubation, 150 plaques are counted from the diluted sample (Figure 4.4). What is the burst size, expressed as PFUs/IC?

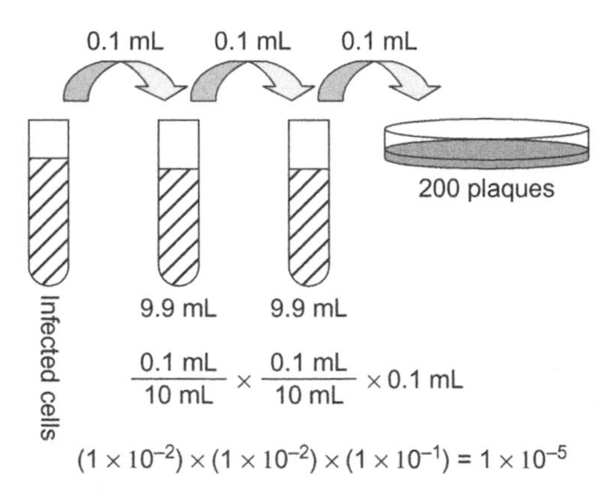

■ **FIGURE 4.3** The dilution series described in Problem 4.13 yields 200 plaques.

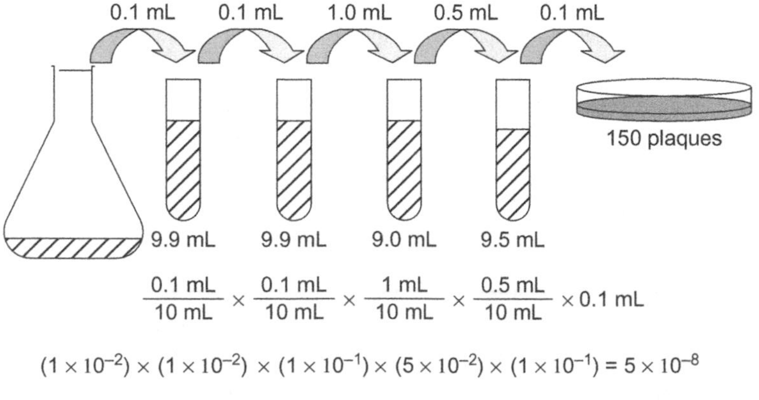

$$\frac{0.1\,\text{mL}}{10\,\text{mL}} \times \frac{0.1\,\text{mL}}{10\,\text{mL}} \times \frac{1\,\text{mL}}{10\,\text{mL}} \times \frac{0.5\,\text{mL}}{10\,\text{mL}} \times 0.1\,\text{mL}$$

$$(1 \times 10^{-2}) \times (1 \times 10^{-2}) \times (1 \times 10^{-1}) \times (5 \times 10^{-2}) \times (1 \times 10^{-1}) = 5 \times 10^{-8}$$

■ **FIGURE 4.4** The dilution series described in Problem 4.13 (a dilution of 5×10^{-8}) yields 150 plaques.

Solution 4.13

The number of ICs is determined by dividing the number of plaques obtained following dilution by the dilution factor. The dilution scheme to obtain ICs is

$$\frac{0.1\,\text{mL}}{10\,\text{mL}} \times \frac{0.1\,\text{mL}}{10\,\text{mL}} \times 0.1\,\text{mL plated}$$

This is equivalent to the expression

$$(1 \times 10^{-2})(1 \times 10^{-2})(1 \times 10^{-1}\,\text{mL}) = 1 \times 10^{-5}\,\text{mL}$$

The number of ICs is then equal to

$$\frac{200\,\text{PFUs}}{1 \times 10^{-5}\,\text{mL}} = 2 \times 10^{7}\,\text{PFUs/mL} = 2 \times 10^{7}\,\text{ICs/mL}$$

The number of PFUs following the 90 min incubation is obtained by diluting the sample in the following manner:

$$\frac{0.1\,\text{mL}}{10\,\text{mL}} \times \frac{0.1\,\text{mL}}{10\,\text{mL}} \times \frac{1\,\text{mL}}{10\,\text{mL}} \times \frac{0.5\,\text{mL}}{10\,\text{mL}} \times 0.1\,\text{mL plated}$$

This is equivalent to 5×10^{-8}, as shown below.

$$(1 \times 10^{-2})(1 \times 10^{-2})(1 \times 10^{-1})(5 \times 10^{-2})(1 \times 10^{-1}\,\text{mL}) = 5 \times 10^{-8}\,\text{mL}$$

Since 150 plaques were obtained from this dilution, the concentration of phage in the culture after 90 min is

$$\frac{150\,\text{PFUs}}{5 \times 10^{-8}\,\text{mL}} = 3 \times 10^{9}\,\text{PFUs/mL}$$

The burst size is then calculated as the concentration of phage following 90 min incubation divided by the concentration of ICs:

$$\frac{3 \times 10^{9}\,\text{PFUs/mL}}{2 \times 10^{7}\,\text{ICs/mL}} = \frac{3 \times 10^{9}\,\text{PFUs}}{\text{mL}} \times \frac{\text{mL}}{2 \times 10^{7}\,\text{ICs}}$$

$$= \frac{3 \times 10^{9}\,\text{PFUs}}{2 \times 10^{7}\,\text{ICs}} = 150\,\text{PFUs/IC}$$

Note: To perform this calculation, a fraction (3×10^{9} PFUs/mL) is divided by a fraction (2×10^{7} ICs/mL). The relationship described below can be used in such a situation.

Dividing a fraction by a fraction is the same as multiplying the numerator fraction by the reciprocal of the denominator fraction. A phrase frequently used to describe this action is 'to invert and multiply.' Therefore,

$$\frac{\frac{1}{a}}{\frac{1}{b}} = \frac{1}{a} \times \frac{b}{1} = \frac{b}{a} \quad \text{and} \quad \frac{\frac{a}{1}}{\frac{b}{1}} = \frac{a}{1} \times \frac{1}{b} = \frac{a}{b}$$

CHAPTER SUMMARY

The ratio of infecting bacteriophage to host cell is called the multiplicity of infection (moi). The probability that any bacterial cell will or will not be infected when a culture of that bacteria is exposed to a certain number of virus particles can be calculated using the Poisson distribution

$$P_r = \frac{e^{-m} m^r}{r!}$$

where P is the probability, r is the number of successes, m is the average number of phage/cell (or infected cells), and e is the base of the natural logarithm. The concentration of bacteriophage per unit volume is called the phage titer and is calculated by determining the number of PFUs in a dilution of the phage stock. Burst size is the number of viable phage released by an infected cell following replication of the virus.

Chapter 5

Nucleic acid quantification

5.1 QUANTIFICATION OF NUCLEIC ACIDS BY ULTRAVIOLET (UV) SPECTROSCOPY

Any experiment requiring manipulation of a nucleic acid most likely also requires its accurate quantification to ensure optimal and reproducible results. The nitrogenous bases positioned along a nucleic acid strand absorb ultraviolet (UV) light at a wavelength of 260 nm; at this wavelength, light absorption is proportional to nucleic acid concentration. This relationship is so well characterized that UV absorption is used to accurately determine the concentration of nucleic acids in solution. The relationship between DNA concentration and absorption is linear up to an absorption at 260 nm (A_{260}) of 2 (Figure 5.1). For measuring the absorption of a nucleic acid solution in a spectrophotometer, most molecular biology laboratories will use quartz cuvettes with a width through which the light beam will travel 1 cm. Therefore, all discussions in this chapter assume a 1 cm light path.

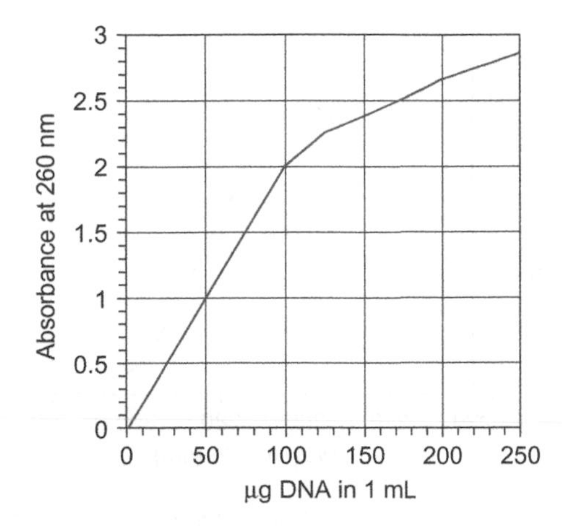

■ **FIGURE 5.1** The concentration of DNA and absorbance at 260 nm is a linear relationship up to an A_{260} value of approximately 2.

Calculations for Molecular Biology and Biotechnology. DOI: 10.1016/B978-0-12-375690-9.00005-X
© 2010 Elsevier Inc. All rights reserved.

5.2 DETERMINING THE CONCENTRATION OF DOUBLE-STRANDED DNA (dsDNA)

Applications requiring quantitation of double-stranded DNAs (dsDNAs) include protocols utilizing plasmids, viruses, or genomes. Quantitation is typically performed by taking absorbance measurements at 260, 280, and 320 nm. Absorbance at 260 nm is used to specifically detect the nucleic acid component of a solution. Absorbance at 280 nm is used to detect the presence of protein (since tryptophan (Trp) residues absorb at this wavelength). Absorbance at 320 nm is used to detect any insoluble light-scattering components. A spectrophotometer capable of providing a scan from 200 to 320 nm will yield maximum relevant information (Figure 5.2).

For nucleic acids purified from a biological source (as opposed to those made synthetically), calculating the ratio of the readings obtained at 260 and 280 nm can give an estimate of protein contamination. Pure DNA free of protein contamination will have an A_{260}/A_{280} ratio close to 1.8. If phenol or protein contamination is present in the DNA prep, the A_{260}/A_{280} ratio will be less than 1.8. If RNA is present in the DNA prep, the A_{260}/A_{280} ratio may be greater than 1.8. Pure RNA preparations will have an A_{260}/A_{280} ratio close to 2.0.

At 260 nm, DNA concentrations as low as 2 mg/mL can be detected. A solution of DNA with a concentration of 50 mg/mL will have an absorbance at 260 nm equal to 1.0. Written as an equation, this relationship is

$$1\ \mathrm{A}_{260}\ \text{unit of dsDNA} = 50\ \mu\text{g DNA/mL}$$

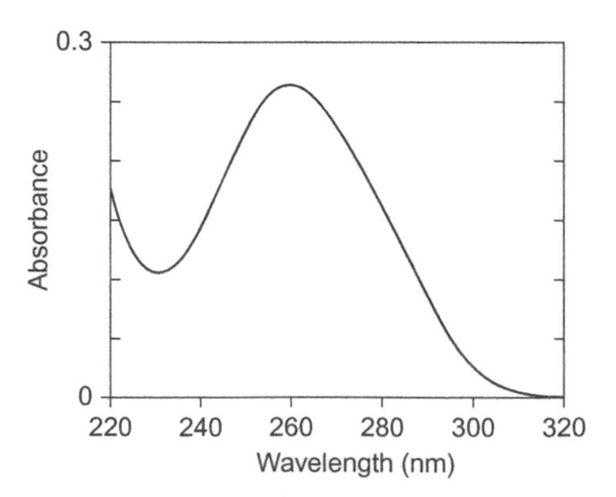

■ **FIGURE 5.2** A typical spectrophotometric scan of double-stranded DNA (dsDNA). Maximum absorbance occurs at 260 nm. Absorbance at 230 nm is an indication of salt in the sample. If the sample is optically clear, it should give a very low reading at 320 nm, which is in the visible wavelength region of the spectrum.

Absorbance and **optical density (OD)** are terms often used interchangeably. The foregoing relationship can also be written as a conversion factor:

$$\frac{50\,\mu\text{g DNA/mL}}{1.0\text{ OD}}$$

The following problems use this relationship.

Problem 5.1 From a small culture, you have purified the DNA of a recombinant plasmid. You have resuspended the DNA in a volume of 50 μL. You dilute 20 μL of the purified DNA sample into a total volume of 1000 μL distilled water. You measure the absorbance of this diluted sample at 260 and 280 nm and obtain the following readings: $A_{260} = 0.550$; $A_{280} = 0.324$.

a) What is the DNA concentration of the 50 μL plasmid prep?

b) How much total DNA was purified by the plasmid prep procedure?

c) What is the A_{260}/A_{280} ratio of the purified DNA?

Solution 5.1(a)

This problem can be solved by setting up the following ratio:

$$\frac{x\,\mu\text{g DNA/mL}}{0.550\text{ OD}} = \frac{50\,\mu\text{g DNA/mL}}{1.0\text{ OD}}$$

$$x\,\mu\text{g DNA/mL} = \frac{(0.550\text{ OD})(50\,\mu\text{g DNA/mL})}{1.0\text{ OD}}$$

$$x = 27.5\,\mu\text{g DNA/mL}$$

This amount represents the concentration of the diluted DNA solution that was used for the spectrophotometer. To determine the concentration of DNA in the original 50 mL plasmid prep, this value must be divided by the dilution factor. A 20 mL sample of the plasmid prep DNA was diluted into water for a total diluted volume of 1000 mL.

$$\frac{27.5\,\mu\text{g DNA/mL}}{\frac{20}{1000}} = 27.5\,\mu\text{g DNA/mL} \times \frac{1000}{20}$$

$$= \frac{27\,500\,\mu\text{g DNA/mL}}{20} = 1375\,\mu\text{g DNA/mL}$$

Therefore, the original 50 μL plasmid prep has a concentration of 1375 μg DNA/mL. To bring this value to an amount of DNA per μL, it can be multiplied by the conversion factor, 1 mL/1000 μL:

$$\frac{1375\,\mu\text{g DNA}}{\text{mL}} \times \frac{1\text{mL}}{1000\,\mu\text{L}} = 1.35\,\mu\text{g DNA}/\mu\text{L}$$

Therefore, the 50 μL plasmid prep has a concentration of 1.35 μg DNA/mL.

Solution 5.1(b)

The total amount of DNA recovered by the plasmid prep procedure can be calculated by multiplying the DNA concentration obtained earlier by the volume containing the recovered DNA:

$$\frac{1375\,\mu\text{g DNA}}{\text{mL}} \times \frac{1\text{ mL}}{1000\ \mu\text{L}} \times 50\,\mu\text{L} = \frac{(1375)(50)}{1000} = 68.75\,\mu\text{g DNA}$$

Therefore, the original 50 μL plasmid prep contained a total of 68.75 μg DNA. However, since 20 μL was used for diluting and reading in the spectrophotometer, only 30 μL of sample prep remains. Therefore, the total remaining amount of DNA is calculated by multiplying the remaining volume by the concentration:

$$30\,\mu\text{L} \times \frac{1.35\,\mu\text{g DNA}}{\mu\text{L}} = 40.5\,\mu\text{g DNA}$$

Therefore, although 68.75 μg of DNA were recovered from the plasmid prep procedure, you used up some of it for spectrophotometry and you now have 40.5 μg DNA remaining.

Solution 5.1(c)

The A_{260}/A_{280} ratio is

$$\frac{0.550}{0.323} = 1.703$$

5.2.1 Using absorbance and an extinction coefficient to calculate double-stranded DNA (dsDNA) concentration

At a neutral pH and assuming a G+C DNA content of 50%, a DNA solution having a concentration of 1 mg/mL will have an absorption in a 1 cm light path at 260 nm (A_{260}) of 20. For most applications in molecular biology, G+C content, unless very heavily skewed, need not be a consideration when quantitating high-molecular-weight dsDNA. The absorption value of 20 for a 1 mg DNA/mL solution is referred to as DNA's **extinction coefficient**. It is represented by the symbol E or e. The term **extinction coefficient** is used interchangeably with **absorption constant** or **absorption coefficient**. The formula that describes the relationship between absorption at 260 nm (A_{260}), concentration (c) (in mg/mL), the light path length (l) of the cuvette (in centimeters), and the extinction coefficient at 260 nm (E_{260}) for a 1 cm light path is

$$A_{260} = E_{260} l c$$

This relationship is known as **Beer's Law**. Since the light path, l, is 1, this equation becomes

$$A_{260} = E_{260} c$$

Rearranging the equation, the concentration of the nucleic acid, c, becomes

$$c = \frac{A_{260}}{E_{260}}$$

Problem 5.2 A DNA solution has an A_{260} value of 0.5. What is the DNA concentration in μg DNA/mL?

Solution 5.2

The answer can be obtained using the equation for Beer's Law.

$$c = \frac{A_{260}}{E_{260}}$$

$$c = \frac{0.5}{20} = 0.025\ \text{mg DNA/mL}$$

$$0.025\ \text{mg DNA/mL} \times \frac{1000\ \mu\text{g}}{\text{mg}} = 25\ \mu\text{g DNA/mL}$$

Therefore, the sample has a concentration of 25 μg DNA/mL.

Problem 5.3 A DNA solution has an A_{260} value of 1.0. What is the DNA concentration in μg DNA/mL?

Solution 5.3

The answer can be obtained using the equation for Beer's Law.

$$c = \frac{A_{260}}{E_{260}}$$

$$c = \frac{1.0}{20} = 0.05 \text{ mg DNA/mL}$$

$$0.05 \text{ mg DNA/mL} \times \frac{1000\ \mu\text{g}}{\text{mg}} = 50\ \mu\text{g DNA/mL}$$

Therefore, the solution has a concentration of 50 μg DNA/mL. Notice that this value is the one described earlier.

$$1\,A_{260} \text{ unit of dsDNA} = 50\ \mu\text{g DNA/mL}$$

5.2.2 Calculating DNA concentration as a millimolar (m*M*) amount

The extinction coefficient (E_{260}) for a 1 m*M* solution of dsDNA is 6.7. This value can be used to calculate the molarity of a solution of DNA.

Problem 5.4 A solution of DNA has an absorbance at 260 nm of 0.212. What is the concentration of the DNA solution expressed as millimolarity?

Solution 5.4

This problem can be solved by setting up a relationship of ratios such that it is read '1 m*M* is to 6.7 OD as *x* m*M* is to 0.212 OD.'

$$\frac{1\text{ m}M}{6.7\text{ OD}} = \frac{x\text{ m}M}{0.212\text{ OD}}$$

$$\frac{(1\text{ m}M)(0.212\text{ OD})}{6.7\text{ OD}} = x\text{ m}M$$

$$0.03\text{ m}M = x$$

Therefore, a DNA solution with an A_{260} of 0.212 has a concentration of 0.03 m*M*.

Problem 5.5 A solution of DNA has an absorbance at 260 nm of 1.00. What is its concentration expressed as millimolarity?

Solution 5.5

This problem can be solved using ratios, with one of those ratios being the relationship of a 1 m*M* solution of dsDNA to the extinction coefficient 6.7.

$$\frac{x \text{ m}M}{1.00 \text{ OD}} = \frac{1 \text{ m}M}{6.7 \text{ OD}}$$

$$x \text{ m}M = \frac{(1 \text{ m}M)(1.00 \text{ OD})}{6.7 \text{ OD}} = 0.15 \text{ m}M$$

Therefore, a solution of dsDNA with an A_{260} of 1.00 has a concentration of 0.15 m*M*. This relationship,

$$1.0\ A_{260} \text{ of dsDNA} = 0.15 \text{ m}M$$

has frequent use in the laboratory.

Problem 5.6 A solution of DNA has a concentration of 0.03 m*M*. What is its concentration expressed as pmol/μL?

Solution 5.6

A 0.03 m*M* solution, by definition, has a concentration of 0.03 millimoles per liter. A series of conversion factors is used to cancel terms and to transform a concentration expressed as millimolarity to one expressed as pmol/μL:

$$\frac{0.03 \text{ mmol}}{\text{L}} \times \frac{1\text{L}}{1 \times 10^6 \ \mu\text{L}} \times \frac{1 \times 10^9 \text{ pmol}}{\text{mmol}} = 30 \text{ pmol}/\mu\text{L}$$

Therefore, a 0.03 m*M* DNA solution has a concentration of 30 pmol DNA/μL.

5.2.3 Using PicoGreen® to determine DNA concentration

Determining DNA concentration by measuring a sample's absorbance at 260 nm, though a common practice and, for many years, the gold standard of methods for DNA quantification, can be prone to inaccuracies that result from the contribution that such contaminants as salt, protein, nucleotides (nts), and RNA can make to the absorbance value. In addition, a solution of DNA having a concentration less than 2 μg/mL cannot be quantified reliably by measuring its 260 nm absorbance. More recently, however, fluorescent dyes have been used as a tool for nucleic acid quantification, the best example of which is PicoGreen® from Life Technologies. PicoGreen offers the advantages that it is specific for dsDNA, only fluoresces when bound to dsDNA, and can detect dsDNA at concentrations as low as 25 pg/mL.

Quantification of dsDNA using PicoGreen requires that a standard curve of known concentrations be made using the same reagents prepared for the unknown sample. The DNA used for the standard curve is typically prepared from bacteriophage λ or calf thymus and is diluted (in TE) in concentrations from 1 ng/mL to 1000 ng/mL. The DNA dilutions are combined with PicoGreen reagent, allowed to incubate for several minutes at room temperature, and then read on a spectrofluorometer that excites the samples at a wavelength of 480 nm and reads their emission intensity at 520 nm. A standard curve presents the samples' fluorescence (y axis) vs. DNA concentration (x axis). An equation describing that curve can then be used to calculate the DNA concentration of an unknown sample based on its fluorescence, as demonstrated in the following problem.

Problem 5.7 A dilution series of calf thymus DNA is assayed for its DNA content using PicoGreen. The following results are obtained. (These values represent the concentrations of DNA in the assay tubes.)

DNA concentration (ng/mL)	Fluorescence
1	5572
2.5	6945
5	9245
10	13 820
25	27 585
50	50 520
100	99 710
250	234 952
500	450 210
750	700 025
1000	920 110

A sample of human DNA using PicoGreen reagent generates fluorescence of 28 795 units. What is its concentration?

Solution 5. 7

We will generate a standard curve using Microsoft Excel, determine the line of best fit's (the regression line's) equation, and then use that equation to calculate the concentration of the unknown human DNA sample. The protocol for using the Excel graphing utility can be found in Appendix A.

In the Excel spreadsheet, enter 'ng/mL (x)' in the column A, row 1 box and 'Fluorescence (y)' in the column B, row 1 box. Fill the data in for both columns. The spreadsheet should look similar to Figure 5.3.

Plot the above data using the 'XY (Scatter)' chart type. When a trendline is added according to the instructions in Appendix A, the chart will appear on the spreadsheet as shown in Figure 5.4.

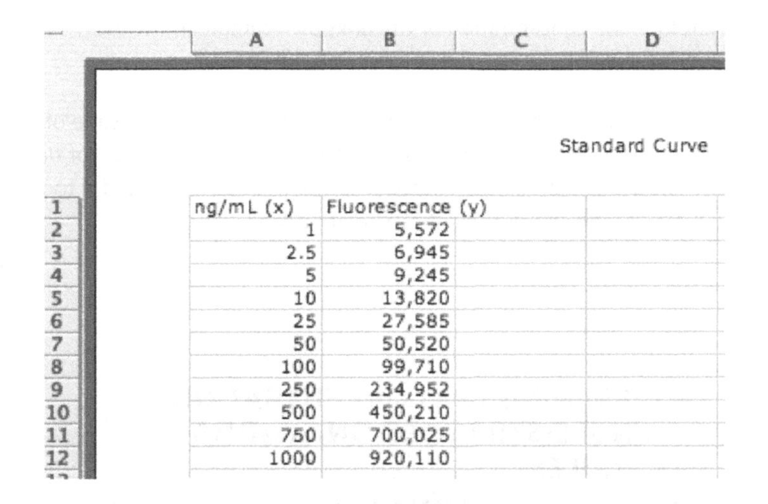

Standard Curve

	A	B	C	D
1	ng/mL (x)	Fluorescence (y)		
2	1	5,572		
3	2.5	6,945		
4	5	9,245		
5	10	13,820		
6	25	27,585		
7	50	50,520		
8	100	99,710		
9	250	234,952		
10	500	450,210		
11	750	700,025		
12	1000	920,110		

■ **FIGURE 5.3** The values of fluorescence for each diluted sample of calf thymus DNA used for the standard curve assay in Problem 5.7, as entered in an Excel spreadsheet.

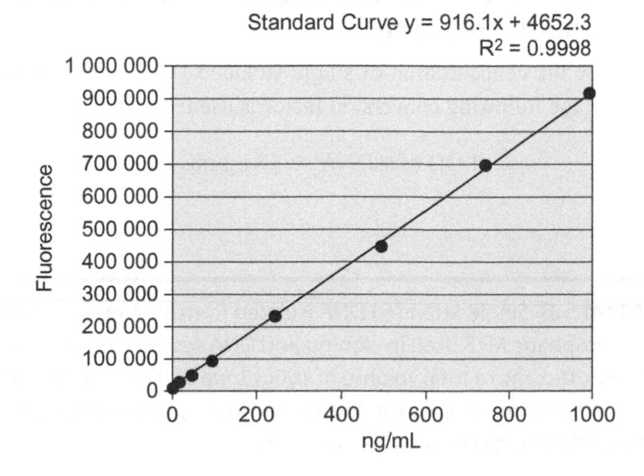

■ **FIGURE 5.4** The regression line and equation for Problem 5.7 data, as calculated in Microsoft Excel.

The regression equation for this line is, therefore, $y = 916.1x + 4652.3$. We now calculate the concentration of the unknown sample using this equation. The unknown sample generated a fluorescence value of 28 795. This is the y value in the equation. We then solve for x to give us the DNA concentration in ng/mL:

$$y = 916.1x + 4652.3$$
$$28\,795 = 916.1x + 4652.3$$
$$916.1x = 28\,795 - 4652.3$$
$$916.1x = 24\,142.7$$
$$x = \frac{24\,142.7}{916.1} = 26.35$$

Therefore, the concentration of the unknown sample in the assay tube is 26.35 ng/mL.

Note: If a fluorescence value for an unknown sample falls outside of that covered by the standard curve and outside of the linear range of detection, the sample should be diluted if too high or concentrated if too low so that a reliable measurement can be obtained.

5.3 DETERMINING THE CONCENTRATION OF SINGLE-STRANDED DNA (ssDNA) MOLECULES

5.3.1 Single-stranded DNA (ssDNA) concentration expressed in μg/mL

To determine the concentration of single-stranded DNA (ssDNA) as a μg/mL amount, the following conversion factor is used:

$$1 \text{ OD of ssDNA} = 33\ \mu\text{g/mL}$$

Problem 5.8 Single-stranded DNA isolated from M13mp18, a derivative of bacteriophage M13 used in cloning and DNA sequencing applications, is diluted 10 μL into a total volume of 1000 μL water. The absorbance of this diluted sample is read at 260 nm and an A_{260} value of 0.325 is obtained. What is its concentration in μg/mL?

Solution 5.8

This problem can be solved by setting up a ratio, with the variable x representing the concentration in μg/mL for the diluted sample. The equation can be read 'x μg/mL is to 0.125 OD as 33 μg/mL is to 1 OD.' Once x is obtained, the concentration of the stock DNA can be determined by multiplying the concentration of the diluted sample by the dilution factor.

$$\frac{x\ \mu g/mL}{0.125\ OD} = \frac{33\ \mu g/mL}{1\ OD}$$

$$x\ \mu g/mL = 0.125 \times 33\ \mu g/mL$$

$$= 4.125\ \mu g/mL$$

Therefore, the concentration of the diluted sample is 4.125 μg/mL. To determine the concentration of the M13mp18 DNA stock solution, this value must be multiplied by the dilution factor:

$$4.125\ \mu g/mL \times \frac{1000\ \mu L}{10\ \mu L} = \frac{4125\ \mu g/mL}{10}$$

$$= 412.5\ \mu g/mL$$

Therefore, the stock of M13mp18 DNA has a concentration of 412.5 μg/mL.

■

5.3.2 Determining the concentration of high-molecular-weight single-stranded DNA (ssDNA) in pmol/μL

The concentration of high-molecular-weight ssDNA can be expressed as a pmol/μL amount by first determining how many micrograms of the ssDNA are equivalent to one pmol. To do this, we use the average molecular weight of a deoxynucleotide in a DNA strand. For ssDNA, it is taken to be 330 daltons. This value is then used as a conversion factor to bring the concentration of ssDNA expressed as μg/mL to a concentration expressed in pmol/μL.

> The unit **dalton** is defined as 1/12 the mass of the carbon-12 atom. It is used interchangeably with 'molecular weight,' a quantity expressed as grams/mole. That is, there are 330 g of nt per mole of nt.

Problem 5.9 A stock of M13mp18 DNA has a concentration of 412.5 μg/mL. What is this concentration expressed in pmol/μL?

Solution 5.9

The cloning vector M13mp18 is 7250 nts in length. To express this as μg/pmol, the following relationship is set up, in which a series of conversion factors is used to cancel terms:

$$7250 \text{ nts} \times \frac{330 \text{ g/mol}}{\text{nt}} \times \frac{1 \times 10^6 \ \mu\text{g}}{\text{g}} \times \frac{\text{mole}}{1 \times 10^{12} \text{ pmol}} = 2.39 \ \mu\text{g/pmol}$$

Therefore, 2.39 μg of a 7250 nt-long ssDNA molecule is equivalent to 1 pmol. This value can now be used to convert μg/mL to pmol/μL:

$$\frac{412.5 \ \mu\text{g}}{\text{mL}} \times \frac{1 \text{pmol}}{2.39 \ \mu\text{g}} \times \frac{1 \text{mL}}{1000 \ \mu\text{L}} = 0.17 \text{pmol}/\mu\text{L}$$

Therefore, the M13mp18 DNA stock has a concentration of 0.17 pmol/μL.

5.3.3 Expressing single-stranded DNA (ssDNA) concentration as a millimolar (m*M*) amount

The extinction coefficient (E_{260}) for a 1 m*M* solution of ssDNA is 8.5. This value can be used to determine the millimolarity concentration of any ssDNA solution from its absorbance.

Problem 5.10 A 1 mL sample of ssDNA has an absorbance of 0.285. What is its m*M* concentration?

Solution 5.10

The following relationship can be used to determine the millimolarity concentration.

$$\frac{x \text{ m}M}{0.285 \text{ OD}} = \frac{1 \text{m}M}{8.5 \text{ OD}}$$

$$x \text{ m}M = \frac{(1 \text{m}M)(0.285 \text{ OD})}{8.5 \text{ OD}} = 0.03 \text{m}M$$

Therefore, a solution of ssDNA with an absorbance of 0.285 has a concentration of 0.03 m*M*.

5.4 OLIGONUCLEOTIDE QUANTIFICATION

5.4.1 Optical density (OD) units

Many laboratories express an amount of an oligonucleotide in terms of **optical density (OD) units**. An **OD unit** is the amount of oligonucleotide dissolved in 1.0 mL giving an A_{260} of 1.00 in a cuvette with a 1 cm light path length. It is calculated by the equation

$$\text{OD units} = (A_{260}) \times (\text{oligonucleotide volume}) \times (\text{dilution factor})$$

Problem 5.11 Following its synthesis, an oligonucleotide is dissolved in 1.5 mL of water. You dilute 50 μL of the oligonucleotide into a total volume of 1000 μL and read the absorbance of the diluted sample at 260 nm. An A_{260} of 0.264 is obtained. How many OD units are present in the 1.5 mL of oligonucleotide stock?

Solution 5.11

Using the formula just given, the number of OD units is

$$\text{OD units} = 0.264 \times 1.5\ \text{mL} \times \frac{1000\ \mu\text{L}}{50\ \mu\text{L}}$$

$$= \frac{396}{50} = 7.92\ \text{OD units}$$

Therefore, the 1.5 mL solution contains 7.92 OD units of oligonucleotide.

5.4.2 Expressing an oligonucleotide's concentration in μg/mL

An A_{260} reading can be converted into a concentration expressed as μg/mL using the extinction coefficient for ssDNA of 1 mL/33 μg for a 1 cm light path. In other words, a solution of ssDNA with an A_{260} value of 1.0 (1.0 OD unit) contains 33 μg of ssDNA per milliliter. Written as an equation, this relationship is

$$1\ \text{OD unit} = 33\ \mu\text{g ssDNA/mL}$$

Problem 5.12 In Problem 5.11, a diluted oligonucleotide gave an A_{260} reading of 0.264. What is the concentration of the oligonucleotide in μg DNA/mL?

Solution 5.12

This problem can be solved using the following ratio.

$$\frac{x\ \mu\text{g ssDNA/mL}}{0.264\ \text{OD}} = \frac{33\ \mu\text{g ssDNA/mL}}{1\ \text{OD}}$$

$$x\ \mu\text{g ssDNA} = \frac{(0.264\ \text{OD})(33\ \mu\text{g ssDNA/mL})}{1\ \text{OD}}$$

$$x = 8.712\ \mu\text{g ssDNA/mL}$$

This value represents the concentration of oligonucleotide in the diluted sample used for spectrophotometry. The concentration of the oligonucleotide stock solution is obtained by multiplying this value by the dilution factor (the inverse of the dilution):

$$8.712\ \mu\text{g ssDNA/mL} \times \frac{1000\ \mu\text{L}}{50\ \mu\text{L}} = 174.24\ \mu\text{g ssDNA/mL}$$

Therefore, the concentration of the oligonucleotide stock solution is 174.24 μg ssDNA/mL.

5.4.3 Oligonucleotide concentration expressed in pmol/μL

Many applications in molecular biology require that a certain number of picomoles (pmol) of oligonucleotide be added to a reaction. This is true in the case of fluorescent DNA sequencing, for example, which, for some protocols, requires 3.2 pmol of oligonucleotide primer per sequencing reaction. An oligonucleotide's concentration can be calculated from an A_{260} value by using the formula

$$C = \left(\frac{A_{260} \times 100}{(1.54 \times n\text{A}) + (0.75 \times n\text{C}) + (1.17 \times n\text{G}) + (0.92 \times n\text{T})} \right) \times \text{dilution factor}$$

In this formula, the concentration, C, is calculated as picomoles per microliter (pmol/μL). The denominator consists of the sum of the extinction coefficients of each base multiplied by the number of times that base appears in the oligonucleotide.

Extinction coefficients for the bases of DNA

The extinction coefficients for the bases are usually expressed in liters per mole or as the absorbance for a 1 *M* solution at the wavelength where the base exhibits maximum absorbance. Either way, the absolute value is the same. The extinction coefficient for dATP, for example, is 15 400 L/mol, or a 1 *M* solution (at pH 7) of dATP will have an absorbance of 15 400 at the wavelength of its maximum absorbance. The amount 15 400 L/mol is equivalent to 0.0154 μL/pmol, as shown by the following conversion:

$$\frac{15\,400\ \text{L}}{\text{mole}} \times \frac{1 \times 10^6\ \mu\text{L}}{1\ \text{L}} \times \frac{1\,M}{1 \times 10^{12}\ \text{pmol}} = \frac{0.0154\ \mu\text{L}}{\text{pmol}}$$

The earlier formula to calculate oligonucleotide concentration uses the value 1.54 as the extinction coefficient for dATP. This value is obtained by multiplying the expression by 100/100 (which is the same as multiplying by 1). This manipulation leaves a '100' in the numerator and makes the equation more manageable.

Problem 5.13 An oligonucleotide with the sequence GAACTACGTTCGATCAAT is suspended in 750 μL of water. A 20 μL aliquot is diluted to a final volume of 1000 μL with water, and the absorbance at 260 nm is determined for the diluted sample. An OD of 0.242 is obtained. What is the concentration of the oligonucleotide stock solution in pmol/μL?

Solution 5.13

The oligonucleotide contains six A residues, four C residues, three G residues, and five T residues. Placing these values into the equation given earlier yields the following result:

$$C = \frac{A_{260} \times 100}{(1.54 \times n\text{A}) + (0.75 \times n\text{C}) + (1.17 \times n\text{G}) + (0.92 \times n\text{T})} \times \text{dilution factor}$$

$$\text{pmol}/\mu\text{L} = \frac{0.242 \times 100}{(1.54 \times 6) + (0.75 \times 4) + (1.17 \times 3) + (0.92 \times 5)} \times \frac{1000\ \mu\text{L}}{20\ \mu\text{L}}$$

$$\text{pmol}/\mu\text{L} = \frac{24.2}{9.24 + 3.00 + 3.51 + 4.60} \times \frac{1000}{20}$$

$$\text{pmol}/\mu\text{L} = \frac{24.2}{20.35} \times \frac{1000}{20} = \frac{24{,}200}{407} = 59.46$$

Therefore, the oligonucleotide stock solution has a concentration of 59.46 pmol/μL.

Problem 5.14 An oligonucleotide stock has a concentration of 60 pmol/μL. You wish to use 3.2 pmol of the oligonucleotide as a primer in a DNA sequencing reaction. How many microliters of oligo stock will give you the desired 3.2 pmol?

Solution 5.14

The answer can be obtained in the following manner:

$$\frac{60\ \text{pmol}}{\mu\text{L}} \times x\ \mu\text{L} = 3.2\ \text{pmol}$$

Setting up the equation in this way allows cancellation of units to give us the number of picomoles. Solving then for x yields

$$x = \frac{3.2\ \text{pmol}}{60\ \text{pmol}/\mu\text{L}} = 0.05\ \mu\text{L}$$

Therefore, to deliver 3.2 pmol of oligonucleotide primer to a reaction, you need to take a 0.05 μL aliquot of the 60 pmol/μL oligonucleotide stock. However, delivering that small a volume with a standard laboratory micropipette is neither practical nor accurate. It is best to dilute the sample to a concentration such that the desired amount of DNA is contained within a volume that can be accurately delivered by your pipettor system. For example, say that you wanted to deliver 3.2 pmol of oligonucleotide in a 2 μL volume. You can then ask, 'If I take a certain specified and convenient volume from the oligonucleotide stock solution, say 5 μL, into what volume of water (or TE dilution buffer) do I need to dilute that 5 μL to give me a concentration such that 3.2 pmol of oligonucleotide are contained in a 2 μL aliquot?' That problem can be set up in the following way:

$$\frac{60\ \text{pmol}}{\mu\text{L}} \times \frac{5\ \mu\text{L}}{x\ \mu\text{L}} = \frac{3.2\ \text{pmol}}{2\ \mu\text{L}}$$

$$\frac{300\ \text{pmol}}{x\ \mu\text{L}} = \frac{3.2\ \text{pmol}}{2\ \mu\text{L}}$$

$$300\ \text{pmol} = \frac{3.2x\ \text{pmol}}{2}$$

$$600\ \text{pmol} = 3.2x\ \text{pmol}$$

$$\frac{600\ \text{pmol}}{3.2\ \text{pmol}} = x$$

$$187.5 = x$$

Therefore, if 5 μL is taken from the 60 pmol/μL oligonucleotide stock solution and diluted to a final volume of 187.5 μL, then 2 μL of this diluted sample will contain 3.2 pmol of oligonucleotide.

5.5 MEASURING RNA CONCENTRATION

The following relationship is used for quantifying RNA:

$$1\ A_{260}\ \text{unit of RNA} = 40\ \mu\text{g/mL}$$

Problem 5.15 Forty microliters of a stock solution of RNA is diluted with water to give a final volume of 1000 μL. The diluted sample has an absorbance at 260 nm of 0.142. What is the concentration of the RNA stock solution in μg/mL?

Solution 5.15

Ratios are set up that read 'x μg RNA/mL is to 0.142 OD as 40 μg RNA/mL is to 1.0 OD.'

$$\frac{x\ \mu\text{g RNA/mL}}{0.142\ \text{OD}} = \frac{40\ \mu\text{g RNA/mL}}{1\ \text{OD}}$$

$$x\ \mu\text{g RNA/mL} = \frac{(0.142\ \text{OD})(40\ \mu\text{g RNA/mL})}{1\ \text{OD}}$$

$$= 5.7\ \mu\text{g RNA/mL}$$

This value represents the concentration of the diluted sample. To obtain the concentration of the RNA stock solution, it must be multiplied by the dilution factor:

$$5.7\ \mu\text{g RNA/mL} \times \frac{1000\ \mu\text{L}}{40\ \mu\text{L}} = 142.5\ \mu\text{g RNA/mL}$$

Therefore, the RNA stock solution has a concentration of 142.5 μg RNA/mL.

5.6 MOLECULAR WEIGHT, MOLARITY, AND NUCLEIC ACID LENGTH

The average molecular weight of a DNA base is approximately 330 daltons (or 330 grams/mole (g/M)). The average molecular weight of a DNA bp is twice this, approximately 660 daltons (or 660 g/M). These values can be used to calculate how much DNA is present in any biological source.

For small ssDNA molecules, such as synthetic oligonucleotides, the molecular weights of the individual nts can be summed to determine the strand's total MW according to the following formula:

$$\text{MW} = (n_A \times 335.2) + (n_C \times 311.2) + (n_G \times 351.2) + (n_T \times 326.2) + P$$

where n_X is the number of nts of A, C, G, or T in the oligonucleotide and P is equal to -101.0 for dephosphorylated (lacking an end phosphate group) or 40.0 for phosphorylated oligonucleotides.

The following problems demonstrate how molecular weight, molarity, and nucleic acid length relate to DNA quantity.

Problem 5.16 How much genomic DNA is present inside a single human diploid nucleated cell? Express the answer in picograms.

Solution 5.16

The first step in solving this problem is to calculate the weight, in grams, of a single bp. This value will then be multiplied by the number of bps in a single diploid cell. The first calculation uses Avogadro's number:

$$\frac{660\text{ g}}{\text{mol}} \times \frac{1\text{mol}}{6.023 \times 10^{23}\text{ bp}} = 1.1 \times 10^{-21}\text{ g/bp}$$

Therefore, a single bp weighs 1.1×10^{-21} g.

There are approximately 6×10^9 bps per diploid human nucleated cell. This can be converted to a picogram amount by using the conversion factor 1×10^{12} picograms/gram, as shown in the following equation:

$$6 \times 10^9\text{ bp} \times \frac{1.1 \times 10^{-21}\text{ g}}{\text{bp}} \times \frac{1 \times 10^{12}\text{ pg}}{\text{g}} = 7\text{ pg}$$

Therefore, a single diploid human cell contains 7 pg of genomic DNA.

Problem 5.17 You have 5 mL of bacteriophage λ stock having a concentration of 2.5×10^{11} phage/mL. You will purify λ DNA using the entire stock. Assuming 80% recovery, how many micrograms of λ DNA will you obtain?

Solution 5.17

First, calculate the total number of phage particles by multiplying the stock's concentration by its total volume:

$$\frac{2.5 \times 10^{11}\,\text{phage}}{\text{mL}} \times 5\,\text{mL} = 1.25 \times 10^{12}\,\text{phage}$$

In the previous problem, we found that one bp weighs 1.1×10^{-21} g. The bacteriophage λ is 48 502 bp in length. With these values, the quantity of DNA contained in the 5 mL of λ phage stock can be calculated:

$$1.25 \times 10^{12}\,\text{phage} \times \frac{48\ 502\,\text{bp}}{\text{phage}} \times \frac{1.1 \times 10^{-21}\,\text{g}}{\text{bp}} \times \frac{1 \times 10^{6}\,\mu\text{g}}{\text{g}} = 66.7\,\mu\text{g DNA}$$

Therefore, a total of 66.7 μg of DNA can be recovered from 12.5×10^{12} phage. Assuming 80% recovery, this value must be multiplied by 0.8:

$$66.7\,\mu\text{g DNA} \times 0.8 = 53.4\,\mu\text{g DNA}$$

Therefore, assuming 80% recovery, 53.4 μg of DNA will be recovered from the 5 mL phage stock.

Problem 5.18 λ DNA is recovered from the 5 mL phage stock described in Problem 5.16. In actuality, 48.5 μg of DNA are recovered. What is the percent recovery?

Solution 5.18

% recovery = (actual yield/expected yield) × 100

$$\frac{48.5\,\mu\text{g}}{66.7\,\mu\text{g}} \times 100 = 72.7\%\ \text{recovery}$$

Problem 5.19 λ DNA is 48 502 bp in length. What is its molecular weight?

Solution 5.19

Since the λ genome is dsDNA, the 660 daltons/bp conversion factor is used.

$$48\,502 \text{ bp} \times \frac{660 \text{ daltons}}{\text{bp}} = 3.2 \times 10^7 \text{ daltons}$$

Problem 5.20 A plasmid cloning vector is 3250 bp in length.

a) How many picomoles of vector are represented by 1 μg of purified DNA?

b) How many molecules does 1 μg of this vector represent?

Solution 5.20(a)

The first step is to calculate the MW of the vector:

$$\text{MW} = 3250 \text{ bp} \times \frac{660 \text{ g/mol}}{\text{bp}} = 2.1 \times 10^6 \text{ g/mol}$$

This value can then be used to calculate how many picomoles of vector are represented by 1 μg of DNA.

$$1\,\mu\text{g vector} \times \frac{1 \text{ mol}}{2.1 \times 10^6 \text{ g}} \times \frac{1 \times 10^{12} \text{ pmol}}{1 \text{ mol}} \times \frac{1 \text{ g}}{1 \times 10^6\,\mu\text{g}} = \frac{1 \times 10^{12} \text{ pmol}}{2.1 \times 10^{12}}$$

$$= 0.48 \text{ pmol}$$

Therefore, 1 μg of the 3250 bp vector is equivalent to 0.48 pmol.

Solution 5.20(b)

To calculate the number of molecules of vector in 1 μg, we need to use the conversion factor relating bps and grams determined in Problem 5.15. The problem is then written as follows:

$$1\,\mu\text{g} \times \frac{1 \text{ molecule}}{3250 \text{ bp}} \times \frac{1 \text{ bp}}{1.1 \times 10^{-21} \text{ g}} \times \frac{1 \text{ g}}{1 \times 10^6\,\mu\text{g}} = \frac{1 \text{ molecule}}{3.6 \times 10^{-12}}$$

$$= 2.8 \times 10^{11} \text{ molecules}$$

Therefore, there are 2.8×10^{11} molecules of the 3250 bp plasmid vector contained in 1 μg.

To calculate picomole quantities or molecule abundance for ssDNA molecules, a similar approach is taken to that used in the preceding problems. However, a conversion factor of 330 g/mol should be used, rather than 660 g/mol as for dsDNA.

Problem 5.21 The ssDNA vector M13mp18 is 7250 bases in length. How many micrograms of DNA does 1 pmol represent?

Solution 5.21

First, calculate the molecular weight of the vector:

$$7250 \text{ bases} \times \frac{330 \text{ g/mol}}{\text{base}} = 2.4 \times 10^{6} \text{ g/mol}$$

This value can then be used as a conversion factor to convert picomoles to micrograms:

$$1 \text{ pmol} \times \frac{2.4 \times 10^{6} \text{ g}}{\text{mol}} \times \frac{1 \times 10^{6} \text{ µg}}{\text{g}} \times \frac{1 \text{ mol}}{1 \times 10^{12} \text{ pmol}} = \frac{2.4 \times 10^{12} \text{ µg}}{1 \times 10^{12}} = 2.4 \text{ µg}$$

Therefore, 1 pmol of M13mp18 is equivalent to 2.4 µg DNA.

Problem 5.22 What is the molecular weight of a dephosphorylated oligonucleotide having the sequence 5′-GGACTTAGCCTTAGTATTGCCG-3′?

Solution 5.22

The oligonucleotide has four As, five Cs, six Gs, and seven Ts. These values are placed into the formula for calculating an oligonucleotide's molecular weight. Since the oligonucleotide is dephosphorylated, *P* in the equation is equal to −101.0.

$$\text{MW} = (n_A \times 335.2) + (n_C \times 311.2) + (n_G \times 351.2) + (n_T \times 326.2) + P$$
$$\text{MW} = (4 \times 335.2) + (5 \times 311.2) + (6 \times 351.2) + (7 \times 326.2) + (-101.0)$$
$$\text{MW} = (1340.8) + (1556) + (2107.2) + (2283.4) - 101.0$$
$$\text{MW} = 7186.4 \text{ daltons}$$

Therefore, the oligonucleotide has a molecular weight of 7186.4 daltons.

5.7 ESTIMATING DNA CONCENTRATION ON AN ETHIDIUM BROMIDE-STAINED GEL

Agarose gel electrophoresis is commonly used to separate DNA fragments following a restriction digest or PCR amplification. Fragments are detected by staining the gel with the intercalating dye, ethidium bromide, followed by visualization/photography under UV light. Ethidium bromide stains DNA in a concentration-dependent manner such that the more DNA that is present in a band on the gel, the more intensely it will stain. This relationship makes it possible to estimate the quantity of DNA present in a band through comparison with another band of known DNA amount. If the intensities of two bands are similar, then they contain similar amounts of DNA. Ethidium bromide stains ssDNA and RNA only very poorly. These forms of nucleic acid will not give reliable quantitation by gel electrophoresis.

Problem 5.23 Five hundred nanograms (0.5 μg) of λ DNA digested with the restriction endonuclease *Hin*dIII is loaded onto an agarose gel as a size marker. A band generated from a DNA amplification experiment has the same intensity upon staining with ethidium bromide as the 564 bp fragment from the λ *Hin*dIII digest. What is the approximate amount of DNA in the amplified fragment?

Solution 5.23

This problem is solved by determining how much DNA is in the 564 bp fragment. Since the amplified DNA fragment has the same intensity after staining as the 564 bp fragment, the two bands contain equivalent amounts of DNA.

Phage λ is 48 502 bp in length. The 564 bp *Hin*dIII fragment is to the total length of the phage λ genome as its amount (in ng) is to the total amount of λ *Hin*dIII marker run on the gel (500 ng). This allows the following relationship:

$$\frac{x \text{ ng}}{500 \text{ ng}} = \frac{564 \text{ bp}}{48\,502 \text{ bp}}$$

$$x \text{ ng} = \frac{(500 \text{ ng})(564 \text{ bp})}{48\,502 \text{ bp}} = 5.8 \text{ ng}$$

Therefore, there are approximately 5.8 ng of DNA in the band of the amplified DNA fragment.

CHAPTER SUMMARY

Nucleic acids maximally absorb UV light at approximately 260 nm. Double-stranded DNA (dsDNA) having a concentration of 50 μg/mL will have an absorbance at 260 nm (an A_{260}) or 1.0. This relationship is used to calculate the DNA concentration of a sample having unknown DNA content. Proteins absorb UV light at approximately 280 nm. A DNA sample's 260 nm/280 nm ratio gives an indication of its purity. Double-stranded DNA having a concentration of 1 mg/mL has an extinction coefficient (E) at 260 nm of 20. This value, along with a DNA sample's measured absorbance at 260 nm, can be used to calculate that sample's DNA concentration, using the relationship

$$c = \frac{A_{260}}{E_{260}}$$

A 1 mM solution of dsDNA has an extinction coefficient at 260 nm of 6.7. Double-stranded DNA having an absorbance at 260 nm of 1.0 has a concentration of 0.15 mM. The concentration of DNA can also be calculated using a standard curve generated by staining known quantities of a DNA control with a DNA-binding dye such as PicoGreen. The regression line describing the standard curve can be used to calculate the concentration of an unknown DNA sample.

Single-stranded DNA (ssDNA) having a concentration of 33 μg/mL will have an absorbance at 260 nm of 1.0. The average molecular weight of a nt in ssDNA is 330 g/m. The average molecular weight of a bp is 660 g/m. A 1 mM solution of ssDNA has an extinction coefficient at 260 nm of 8.5. A solution of ssDNA with an A_{260} of 1.0 has a concentration of 33 μg/mL.

Oligonucleotide concentration is often expressed in OD units, which is calculated using the relationship

$$\text{OD units} = (A_{260}) \times (\text{oligonucleotide volume}) \times (\text{dilution factor})$$

An oligonucleotide's concentration can also be determined using the sum of the extinction coefficients of its composite nts, as expressed in the following relationship:

$$C = \left(\frac{A_{260} \times 100}{(1.54 \times n\text{A}) + (0.75 \times n\text{C}) + (1.17 \times n\text{G}) + (0.92 \times n\text{T})}\right) \times \text{dilution factor}$$

A sample of RNA with an absorbance at 260 nm of 1.0 has a concentration of 40 μg/mL.

The molecular weight of DNA can be calculated by adding the molecular weights of its constituent bases using the relationship

$$\mathrm{MW} = (n_{\mathrm{A}} \times 335.2) + (n_{\mathrm{C}} \times 311.2) + (n_{\mathrm{G}} \times 351.2) + (n_{\mathrm{T}} \times 326.2) + P$$

where n_X is the number of nts of A, C, G, or T in the oligonucleotide and P is equal to -101.0 for dephosphorylated (lacking an end phosphate group) or 40.0 for phosphorylated oligonucleotides. A single bp weighs roughly 1.1×10^{-21} g.

DNA concentration of a band on an ethidium bromide-stained gel can be estimated by comparing its intensity to fragments of known length and of known amount.

Labeling nucleic acids with radioisotopes

■ INTRODUCTION

Sequencing DNA, quantitating gene expression, detecting genes or recombinant clones by probe hybridization, monitoring cell replication, measuring the rate of DNA synthesis – there are a number of reasons why researchers need to label nucleic acids. Radioisotopes have found wide use as nucleic acid tags because of several unique features. They are commercially available, sparing the researcher from having to perform a great deal of up-front chemistry. They can be attached to individual nts and subsequently incorporated into DNA or RNA through enzyme action. Nucleic acids labeled with radioisotopes can be detected by several methods, including liquid scintillation counting (providing information on nucleic acid quantity) and exposure to X-ray film (providing information on nucleic acid size or clone location). This section discusses the mathematics involved in the use of radioisotopes in nucleic acid research.

6.1 UNITS OF RADIOACTIVITY – THE CURIE (Ci)

The basic unit characterizing radioactive decay is the **curie** (**Ci**). It is defined as the quantity of radioactive substance having a decay rate of 3.7×10^{10} disintegrations per second (2.22×10^{12} **disintegrations per minute** (**dpm**)). Most instruments used to detect radioactive decay are less than 100% efficient. Consequently, a defined number of curies of some radioactive substance will yield fewer counts than theoretically expected.

The disintegrations actually detected by an instrument, such as by a liquid scintillation counter, are referred to as **counts per minute** (**cpm**).

Calculations for Molecular Biology and Biotechnology. DOI: 10.1016/B978-0-12-375690-9.00006-1
© 2010 Elsevier Inc. All rights reserved.

Problem 6.1 How many dpm are associated with 1 μCi of radioactive material?

Solution 6.1

The answer is obtained by using conversion factors to cancel terms, as shown in the following equation:

$$1\,\mu\text{Ci} \times \frac{1\,\text{Ci}}{1 \times 10^{6}\,\mu\text{Ci}} \times \frac{2.22 \times 10^{12}\,\text{dpm}}{\text{Ci}} = 2.22 \times 10^{6}\,\text{dpm}$$

Therefore, 1 μCi is equivalent to 2.22×10^{6} dpm.

Problem 6.2 If the instrument used for counting radioactive material is 25% efficient at detecting disintegration events, how many cpm will 1 μCi yield?

Solution 6.2

In Problem 6.1, it was determined that 1 μCi is equivalent to 2.22×10^{6} dpm. To calculate the cpm for 1 μCi, the dpm value should be multiplied by 0.25 (25% detection efficiency):

$$2.22 \times 10^{6}\,\text{dpm} \times 0.25 = 5.55 \times 10^{5}\,\text{cpm}$$

Therefore, an instrument counting radioactive decay with 25% efficiency should detect 5.55×10^{5} cpm for a 1 μCi sample.

6.2 ESTIMATING PLASMID COPY NUMBER

Plasmids – autonomously replicating, circular DNA elements used for cloning of recombinant genes – exist within a cell in one to several hundred copies. In one technique for determining how many copies of a plasmid are present within an individual cell, a culture of the plasmid-carrying strain is grown in the presence of [^{3}H]-thymine or [^{3}H]-thymidine (depending on the needs of the strain). After growth to log phase, the cells are harvested and lysed, and total DNA is isolated. A sample of the isolated DNA is centrifuged in an ethidium bromide–cesium chloride gradient to separate plasmid from chromosomal DNA. Following centrifugation, the

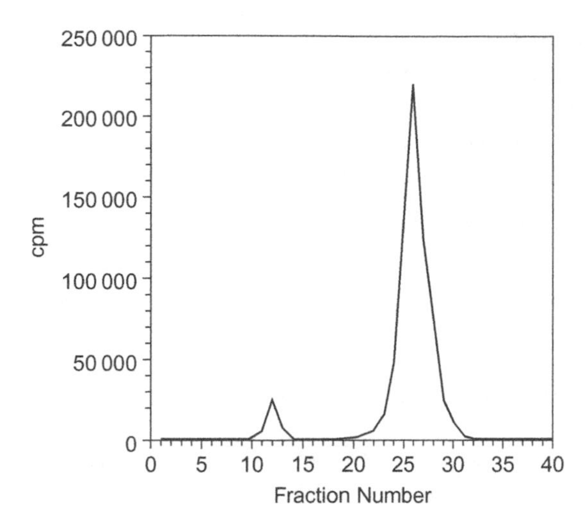

■ **FIGURE 6.1** Fractions collected from an ethidium bromide–cesium chloride gradient separating plasmid from chromosomal DNA. Cells are grown in the presence of radioactive thymine or thymidine so that replicating DNA is labeled with the tritium isotope. Supercoiled plasmid DNA, centered around fraction 12, sediments at a higher density than sheared chromosomal DNA, centered around fraction 26.

centrifuge tube is punctured at its bottom, fractions are collected onto filter discs, and the DNA is precipitated on the filters by treatment with trichloroacetic acid (TCA) and ethanol. The filters are dried and counted in a scintillation spectrometer. A graph of fraction number vs. cpm might produce a profile such as that shown in Figure 6.1. The smaller peak, centered at fraction number 12, represents plasmid DNA, which, in a supercoiled form, is denser than the sheared chromosomal DNA centered at fraction number 26.

Plasmid copy number is calculated using the following relationship:

$$\text{plasmid copy number} = \frac{\text{cpm of plasmid peak}}{\text{cpm of chromosome peak}} \times \frac{\text{chromosome MW}}{\text{plasmid MW}}$$

Problem 6.3 An experiment to determine the copy number of a 6000 bp plasmid in *E. coli* is performed as described earlier. The major plasmid-containing fraction is found to contain 25 000 cpm. The chromosomal peak fraction contains 220 000 cpm. What is the plasmid copy number?

Solution 6.3

The first step in solving this problem is to calculate the molecular weights of the plasmid and the *E. coli* chromosome. The plasmid is 6000 bp. The *E. coli* chromosome is approximately 4.6 million bp in length. Their molecular weights are calculated using the 660 daltons/bp conversion factor.

The molecular weight of the plasmid is

$$6000\,\text{bp} \times \frac{660\ \text{daltons}}{\text{bp}} = 39\,60\,000\ \text{daltons}$$

The molecular weight of the *E. coli* chromosome is

$$46\,00\,000\,\text{bp} \times \frac{660\ \text{daltons}}{\text{bp}} = 3.0 \times 10^{9}\ \text{daltons}$$

These values can then be incorporated into the equation (preceding this problem) for calculating copy number:

$$\frac{25\,000\ \text{cpm}}{2\,20\,000\ \text{cpm}} \times \frac{3.0 \times 10^{9}\ \text{daltons}}{3.96 \times 10^{6}\ \text{daltons}} = \frac{7.5 \times 10^{13}}{8.7 \times 10^{11}} = 86\ \text{plasmid copies}$$

6.3 LABELING DNA BY NICK TRANSLATION

Nick translation is a technique for radioactively labeling dsDNA, making it suitable as a hybridization probe for detecting specific genomic sequences. The endonuclease DNase I is used to create nicks in a DNA probe fragment. Following DNase I treatment, DNA polymerase I is used to add nt residues to the free 3′-hydroxyl ends created during the DNase I nicking process. As the DNA polymerase I extends the 3′ ends, the 5′ to 3′ exonuclease activity of the enzyme removes bases from the 5′-phosphoryl terminus of the nick. The sequential addition of bases onto the 3′ end with the simultaneous removal of bases from the 5′ end of the downstream annealed strand results in translation of the nick along the DNA molecule. When performed in the presence of a radioactive deoxynucleoside triphosphate (such as [α-^{32}P]dCTP), the newly synthesized strand becomes labeled.

When measuring cpm by liquid scintillation, a blank sample should be measured containing only scintillation fluid and a filter (if a filter is used for spotting and counting of a labeling reaction). This sample is used to detect background cpm inherent in the scintillation fluid, the filter, and the environment. The background cpm should be subtracted from any sample taken from a labeling experiment.

6.3.1 Determining percent incorporation of radioactive label from nick translation

To determine the percent of radioactive label incorporated into DNA by nick translation, a sample is taken at some time after addition of the radioactive dNTP but prior to DNA polymerase I treatment. This sample represents the total cpm in the reaction. Following the nick translation reaction, another sample is withdrawn. Both samples are spotted onto filter discs. The disc containing the sample representing the reaction after nick translation is treated with TCA to precipitate DNA fragments. The TCA treatment precipitates polynucleotides onto the filter disc but allows unincorporated radioactive dNTPs to pass through. The percent incorporation is then calculated using the equation

$$\frac{\text{cpm in TCA-precipitated sample}}{\text{cpm added to reaction}} \times 100 = \text{percent incorporation}$$

Problem 6.4 One μg of bacteriophage λ DNA is used for a nick translation reaction. Twenty-five μCi of labeled [α-^{32}P]dCTP is used in a 50 μL reaction. Following treatment with DNA polymerase I, a 1 μL sample is withdrawn and diluted into a total volume of 100 μL with TE buffer. Five μL of this dilution are spotted onto a glass fiber filter disc to determine total cpm in the reaction. Another 5 μL are used in a TCA-precipitation procedure to determine the amount of radioisotope incorporation. The TCA-precipitated sample is collected on a glass fiber filter disc and washed with TCA and ethanol. Both filters are dried and counted by liquid scintillation. The filter disc representing the total cpm gives 19 000 cpm. The TCA-precipitated sample gives 11 600 cpm. What is the percent incorporation?

Solution 6.4

The percent incorporation is calculated using the formula described earlier. The calculation goes as follows:

$$\frac{1.16 \times 10^4 \text{ cpm}}{1.9 \times 10^4 \text{ cpm}} \times 100 = 61\% \text{ incorporation}$$

6.3.2 Calculating specific radioactivity of a nick translation product

The **specific radioactivity** (or **specific activity**) of a labeled product is the amount of isotope incorporated, measured as cpm, per quantity of nucleic acid. Incorporation of isotope can be measured as TCA-precipitable cpm by the procedure described earlier. Using TCA-precipitation to measure incorporation, specific radioactivity is calculated using the following formula:

$$\text{cpm}/\mu\text{g DNA} = \frac{\text{cpm in TCA precipitate}}{\text{volume used to TCA precipitate}} \times \text{dilution factor} \times \frac{\text{total reaction volume}}{\text{total } \mu\text{g DNA in reaction}}$$

Problem 6.5 In Problem 6.4, 1.16×10^4 cpm were counted as TCA-precipitable counts in a 5 μL sample taken from a 1/100 dilution of a 50 μL nick translation reaction using 1 μg of λ DNA. What is the specific activity of the labeled product?

Solution 6.5

The incorporated radioactivity is represented by the TCA-precipitable counts. The reaction was diluted 1/100 to obtain a sample (5 μL) to TCA-precipitate. The dilution factor is the inverse of the dilution. Placing these values into the preceding equation yields

$$\frac{1.16 \times 10^4 \text{ cpm}}{5\,\mu\text{L}} \times \frac{100\,\mu\text{L}}{1\,\mu\text{L}} \times \frac{50\,\mu\text{L}}{1\,\mu\text{g DNA}} = \frac{5.8 \times 10^7 \text{ cpm}}{5\,\mu\text{g}}$$

$$= 1.16 \times 10^7 \text{ cpm}/\mu\text{g}$$

Therefore, the nick translation product has a specific activity of 1.16×10^7 cpm/μg.

6.4 RANDOM PRIMER LABELING OF DNA

An oligonucleotide six bases long is called a **hexamer** or **hexanucleotide**. If synthesis of a hexamer is carried out in such a way that A, C, G, and T nts have an equal probability of coupling at each base addition step, then the resulting product will be a mixture of oligonucleotides having many different random sequences. Such an oligonucleotide is called a **random hexamer**. For any six contiguous nts on DNA derived from a biological source, there should be a hexamer within the random hexamer pool that will be complementary. Under the proper conditions of temperature and salt, the annealed hexamer can then serve as a primer for *in vitro* DNA synthesis. By using a radioactively labeled dNTP (such as [α-^{32}P]dCTP) in the DNA synthesis reaction, the synthesis products become radioactively labeled. The labeled DNA can then be used as a probe to identify specific sequences by any of a number of gene detection protocols using hybridization.

Several mathematical calculations are necessary to assess the quality of the reaction, and these are outlined in the following sections.

6.4.1 Random primer labeling – percent incorporation

The percent incorporation for a random-primer-labeling experiment is determined in the same manner as that used in the nick translation procedure. It is calculated by dividing the cpm incorporated into DNA (as TCA-precipitable cpm) by the total cpm in the reaction and then multiplying this value by 100:

$$\frac{\text{cpm incorporated into DNA}}{\text{total cpm in reaction}} \times 100 = \%\ \text{incorporation}$$

Problem 6.6 In a random-primer-labeling experiment, random hexamers, 90 ng of denatured DNA template, DNA polymerase Klenow enzyme, enzyme buffer, 100 μCi [α-^{32}P]dCTP (2000 Ci/m*M*), and a mixture of the three nonisotopically labeled dNTPs (at molar excess to labeled dCTP) are incubated in a 70 μL reaction at room temperature for 75 minutes to allow DNA synthesis by extension of the annealed primers. The reaction is terminated by heating. A 1 μL aliquot of the reaction is diluted into 99 μL of TE buffer. Five μL of this dilution are spotted onto a filter disc, dried, and counted by liquid scintillation to measure total cpm in the reaction. Another 5 μL of the 1/100 dilution are used in a TCA-precipitation procedure to measure the amount of label incorporated into polynucleotides. The filter disc used to measure total cpm gives 6.2×10^4 cpm. The TCA-precipitated sample gives 5.1×10^4 cpm. What is the percent incorporation?

Solution 6.6

Using the formula for calculating percent incorporation given earlier, the solution is calculated as follows:

$$\frac{5.1 \times 10^4\ \text{cpm}}{6.2 \times 10^4\ \text{cpm}} \times 100 = 82\%\ \text{incorporation}$$

Therefore, the reaction achieved 82% incorporation of the labeled base.

6.4.2 Random primer labeling – calculating theoretical yield

Nick translation replaces existing DNA with radioactively labeled strands with no net addition of DNA quantity. Random primer labeling, on the other hand, actually results in the production of completely new and additional DNA strands. To calculate theoretical yield, it is assumed that the reaction achieves 100% incorporation of the labeled dNTP. In the random-primer-labeling experiment, the unlabeled dNTPs are placed into the reaction in molar excess to the labeled dNTP. The unlabeled dNTPs can be at a concentration 10- to 100-fold higher than that of the radioactively labeled dNTP.

Theoretical yield is calculated using the following formula:

$$\text{grams DNA} = \frac{\mu\text{Ci dNTP added to reaction} \times 4\text{ dNTPs} \times \dfrac{330\text{ g/mol}}{\text{dNTP}}}{\text{dNTP specific activity in }\mu\text{Ci/mol}}$$

In this formula, all four dNTPs are assumed to have the same concentration as that of the labeled dNTP, since it is the labeled dNTP that is in a limiting amount.

In a typical random-primer-labeling experiment, nanogram amounts of DNA are synthesized. The preceding equation can be converted to a form that more readily yields an answer in nanograms of DNA. The average molecular weight of a nucleotide is 330 g/mole. This is equivalent to 330 ng/nmole, as shown in the following equation:

$$\frac{330\text{ g}}{\text{mol}} \times \frac{1 \times 10^{9}\text{ ng}}{\text{g}} \times \frac{1\text{ mol}}{1 \times 10^{9}\text{ nmol}} = \frac{330\text{ ng}}{\text{nmol}}$$

The specific activity of a radioactive dNTP, such as $[\alpha\text{-}^{32}\text{P}]$dCTP, from most suppliers, is expressed in units of Ci/mmol. This unit is equivalent to μCi/nmol:

$$\frac{\text{Ci}}{\text{mmol}} \times \frac{1 \times 10^{6}\,\mu\text{Ci}}{\text{Ci}} \times \frac{\text{mmol}}{1 \times 10^{6}\text{ nmol}} = \frac{\mu\text{Ci}}{\text{nmol}}$$

Placing these new conversion factors into the equation for calculating theoretical yield gives

$$\text{ng DNA} = \frac{\mu\text{Ci dNTP added} \times 4\text{ dNTPs} \times \dfrac{330\text{ ng/nmol}}{\text{dNTP}}}{\text{dNTP specific activity in }\mu\text{Ci/nmol}}$$

Problem 6.7 In Problem 6.6, 100 μCi of [α-^{32}P]dCTP having a specific activity of 2000 Ci/mmol was used in a random-primer-labeling experiment. Assuming 100% efficient primer extension, what is the theoretical yield?

Solution 6.7

As shown earlier, a dNTP-specific activity of 2000 Ci/mmol is equivalent to 2000 μCi/nmol. The equation for calculating yield is then as follows:

$$\text{ng DNA} = \frac{100\,\mu\text{Ci} \times 4\text{ dNTPs} \times \dfrac{330\text{ ng/nmol}}{\text{dNTP}}}{2000\,\mu\text{Ci dNTP/nmol}}$$

$$= \frac{132\,000\text{ ng}}{2000} = 66\text{ ng DNA theoretical yield}$$

Therefore, in the random-primer-labeling experiment described in Problem 6.6, 66 ng of DNA should be synthesized if the reaction is 100% efficient.

6.4.3 Random primer labeling – calculating actual yield

The actual yield of DNA synthesized by random primer labeling is calculated by multiplying the theoretical yield by the percent incorporation (expressed as a decimal):

$$\text{theoretical yield} \times \%\text{ incorporation} = \text{ng DNA synthesized}$$

Problem 6.8 In Problem 6.6, it was calculated that the described reaction achieved 82% incorporation of the labeled dNTP. In Problem 6.7, it was calculated that the random-primer-labeling experiment described in Problem 6.6 would theoretically yield 66 ng of newly synthesized DNA. What is the actual yield?

Solution 6.8

The actual yield is obtained by multiplying the percent incorporation by the theoretical yield. Expressed as a decimal, 82% (the percent incorporation) is 0.82. Using this value for calculating actual yield gives the following equation:

$$0.82 \times 66\text{ ng DNA} = 54.1\text{ ng DNA}$$

Therefore, under the conditions of the random-primer-labeling experiment described in Problem 6.6, an actual yield of 54.1 ng of newly synthesized DNA is obtained.

6.4.4 Random primer labeling – calculating specific activity of the product

The specific activity of a DNA product generated by random primer labeling is expressed as cpm/μg DNA. It is calculated using the following formula:

$$\frac{\dfrac{\text{cpm incorporated}}{\text{mL used to measure incorporation}} \times \begin{pmatrix}\text{dilution}\\ \text{factor}\end{pmatrix} \times \begin{pmatrix}\text{reaction}\\ \text{volume}\end{pmatrix}}{(\text{ng DNA actual yield} + \text{ng input DNA}) \times \dfrac{0.001\,\mu\text{g}}{\text{ng}}} = \text{cpm}/\mu\text{g}$$

Problem 6.9 Problem 6.6 describes a random-primer-labeling experiment in which 90 ng of denatured template DNA are used in a 70 μL reaction. To measure incorporation of the labeled dNTP, 5 μL of a 1/100 dilution taken from the 70 μL reaction are TCA-precipitated on a filter and counted. This 5 μL sample gives 5.1×10^4 cpm. In Problem 6.8, an actual yield of 54.1 ng of DNA was calculated for this labeling experiment. What is the specific activity of the DNA synthesis product?

Solution 6.9

Substituting the given values into the preceding equation yields the following result. Note the use of the dilution factor when calculating the cpm incorporated.

$$\frac{\dfrac{5.1 \times 10^4\ \text{cpm}}{5\,\mu\text{L}} \times \dfrac{100\,\mu\text{L}}{1\,\mu\text{L}} \times 70\,\mu\text{L}}{(54.1\text{ng DNA} + 90\,\text{ng DNA}) \times \dfrac{0.001\,\mu\text{g}}{\text{ng}}} = \frac{7.1 \times 10^7\ \text{cpm}}{0.144\,\mu\text{g}}$$

$$= 4.9 \times 10^8\ \text{cpm}/\mu\text{g DNA}$$

Therefore, the newly synthesized DNA in this random-primer-labeling experiment has a specific activity of 4.9×10^8 cpm/μg.

6.5 LABELING 3′ TERMINI WITH TERMINAL TRANSFERASE

The enzyme **terminal deoxynucleotidyl transferase** adds mononucleotides onto 3′ hydroxyl termini of either ssDNA or dsDNA. The mononucleotides are donated by a dNTP. The addition of each mononucleotide releases inorganic phosphate. In a procedure referred to as **homopolymeric tailing** (or simply **tailing**), terminal transferase is used to add a series of bases onto the 3′ end of a DNA fragment. The reaction can also be manipulated so that only one mononucleotide is added onto the 3′ end. This is accomplished by supplying the reaction exclusively with either dideoxynucleoside 5′-triphosphate (ddNTP) or cordycepin-5′-triphosphate. Both of these analogs lack a free 3′ hydroxyl group. Once one residue has been added, polymerization stops, since the added base lacks the 3′ hydroxyl required for coupling of a subsequent mononucleotide. By either approach, use of a labeled dNTP donor provides another method of preparing a labeled probe suitable for use in gene detection hybridization assays.

6.5.1 3′-end labeling with terminal transferase – percent incorporation

As with the other labeling methods described in this chapter, TCA-precipitation (described in Section 6.3.1) can be used to measure incorporation of a labeled dNTP. Percent incorporation is calculated using the following equation:

$$\frac{\text{cpm incorporated}}{\text{total cpm}} \times 100 = \%\ \text{incorporation}$$

Problem 6.10 A 3′-end-labeling experiment is performed in which 60 μg of DNA fragment, terminal transferase, buffer, and 100 μCi of [α-^{32}P] cordycepin-5′-triphosphate (2000 Ci/mmol) are combined in a 50 μL reaction. The mixture is incubated at 37°C for 20 min and stopped by heating at 72°C for 10 min. One μL of the reaction is diluted into 99 μL of TE buffer. Five μL of this dilution are spotted onto a filter; the filter is dried and then counted in a liquid scintillation counter. This sample, representing total cpm, gives 90 000 cpm. Another 5 μL of the 1/100 dilution are treated with TCA to measure the amount of label incorporated into DNA. The filter from the TCA procedure gives 76 500 cpm. What is the percent incorporation?

Solution 6.10

The total cpm was measured to be 90 000 cpm. By TCA-precipitation, it was found that 76 500 cpm are incorporated into DNA. Using these values and the equation shown earlier, the calculation goes as follows:

$$\frac{76\,500\text{ cpm}}{90\,000\text{ cpm}} \times 100 = 85\%\text{ incorporation}$$

Therefore, in this 3′-end-labeling experiment, 85% of the label was incorporated into DNA.

6.5.2 3′-end labeling with terminal transferase – specific activity of the product

The specific activity of the 3′-end-labeled product, expressed as cpm/μg DNA, is given by the equation

$$\text{cpm}/\mu\text{g} = \frac{\%\text{ incorp.} \times \dfrac{\text{total cpm}}{\begin{pmatrix}\mu\text{L used}\\ \text{to measure}\\ \text{total cpm}\end{pmatrix}} \times \begin{pmatrix}\text{dilution}\\ \text{factor}\end{pmatrix} \times \begin{pmatrix}\text{reaction}\\ \text{volume}\end{pmatrix}}{\mu\text{g of DNA fragment in reaction}}$$

Problem 6.11 In Problem 6.10, a 3′-end-labeling reaction is described in which 60 μg of DNA fragment are labeled with [α-^{32}P] cordycepin-5′-triphosphate in a 50 μL reaction. A 1 μL aliquot of the reaction is diluted 1 μL/100 μL, and 5 μL are counted, giving 90 000 cpm. In Problem 6.10, it was shown that 85% of the radioactive label was incorporated into DNA. What is the specific activity of the product?

Solution 6.11

Using the preceding equation yields

$$\frac{0.85 \times \dfrac{90\,000\text{ cpm}}{5\,\mu\text{L}} \times \dfrac{100\,\mu\text{L}}{1\,\mu\text{L}} \times 50\,\mu\text{L}}{60\,\mu\text{g DNA}} = \frac{7.65 \times 10^{7}\text{ cpm}}{60\,\mu\text{g DNA}}$$

$$= 1.3 \times 10^{6}\text{ cpm}/\mu\text{g DNA}$$

Therefore, the product of this 3′-end-labeling experiment has a specific activity of 1.3×10^{6} cpm/μg.

6.6 COMPLEMENTARY DNA (cDNA) SYNTHESIS

RNA is a fragile molecule and is easily degraded when released from a cell during an extraction protocol. This can make the study of gene expression a challenging task. One method to make the study of RNA more accessible has been to copy RNA into DNA. Double-stranded DNA is far more stable and amenable to manipulation by recombinant DNA techniques than is RNA.

The retroviral enzyme **reverse transcriptase** has the unique ability to synthesize a complementary DNA strand from an RNA template. DNA produced in this manner is termed **cDNA** (for complementary DNA). Once made, a DNA polymerase enzyme can be used in a second step to synthesize a complete double-stranded cDNA molecule using the first cDNA strand as a template. The cloning and sequencing of cDNAs have provided molecular biologists with valuable insights into gene structure and function.

6.6.1 First strand cDNA synthesis

Since most eukaryotic mRNAs can be isolated having intact poly (A) tails, an oligo(dT) primer designed to anneal to homopolymer tails can be used to initiate synthesis of the first cDNA strand. When long RNA molecules are to be reverse transcribed, however, using random hexanucleotide primers is a more favored approach. Other reaction components include reverse transcriptase, a ribonuclease inhibitor, the four dNTPs, dithiothreitol (a reducing agent), and a reaction buffer. RNA is usually added to the reaction at a concentration of roughly 0.1 μg/μL. When all reaction components have been combined, a small aliquot of the reaction is transferred to a separate tube containing a small amount of [α-^{32}P]dCTP to monitor first strand synthesis. The original reaction mixture (with no radioisotope) is carried through synthesis of the second strand.

Incorporation of label into polynucleotide can be followed by TCA-precipitation, as described previously for other labeling reactions (see Section 6.3.1). Calculating percent incorporation of label is performed as described previously in this chapter. The purpose of assessing the first strand cDNA synthesis reaction is to determine the quantity of cDNA synthesized and the efficiency with which mRNA was converted to cDNA. These calculations require the following determinations:

a) The moles of dNTP per reaction, given by the equation

$$\text{moles of dNTP/reaction} = \begin{pmatrix}\text{molar concentration}\\ \text{of dNTP}\end{pmatrix} \times \begin{pmatrix}\text{reaction}\\ \text{volume}\end{pmatrix}$$

b) The percent incorporation of labeled dNTP, given by the equation

$$\%\text{ incorporation} = \frac{\text{TCA precipitable cpm}}{\text{total cpm}} \times 100$$

c) Moles of dNTP incorporated into polynucleotide, given by the equation

$$\begin{pmatrix}\text{moles dNTP}\\\text{incorporated}\end{pmatrix} = \frac{\text{moles dNTP}}{\mu\text{L}} \times \begin{pmatrix}\text{reaction}\\\text{volume}\end{pmatrix} \times \frac{\%\text{ incorporation}}{100}$$

d) Grams of first strand cDNA synthesized, given by the equation

$$\text{g cDNA synthesized} = \text{moles dNTP incorporated} \times \frac{330\text{ g}}{\text{moles dNTP}}$$

e) The number of grams of first strand cDNA carried through to second strand synthesis, given by the equation

$$\begin{pmatrix}\text{g cDNA available for}\\\text{second strand synthesis}\end{pmatrix} = \begin{pmatrix}\text{g first}\\\text{strand}\\\text{cDNA}\end{pmatrix} \times \frac{\begin{pmatrix}\text{reaction volume after sample}\\\text{removed for labeling}\end{pmatrix}}{\text{initial reaction volume}}$$

Problem 6.12 A first strand cDNA synthesis experiment is performed in which 5 μg of poly (A) mRNA is combined with reverse transcriptase, 2 m*M* each dNTP, and other necessary reagents in a total volume of 55 μL. After all components have been mixed, 5 μL of the reaction is transferred to a different tube containing 0.5 μL of [α-^{32}P]dCTP (1000 Ci/m*M*; 10 μCi/μL). In the small side reaction with the [α-^{32}P]dCTP tracer, total cpm and incorporation of label are measured by TCA-precipitation in the following manner: Two μL of the side reaction is diluted into 100 μL of TE buffer. Five μL of this dilution is spotted onto a filter and counted (to measure total cpm). Another 5 μL of the 2/100 dilution of the side reaction are precipitated with TCA and counted. The sample prepared to measure total cpm gives 220 000 cpm. The TCA-precipitated sample gives 2000 cpm.

a) How much cDNA was synthesized in this reaction?

b) How much of the input mRNA was converted to cDNA?

Solution 6.12(a)

This problem can be solved using the equations shown earlier. The first step is to determine the concentration of dNTPs (expressed as nmol/μL). The reaction contains 2 m*M* dATP, 2 m*M* dCTP, 2 m*M* dGTP, and 2 m*M* dTTP, or a total dNTP concentration of 8 m*M*. An 8 m*M* dNTP concentration is equivalent to 8 mmol/L. This can be converted to nmol/μL using the following relationship:

$$\frac{8\ \text{mmol dNTP}}{\text{L}} \times \frac{1\text{L}}{1 \times 10^{6}\ \mu\text{L}} \times \frac{1 \times 10^{6}\ \text{nmol}}{\text{mmol}} = \frac{8\ \text{nmol dNTP}}{\mu\text{L}}$$

The percent incorporation of [α-^{32}P]dCTP can be calculated using the data provided:

$$\frac{2000\ \text{incorporated cpm}}{220\,000\ \text{total cpm}} \times 100 = 0.9\%\ \text{incorporation}$$

The number of nanomoles of dNTP incorporated into cDNA can now be calculated. To perform this step, the concentration of dNTPs in the reaction (expressed in nmol dNTP/μL), the total volume of the reaction prior to sample aliquoting into a separate tracer reaction, and the percent incorporation (expressed as a decimal) are used as follows:

$$\frac{8\ \text{nmol dNTP}}{\mu\text{L}} \times 55\ \mu\text{L} \times 0.009 = 4\ \text{nmol dNTP incorporated}$$

Using the number of nanomoles of dNTP incorporated into nucleic acid and the average molecular weight of a dNTP (330 g/mol), the amount of cDNA synthesized can be calculated by multiplying these two values. Conversion factors must be applied in this calculation, since we have determined nanomoles of incorporated dNTP and the molecular weight of a dNTP is given in terms of grams/mole. The calculation goes as follows:

$$4\ \text{nmol dNTP} \times \frac{330\ \text{g dNTP}}{\text{mol}} \times \frac{1 \times 10^{9}\ \text{ng}}{\text{g}} \times \frac{\text{mol}}{1 \times 10^{9}\ \text{nmol}} = 1320\ \text{ng cDNA}$$

Therefore, in this reaction, 1320 ng of cDNA were synthesized.

Solution 6.12(b)

The percent mRNA converted to cDNA is calculated by dividing the amount of cDNA synthesized by the amount of mRNA put into the reaction. The quotient is then multiplied by 100 to give a percent value. In this example, 1320 ng of cDNA were synthesized. Five μg of mRNA were put into the reaction. Since these nucleic acid quantities are expressed in different units, they first must be brought to equivalent units. Converting μg mRNA to ng mRNA is performed as follows:

$$5\,\mu g\ \text{mRNA} \times \frac{1000\ \text{ng}}{\mu g} = 5000\ \text{ng mRNA}$$

Percent mRNA converted to cDNA can now be calculated:

$$\frac{1320\ \text{ng cDNA synthesized}}{5000\ \text{ng mRNA in reaction}} \times 100 = 26.4\%\ \text{mRNA converted to cDNA}$$

Therefore, 26.4% of the mRNA put into the reaction was converted to cDNA.

Problem 6.13 How much of the cDNA synthesized in Problem 6.12 will be carried into the second strand synthesis reaction?

Solution 6.13

In Problem 6.12, 5 μL of a 55 μL reaction were removed to monitor incorporation of [α-^{32}P]dCTP. Therefore, 50 μL of 55 μL remained in the reaction. Multiplying this ratio by the amount of cDNA synthesized (1320 ng) gives the following result:

$$1320\ \text{ng} \times \frac{50\,\mu L}{55\,\mu L} = 1200\ \text{ng cDNA carried to second strand synthesis}$$

Problem 6.14 For the cDNA synthesis reaction described in Problem 6.12, how many nanomoles of dNTP were incorporated into cDNA in the main reaction?

Solution 6.14

In Problem 6.12, a total of 4 nmol of dNTP was incorporated into nucleic acid. Multiplying this amount by the volume fraction left in the main reaction will give the number of nanomoles of dNTP incorporated into cDNA that will be carried into the second strand synthesis reaction:

$$4\ \text{nmol dNTP incorporated} \times \frac{50\,\mu L}{55\,\mu L} = 3.6\ \text{nmol dNTP incorporated}$$

Therefore, 3.6 n*mol* of dNTP were incorporated into cDNA in the main reaction.

6.6.2 Second strand cDNA synthesis

Synthesis of the cDNA second strand can be accomplished by a number of different methods. As a first step, the RNA in the RNA–cDNA hybrid formed during the first strand synthesis reaction can be removed by treatment with RNase A. Priming of second strand synthesis can then be performed using random primers, a gene-specific primer, or the hairpin loop structure typically found at the 3′-end of the first strand cDNA. Alternatively, RNase H can be used to nick the RNA strand of the RNA–cDNA hybrid molecule. DNA polymerase then utilizes the nicked RNA as primer to replace the RNA strand in a reaction similar to nick translation.

To perform second strand cDNA synthesis, the first strand reaction is diluted directly into a new mixture carrying all the components necessary for successful DNA replication. Typically, no new dNTPs are added. They are carried into the second strand reaction from the first strand reaction. As with first strand cDNA synthesis, a separate tracer reaction including [α-^{32}P]dCTP can be used to monitor incorporation of dNTPs and formation of the DNA second strand.

Several calculations are used to determine the quality of second strand cDNA synthesis:

a) Percent incorporation of label into the second strand is calculated using the following formula:

$$\frac{\text{TCA precipitable cpm}}{\text{total cpm}} \times 100 = \%\ \text{incorporation into second strand}$$

b) Nanomoles of dNTP incorporated into cDNA is given by the following equation:

$$\begin{pmatrix}\text{nmol dNTP}\\ \text{incorporated}\end{pmatrix} = \left[\begin{pmatrix}\text{nmol}\\ \text{dNTP}/\mu\text{L}\end{pmatrix} \times \begin{pmatrix}\text{reaction}\\ \text{volume}\end{pmatrix} - \begin{pmatrix}\text{nmol dNTP}\\ \text{incorporated}\\ \text{in first cDNA}\\ \text{strand}\end{pmatrix}\right] \times \%\ \text{incorporation}$$

(In this equation, percent incorporation should be expressed as a decimal.)

c) The amount of second strand cDNA synthesized (in ng) is given by the following equation:

$$\text{ng second strand cDNA} = \text{nmol dNTP incorporated} \times \frac{330\ \text{ng dNTP}}{\text{nmol}}$$

d) The percent of first strand cDNA converted to double-stranded cDNA is given by the following equation:

$$\left(\begin{array}{l}\text{\% conversion to} \\ \text{double-stranded cDNA}\end{array}\right) = \frac{\text{ng second strand cDNA}}{\text{ng first strand cDNA}} \times 100$$

Problem 6.15 The 50 μL first strand cDNA synthesis reaction described in Problem 6.12 is diluted into a 250 μL second strand synthesis reaction containing all the necessary reagents. To prepare a tracer reaction, 10 μL of the 250 μL main reaction are transferred to a separate tube containing 5 μCi [α-^{32}P]dCTP. Following incubation of the reactions, 90 μL of an EDTA solution is added to the tracer reaction. A heating step is used to stop the reaction. Two μL of the 100 μL tracer reaction is spotted on a filter and counted. This sample gives 90 000 cpm. Another 2 μL sample from the diluted tracer reaction is treated with TCA to measure incorporation of [α-^{32}P]dCTP. This sample gives 720 cpm.

a) How many nanograms of second strand cDNA were synthesized?

b) What percentage of the first strand was converted into second strand cDNA?

Solution 6.15(a)

The first step in solving this problem is to calculate the percent incorporation of [α-^{32}P]dCTP:

$$\frac{\text{720 TCA precipitable cpm}}{\text{90 000 total cpm}} \times 100 = 0.8\% \text{ incorporation}$$

The nanomoles of dNTP incorporated can now be calculated. The original first strand cDNA synthesis reaction had a dNTP concentration of 8 nmol/μL (see Solution 6.12a). The 50 μL first strand main reaction was diluted into a total volume of 250 μL. Assuming no incorporation of dNTP in the first strand reaction, the new dNTP concentration in the second strand reaction is

$$\frac{\text{8 nmol dNTP}}{\mu\text{L}} \times \frac{50\,\mu\text{L}}{250\,\mu\text{L}} = \frac{\text{1.6 nmol dNTP}}{\mu\text{L}}$$

The first strand of cDNA incorporated 3.6 nmol of dNTP in the main reaction (Problem 6.14). Since this amount of dNTP is no longer available in the second strand synthesis reaction for incorporation into nucleic acid, it must be subtracted from the total dNTP concentration. Using the percent incorporation calculated earlier, the amount of dNTP incorporated can be calculated from the following equation:

$$\left[\left(\frac{1.6\,\text{nmol}}{\mu\text{L}} \times 250\,\mu\text{L}\right) - \left(\begin{array}{l}3.6\,\text{nmol dNTP}\\ \text{incorporarated}\\ \text{in first strand}\end{array}\right)\right] \times 0.008 = \left(\begin{array}{l}3.2\,\text{nmol dNTP}\\ \text{incorporated}\end{array}\right)$$

Therefore, 3.2 nmol of dNTP was incorporated into cDNA during synthesis of the second strand.

The amount of cDNA produced during second strand synthesis can now be calculated by multiplying the nanomoles of dNTP incorporated by the average molecular weight of a dNTP (330 ng/nmol):

$$3.2\,\text{nmol dNTP} \times \frac{330\,\text{ng}}{\text{nmol}} = 1056\,\text{ng second strand cDNA}$$

Therefore, 1056 ng of cDNA was synthesized in the second strand cDNA synthesis reaction.

Solution 6.15(b)

In Problem 6.13, it was calculated that 1200 ng of cDNA synthesized in the first strand synthesis reaction was carried into the second strand cDNA synthesis reaction. The percent conversion of input nucleic acid to second strand cDNA is calculated by dividing the amount of second strand cDNA synthesized by the amount of first strand cDNA carried into the second strand reaction and multiplying this value by 100:

$$\frac{1056\,\text{ng second strand cDNA}}{1200\,\text{ng first strand cDNA}} \times 100 = \left(\begin{array}{l}88\%\ \text{conversion}\\ \text{to double}\end{array}\right) - \left(\begin{array}{l}\text{stranded}\\ \text{cDNA}\end{array}\right)$$

Therefore, 88% of the input first strand cDNA was converted to double-stranded cDNA in the second strand synthesis reaction.

■

6.7 HOMOPOLYMERIC TAILING

Homopolymeric tailing (or simply **tailing**) is the process by which the enzyme **terminal transferase** is used to add a series of residues (all of the same base) onto the 3′ termini of dsDNA. It has found particular use in facilitating the cloning of cDNA into plasmid cloning vectors. In this technique, a double-stranded cDNA fragment is tailed with a series of either dC or dG residues onto the two 3′ termini. The vector into which the cDNA fragment is to be cloned is tailed with the complementary base (dC if dG was used to tail the cDNA or dG if dC was used to tail the cDNA). By tailing the cDNA and the vector with complementary bases, effective annealing between the two DNAs is promoted. Ligation of the cDNA fragment to the vector is then more efficient:

It is possible to determine the number of residues that terminal transferase has added onto a 3′ end by determining the number of picomoles of 3′ ends available as substrate and the concentration (in picomoles) of dG (or dC) incorporated into DNA. As to the first part, the number of picomoles of 3′ ends is calculated by using the DNA substrate's size in the following formula:

$$\text{pmoles } 3' \text{ ends} = \frac{\mu\text{g DNA substrate} \times \dfrac{1\text{ g}}{1 \times 10^{6}\ \mu\text{g}}}{\dfrac{\text{number of bp}}{\text{molecule}} \times \dfrac{660\text{ g/mol}}{\text{bp}}} \times \frac{1 \times 10^{12}\text{ pmol}}{\text{mol}} \times \frac{\text{number of } 3' \text{ ends}}{\text{molecule}}$$

Problem 6.16 If 2.5 μg of a 3200 bp plasmid cloning vector are cut once with a restriction enzyme, how many picomoles of 3′ ends are available as substrate for terminal transferase?

Solution 6.16

A plasmid is a circular dsDNA molecule. If it is cut with a restriction endonuclease at one site, it will become linearized. As a linear molecule, it will have two 3′ ends. The equation for calculating the number of 3′ ends (see earlier) becomes

$$\text{pmol } 3' \text{ ends} = \frac{2.5\ \mu\text{g DNA} \times \dfrac{1\text{g}}{1 \times 10^{6}\ \mu\text{g}}}{\dfrac{3200\text{ bp}}{\text{molecule}} \times \dfrac{660\text{ g/mol}}{\text{bp}}} \times \frac{1 \times 10^{12}\text{ pmol}}{\text{mol}} \times \frac{2\ 3' \text{ ends}}{\text{molecule}}$$

$$= \frac{5 \times 10^{6}\text{ pmol } 3' \text{ ends}}{2.1 \times 10^{6}} = 2.4\text{ pmol } 3' \text{ ends}$$

Therefore, 2.5 μg of a linear DNA fragment 3200 bp in length is equivalent to 2.4 pmol of 3′ ends.

To estimate the number of dG residues added onto a 3′ end of a DNA fragment, it is necessary to have a radioactive molecule present in the reaction that can be detected and quantitated. The molecule used to monitor the progress of the reaction should be a substrate of the enzyme. In a dG tailing protocol, therefore, tritiated dGTP (^{3}H-dGTP) can serve this purpose. The next step in calculating the number of residues added to a 3′ end is to calculate the specific activity of the dNTP. Specific activity can be expressed as cpm/pmol dNTP. The general formula for calculating the specific activity of tailed DNA is

$$\text{cpm/pmol dNTP} = \frac{\text{total cpm in counted sample}}{\text{pmol dNTP}}$$

To perform this calculation, the number of picomoles of dNTP in the reaction must be determined. Let's assume that the concentration of the starting dNTP material, as supplied by the manufacturer, is expressed in a micromolar (μM) amount. The micromolar concentration of dNTP in the terminal transferase tailing reaction mix is determined by the following formula:

$$\mu M \text{ dNTP} = \frac{(\mu M \text{ concentration of dNTP stock}) \times (\text{volume dNTP added})}{\text{total reaction volume}}$$

Problem 6.17 A 3200 bp vector is being tailed with dG residues using dGTP and terminal transferase. The stock solution of dGTP being used as substrate has a concentration of 200 μM. Four μL of this stock solution are added to a 100 μL reaction (final volume). What is the micromolar concentration of dGTP in the reaction?

Solution 6.17

Using the equation just given yields the following relationship:

$$\mu M \text{ dGTP} = \frac{(200\,\mu M) \times (4\,\mu\text{L})}{100\,\mu\text{L}} = 8\,\mu M$$

Therefore, dGTP in the reaction mix has a concentration of 8 μM.

The dGTP in the reaction is usually used in large molar excess to the ^{3}H-dGTP added. The added ^{3}H-dGTP, therefore, does not contribute significantly to the overall dGTP concentration. When an aliquot of the radioactive dGTP is added to the terminal transferase tailing reaction, the total cpm can be determined by spotting a small amount of the reaction mix on a filter, allowing the filter to dry, and then counting by liquid scintillation. The number of picomoles of dGTP spotted on the filter is given by the following equation:

$$\text{pmol dNTP} = \frac{\text{mmol dNTP}}{\text{L}} \times \frac{\text{L}}{1 \times 10^6\ \mu\text{L}} \times \frac{1 \times 10^6\ \text{pmol}}{\mu\text{mol}} \times \left(\begin{array}{l}\mu\text{L spotted}\\ \text{on filter}\end{array}\right)$$

It must be remembered that μM is equivalent to micromoles/liter (see Chapter 2). This gives us the first term of this equation. For example, an $8\,\mu M$ solution of dGTP is equivalent to $8\,\mu$mol dGTP/L.

Problem 6.18 The terminal transferase tailing reaction just described has a dGTP concentration of $8\,\mu M$. Five μCi of ^{3}H-dGTP are added to the reaction as a reporter molecule. Two microliters of the reaction mix are then spotted on a filter and counted by liquid scintillation to determine total cpm in the reaction. How many picomoles of dGTP are on the filter?

Solution 6.18

The ^{3}H-dGTP added to the reaction is such a small molar amount that it does not significantly change the overall dGTP concentration of $8\,\mu M$. The earlier equation yields the following relationship:

$$\text{pmol dGTP} = \frac{8\ \mu\text{mol}}{\text{L}} \times \frac{\text{L}}{1 \times 10^6\ \mu\text{L}} \times \frac{1 \times 10^6\ \text{pmol}}{\mu\text{mol}} \times 2\ \mu\text{L} = 16\ \text{pmol dGTP}$$

Therefore, the $2\,\mu$L of reaction spotted on the filter to measure total cpm contains 16 pmol of dGTP.

A specific activity can be calculated by measuring the cpm of the spotted material and by use of the following formula:

$$\text{specific activity (in cpm/pmol)} = \frac{\text{spotted filter cpm}}{\text{pmol of dNTP in spotted material}}$$

Problem 6.19 The $2\,\mu$L sample spotted on the filter (referred to in Problem 6.18) was counted by liquid scintillation and gave 90 000 cpm. What is the specific activity, in cpm/pmol dNTP, of the spotted sample?

Solution 6.19

In Problem 6.18, it was calculated that a 2 μL sample of the tailing reaction contains 16 pmol of dGTP. Placing this value in the preceding equation gives the following relationship:

$$\text{specific activity} = \frac{90\,000 \text{ cpm}}{16 \text{ pmol dGTP}} = 5625 \text{ cpm/pmol dGTP}$$

Therefore, the tailing reaction has a specific activity of 5625 cpm/pmol dGTP.

Incorporation of label into a DNA strand can be determined by TCA-precipitation. Knowing the specific activity of the tailing dNTP allows for the determination of the total number of pmoles of base incorporated into a homopolymeric tail. This calculation is done using the following formula:

$$\begin{pmatrix}\text{total pmol dNTP} \\ \text{incorporated}\end{pmatrix} = \frac{\dfrac{\text{TCA} - \text{precipitable cpm}}{\mu\text{L used for TCA precipitation}} \times \text{total } \mu\text{L in reaction}}{\text{specific activity of dNTP (in cpm/pmol dNTP)}}$$

Problem 6.20 Following incubation with terminal transferase, 5 μL of the reaction mixture is treated with TCA and ethanol to measure incorporation of dGTP into polynucleotide. The filter from this assay gives 14 175 cpm by liquid scintillation. How many picomoles of dGTP are incorporated into homopolymeric tails?

Solution 6.20

It was calculated in Problem 6.19 that the dGTP in the tailing reaction has a specific activity of 5625 cpm/pmol. The reaction volume is 100 μL. Placing these values into the preceding equation yields the following relationship:

$$\text{total pmol dGTP incorporated} = \frac{\dfrac{14\,175 \text{ TCA-precipitable cpm}}{5\,\mu\text{L}} \times 100\,\mu\text{L}}{5625 \text{ cpm/pmol dGTP}}$$

$$= 50.4 \text{ pmol dGTP}$$

Therefore, a total of 50.4 picomoles of dGTP were incorporated into homopolymeric tails in this reaction.

The number of nt residues added per 3′ end can be calculated using the following formula:

$$\text{nucleotide residues/3}'\text{end} = \frac{\text{pmol dNTP incorporated}}{\text{pmol } 3' \text{ ends}}$$

It should be remembered that the term 'picomoles' represents a number, a quantity. Picomoles of different molecules are related by Avogadro's number (see Chapter 2). One mole of anything is equivalent to 6.023×10^{23} units. One picomole of anything, therefore, consists of 6.023×10^{11} units, as shown here:

$$\frac{6.023 \times 10^{23} \text{ molecules}}{\text{mol}} \times \frac{\text{mole}}{1 \times 10^{12} \text{ pmol}} = \frac{6.023 \times 10^{11} \text{ molecules}}{\text{pmol}}$$

Let's assume we find that 20 pmol dGTP have been incorporated into polynucleotides in a terminal transferase tailing reaction. Assume further that there are 2 pmol 3′ ends onto which tails can be added. Twenty picomoles of dGTP is equivalent to 1.2×10^{13} molecules of dGTP, as shown in this equation:

$$\frac{6.023 \times 10^{11} \text{ dGTP molecules}}{\text{pmol dGTP}} \times 20 \text{ pmol dGTP} = 1.2 \times 10^{13} \text{ molecules dGTP}$$

Two picomoles of 3′ ends is equivalent to 1.2×10^{12} 3′ ends:

$$\frac{6.023 \times 10^{11} \; 3' \text{ ends}}{\text{pmol } 3' \text{ ends}} \times 2 \text{ pmol } 3' \text{ ends} = 1.2 \times 10^{12} \; 3' \text{ ends}$$

Whether the number of residues added to the available 3′ ends is calculated using picomoles or using molecules, the ratios are the same, in this case 10:

$$\text{dG residues added} = \frac{20 \text{ pmol dGTP incorporated}}{2 \text{ pmol } 3' \text{ ends}} = 10$$

or

$$\text{dG residues added} = \frac{1.2 \times 10^{13} \text{ molecules dGTP incorporated}}{1.2 \times 10^{12} \; 3' \text{ ends}} = 10$$

Keeping quantities expressed as picomoles saves a step.

Problem 6.21 For the tailing reaction experiment described in this section, how long are the dG tails?

Solution 6.2

In Problem 6.20, it was calculated that 50.4 pmol dGTP were incorporated into tails. In Problem 6.16, it was calculated that the reaction contains 2.4 pmol 3′ ends. Using these values in the equation for calculating homopolymeric tail length gives the following relationship:

$$\text{dG tail length} = \frac{50.4\ \text{pmol dGTP incorporated}}{2.4\ \text{pmol } 3'\ \text{ends}} = 21\ \text{residues}$$

Therefore, 21 dG residues are added onto each 3′ end.

6.8 *IN VITRO* TRANSCRIPTION

In vitro RNA transcripts are synthesized for use as hybridization probes, as substrates for *in vitro* translation, and as antisense strands to inhibit gene expression. To monitor the efficiency of RNA synthesis, labeled CTP or UTP can be added to the reaction. Incorporation of labeled nt into RNA strands is determined by the general procedure of TCA-precipitation, as described in Section 6.3.1. Percent incorporation can then be calculated by the following formula:

$$\%\ \text{incorporation} = \frac{\text{TCA precipitable cpm}}{\text{total cpm}} \times 100$$

Other calculations for the determination of quantity of RNA synthesized and the specific activity of the RNA product are similar to those described previously in this section and are demonstrated by the following problems.

Problem 6.22 Five hundred μCi of [α-^{35}S]UTP (1500 Ci/m*M*) are added to a 20 μL *in vitro* transcription reaction containing 500 μM each of ATP, GTP, and CTP, 20 units of T7 RNA polymerase, and 1 μg of DNA. Following incubation of the reaction, 1 μL is diluted into a total volume of 100 μL TE buffer. Five μL of this dilution are spotted onto a filter and counted to determine total cpm. Another 5 μL sample of the dilution is precipitated with TCA and counted to determine incorporation of label. The sample prepared to determine total cpm gives 1.1×10^6 cpm. The

TCA-precipitated sample gives 7.7×10^5 cpm. The RNA synthesized in this reaction is purified free of unincorporated label prior to use.

a) How much RNA is synthesized?

b) What is its specific activity?

Solution 6.22(a)

The first step in solving this problem is to calculate the percent of labeled nt incorporated. This is done as follows:

$$\frac{7.7 \times 10^5 \text{ TCA-precipitated cpm}}{1.1 \times 10^6 \text{ total cpm}} \times 100 = 70\% \text{ incorporation of label}$$

Therefore, 70% of the [α-^{35}S]UTP is incorporated into RNA.

The reaction contained 500 μCi of [α-^{35}S]UTP having a specific activity of 1500 Ci/mmole. This information can be used to calculate the number of nanomoles of [α-^{35}S]UTP added to the reaction, as follows:

$$500\,\mu\text{Ci}\,[\alpha\text{-}^{35}\text{S}]\text{UTP} \times \frac{\text{mmol}}{1500\text{ Ci}} \times \frac{\text{Ci}}{1 \times 10^6\,\mu\text{Ci}} \times \frac{1 \times 10^6\text{ nmol}}{\text{mmol}} = 0.3\text{ nmol UTP}$$

Therefore, the reaction contains 0.3 nmol UTP.

Since 70% of the UTP is incorporated into RNA, the number of nanomoles of UTP incorporated into RNA is given by the following equation:

$$0.3\text{ nmol UTP} \times 0.7 = 0.2\text{ nmol UTP incorporated}$$

Therefore, 0.2 nmol of UTP are incorporated into RNA.

Assuming that there are equal numbers of As, Cs, Gs, and Us in the RNA transcript, then 0.2 nmol (as just calculated) represents 25% of the bases incorporated. Therefore, to calculate the total number of nanomoles of NTP incorporated, 0.2 nmol should be multiplied by 4.

$$0.2\text{ n}M \times 4\text{ nucleotides} = 0.8\text{ n}M\text{ nucleotide incorporated}$$

Using the average molecular weight of a nucleotide (330 ng/nmol), the amount of RNA synthesized can be calculated as follows:

$$0.8\text{ nmol NTP} \times \frac{330\text{ ng}}{\text{nmol NTP}} = 264\text{ ng RNA sythesized}$$

Therefore, 264 ng of RNA is synthesized in the *in vitro* transcription reaction.

Solution 6.22(b)

Specific activity is expressed in cpm/μg RNA. In part (a) of this problem, it was determined that a total of 264 ng of RNA are synthesized. To determine a specific activity, the number of cpm incorporated for the entire reaction must be calculated. A 1 μL sample of the reaction is diluted 1 μL/100 μL, and then 5 μL of this dilution is used to measure TCA-precipitable cpm. The TCA-precipitated sample gives 7.7×10^5 cpm. To calculate the incorporated cpm in the actual reaction, 7.7×10^5 cpm must be multiplied by both the dilution factor (100 μL/1 μL) and the reaction volume:

$$\frac{7.7 \times 10^5 \text{ cpm}}{5\,\mu\text{L}} \times \frac{100\,\mu\text{L}}{1\,\mu\text{L}} \times 20\,\mu\text{L} = 3.1 \times 10^8 \text{ cpm}$$

Therefore, there are a total of 3.1×10^8 cpm incorporated in the *in vitro* transcription reaction. Dividing this value by the total RNA synthesized and multiplying by a conversion factor to bring ng to μg gives the following equation:

$$\frac{3.1 \times 10^8 \text{ cpm}}{264 \text{ ng RNA}} \times \frac{1000 \text{ ng}}{\mu\text{g}} = \frac{1.2 \times 10^9 \text{ cpm}}{\mu\text{g RNA}}$$

Therefore, the RNA synthesized in this reaction has a specific activity of 1.2×10^9 cpm/μg RNA.

CHAPTER SUMMARY

Nucleic acids can be labeled with radioisotopes as a means to monitor their synthesis, location, and relative amount. The basic unit of radioactive decay is the **curie (Ci)**, defined as that amount of radioactive material that generates 2.22×10^{12} **disintegrations per minute (dpm)**. The disintegrations of radioactive material detected by an instrument are referred to as **counts per minute (cpm)**.

Plasmid copy number can be estimated by growing the plasmid-carrying strain of bacteria in the presence of a nt labeled with [^{3}H]. The plasmid DNA is fractionated away from the bacterium's chromosomal DNA by a cesium chloride gradient in the presence of ethidium bromide. The plasmid copy number is then calculated as

$$\text{plasmid copy number} = \frac{\text{cpm of plasmid peak}}{\text{cpm of chromosome peak}} \times \frac{\text{chromosome MW}}{\text{plasmid MW}}$$

In any type of DNA synthesis reaction where nts are incorporated into a chain, whether by nick translation, random primer labeling, or end labeling using terminal transferase, the cpm incorporated into DNA strands can be assessed by collecting TCA-precipitable material on a glass fiber filter and comparing it to the total cpm (a non-TCA-precipitated sample collected on a separate filter). The percent label incorporated into DNA is then calculated as

$$\%\,\text{incorporation} = \frac{\text{cpm incorporated into DNA}}{\text{total cpm in reaction}} \times 100$$

The specific radioactivity (the cpm per weight of nucleic acid) of a nick translation product is calculated as

$$\text{cpm}/\mu\text{g DNA} = \frac{\text{cpm in TCA precipitate}}{\text{volume used to TCA precipitate}} \times \text{dilution factor} \times \frac{\text{total reaction volume}}{\text{total }\mu\text{g DNA in reaction}}$$

DNA can be radioactively labeled by extending short, random sequence oligonucleotides with DNA polymerase. The theoretical yield of a random primer labeling experiment is calculated as

$$\text{grams DNA} = \frac{\mu\text{Ci dNTP added to reaction} \times 4\text{ dNTPs} \times \frac{330\text{ g/mol}}{\text{dNTP}}}{\text{dNTP specific activity in }\mu\text{Ci/mol}}$$

The actual yield in a random primer labeling experiment is calculated as

$$\text{ng DNA synthesized} = \text{theoretical yield} \times \%\,\text{incorporation}$$

The specific radioactivity of a product labeled by random priming is calculated at

$$\text{cpm}/\mu\text{g} = \frac{\frac{\text{cpm incorporated}}{\text{mL used to measure incorporation}} \times \begin{pmatrix}\text{dilution}\\ \text{factor}\end{pmatrix} \times \begin{pmatrix}\text{reaction}\\ \text{volume}\end{pmatrix}}{(\text{ng DNA actual yield} + \text{ng input DNA}) \times \frac{0.001\,\mu\text{g}}{\text{ng}}}$$

DNA can be labeled at its 3′ end using the enzyme terminal transferase. The specific radioactivity of a 3′-end labeled product can be calculated as

$$\text{cpm}/\mu\text{g} = \frac{\%\ \text{incorp.} \times \dfrac{\text{total cpm}}{\mu\text{L used to measure total cpm}} \times \begin{pmatrix}\text{dilution}\\ \text{factor}\end{pmatrix} \times \begin{pmatrix}\text{reaction}\\ \text{volume}\end{pmatrix}}{\mu\text{g of DNA fragment in reaction}}$$

The synthesis of cDNA can be monitored by side reactions having all the components of the main cDNA reaction but with the addition of a radioactive tracer. The number of moles of dNTP incorporated into the first cDNA strand can be calculated as

$$\text{moles dNTP incorporated} = \frac{\text{moles dNTP}}{\mu\text{L}} \times \begin{pmatrix}\text{reaction}\\ \text{volume}\end{pmatrix} \times \frac{\%\ \text{incorporation}}{100}$$

The amount of first strand cDNA synthesized can be calculated as

$$\text{g cDNA synthesized} = \text{moles dNTP incorporated} \times \frac{330\ \text{g}}{\text{moles dNTP}}$$

The amount of first strand cDNA carried into the second strand synthesis reaction is calculated as

$$\begin{pmatrix}\text{g cDNA available}\\ \text{for second strand}\\ \text{synthesis}\end{pmatrix} = \text{g first strand cDNA} \times \frac{\text{reaction volume after sample removed for labeling}}{\text{initial reaction volume}}$$

The quality of the second strand cDNA synthesis reaction can be assessed at several levels. The amount of dNTP incorporated into cDNA is calculated as

$$\begin{pmatrix}\text{nmol dNTP}\\ \text{incorporated}\end{pmatrix} = \left[\begin{pmatrix}\text{nmol}\\ \text{dNTP}/\mu\text{L}\end{pmatrix} \times \begin{pmatrix}\text{reaction}\\ \text{volume}\end{pmatrix} - \begin{pmatrix}\text{nmol dNTP incorporated}\\ \text{in first cDNA strand}\end{pmatrix}\right] \times \%\ \text{incorporation}$$

The amount of second strand cDNA synthesized is given by the equation

$$\text{ng second strand cDNA} = \text{nmol dNTP incorporated} \times \frac{330\ \text{ng dNTP}}{\text{nmol}}$$

The amount of first stand cDNA converted to double-stranded cDNA can be calculated as

$$\%\text{ conversion to double-stranded cDNA} = \frac{\text{ng second strand cDNA}}{\text{ng first strand cDNA}} \times 100$$

Terminal transferase can be used to add a string of a single nt onto the 3′ end of a DNA strand (in a process called **tailing**). The number of picomoles of 3′ ends available for tailing is given by the equation

$$\text{pmoles } 3'\text{ ends} = \frac{\mu\text{g DNA substrate} \times \dfrac{1\text{ g}}{1 \times 10^{6}\ \mu\text{g}}}{\dfrac{\text{number of bp}}{\text{molecule}} \times \dfrac{660\text{ g/mol}}{\text{bp}}} \times \frac{1 \times 10^{12}\text{ pmol}}{\text{mol}} \times \frac{\text{number of } 3'\text{ ends}}{\text{molecule}}$$

The specific activity of tailed DNA is calculated as

$$\text{cpm/pmol dNTP} = \frac{\text{total cpm in counted sample}}{\text{pmol dNTP}}$$

The concentration of dNTP in the tailing reaction (needed to calculate specific activity) is determined as

$$\mu M\text{ dNTP} = \frac{(\mu M\text{ concentration of dNTP stock}) \times (\text{volume dNTP added})}{\text{total reaction volume}}$$

Assessing the quality of the tailing reaction requires knowing the number of picomoles of deoxynucleotide spotted on the filter used to obtain incorporated cpm. It is calculated as

$$\text{pmol dNTP} = \frac{\text{mmol dNTP}}{\text{L}} \times \frac{\text{L}}{1 \times 10^{6}\ \mu\text{L}} \times \frac{1 \times 10^{6}\text{ pmol}}{\mu\text{mol}} \times \left(\begin{array}{l}\mu\text{L spotted}\\ \text{on filter}\end{array}\right)$$

The specific activity of the tailed product is then

$$\text{specific activity (in cpm/pmol)} = \frac{\text{spotted filter cpm}}{\text{pmol of dNTP in spotted material}}$$

The amount of label incorporated in the tailing reaction is calculated as

$$\begin{pmatrix}\text{total pmol dNTP} \\ \text{incorporated}\end{pmatrix} = \frac{\dfrac{\text{TCA} - \text{precipitable cpm}}{\mu\text{L used for TCA precipitation}} \times \text{total } \mu\text{L in reaction}}{\text{specific activity of dNTP (in cpm/pmol dNTP)}}$$

The number of nt residues incorporated into a tail by the terminal transferase reaction is calculated as

$$\text{nucleotide residues}/3'\text{end} = \frac{\text{pmol dNTP incorporated}}{\text{pmol } 3' \text{ ends}}$$

Radioactive labeling can also be used to monitor *in vitro* transcription. Knowing how much radioactive material is placed into the reaction (as a labeled RNA base) and knowing the specific activity of that material (in Ci/mmol) allows the calculation of the amount of RNA synthesized in TCA-precipitable cpm.

Chapter 7

Oligonucleotide synthesis

INTRODUCTION

Oligonucleotides (**oligos**), short single-stranded nucleic acids, are used for a number of applications in molecular biology and biotechnology. Oligonucleotides are used as primers for DNA sequencing and PCRs, as probes for hybridization assays in gene detection, as building blocks for the construction of synthetic genes, and as antisense molecules for the control of gene expression. With such a central role in these applications as well as in many others, it is understandable that many laboratories have acquired the instrumentation for and capability of synthesizing oligonucleotides for their own purposes. It is just as understandable that an entire industry has flourished to provide custom oligos to the general academic and biotechnology marketplace.

A number of methods have been described for the chemical synthesis of nucleic acids. However, almost all DNA synthesis instruments available commercially are designed to support the phosphoramidite method. In this chemistry, a phosphoramidite (a nucleoside with side protecting groups that preserve the integrity of the sugar, the phosphodiester linkage, and the base during chain extension steps) is coupled through its reactive 3′ phosphorous group to the 5′ hydroxyl group of a nucleoside immobilized on a solid support column. The steps of oligonucleotide synthesis include the following: (1) **Detritylation**, in which the dimethoxytrityl (DMT or 'trityl') group on the 5′ hydroxyl of the support nucleoside is removed by treatment with TCA. (2) In the **coupling step**, a phosphoramidite, made reactive by tetrazole (a weak acid), is chemically coupled to the last base added to the column support material. (3) In the **capping step**, any free 5′ hydroxyl groups of unreacted column nts are acetylated by treatment with acetic anhydride and *N*-methylimidazole. (4) In the final step, called **oxidation**, the unstable internucleotide phosphite linkage between the previously coupled base and the most recently added base is oxidized by treatment with iodine and water to a more stable phosphotriester linkage. Following coupling of all bases in the oligonucleotide's sequence, the completed nucleic acid chain is cleaved from the column by treatment with ammonium hydroxide, and the base protecting groups are removed by heating in the ammonium hydroxide solution.

Calculations for Molecular Biology and Biotechnology. DOI: 10.1016/B978-0-12-375690-9.00007-3
© 2010 Elsevier Inc. All rights reserved.

7.1 SYNTHESIS YIELD

Nucleic acid synthesis columns come in several scales, which are functions of the amount of the 3′ nucleoside coupled to the support material. Popular column sizes range from 40 nmol to 10 μmol. Popular support materials include controlled-pore glass (CPG) and polystyrene. The CPG columns are typically supplied having pore sizes of either 500 or 1000 Å. The scale of column used for a synthesis depends on both the quantity and the length of oligonucleotide desired. The greater the amount of 3′ nucleoside attached to the column support, the greater the yield of oligo. Polystyrene or large-pore CPG columns are recommended for oligonucleotides greater than 50 bases in length.

A 0.2 μmol CPG column contains 0.2 μmol of starting 3′ base. If the second base in the synthesis is coupled with 100% efficiency, then 0.2 μmol of that second base will couple, for a total of 0.4 μmol of base now on the column (0.2 μmol 3′ base + 0.2 μmol coupled base = 0.4 μmol of bases). The number of micromoles of an oligonucleotide synthesized on a column, therefore, can be calculated by multiplying the length of the oligonucleotide by the column scale.

Problem 7.1 An oligo 22 nts in length (a 22-mer) is synthesized on a 0.2 μmol column. Assuming 100% synthesis efficiency, how many micromoles of oligo are made?

Solution 7.1

Multiplying the column scale by the length of the oligo yields the following equation:

$$22 \text{ base additions} \times \frac{0.2\,\mu\text{mol}}{\text{base addition}} = 4.4\,\mu\text{mol}$$

Therefore, assuming 100% coupling efficiency, 4.4 μmol of oligo are synthesized.

If bases are coupled at 100% efficiency, then a theoretical yield for a synthesis can be determined by using the following relationship: 1 μmol of DNA oligonucleotide contains about 10 OD units per base. (An **OD unit** is the amount of oligonucleotide dissolved in 1.0 mL, which gives an A_{260} of 1.00 in a cuvette with a 1 cm-length light path.) The theoretical yield of crude oligonucleotide is calculated by multiplying the synthesis scale by the oligo length and then multiplying the product by 10, as shown in the following problem.

Problem 7.2 A 22-mer is synthesized on a 0.2 μmol-scale column. What is the maximum theoretical yield of crude oligo in OD units?

Solution 7.2

The theoretical maximum yield of crude oligo is given by the following equation:

$$22 \text{ base additions} \times \frac{0.2\,\mu\text{mol base}}{\text{base addition}} \times \frac{10\text{ OD units}}{\mu\text{mol base}} = 44\text{ OD units}$$

Therefore, 44 OD units of crude oligonucleotide will be synthesized.

Nucleic acid synthesis chemistry, however, is not 100% efficient. No instrument, no matter how well engineered or how pure the reagents it uses, can make oligonucleotides at the theoretical yield. The actual expected yield of synthesis product will be less. Table 7.1 lists the yield of crude oligonucleotide that can be expected from several different columns when synthesis is optimized.

Table 7.1 The approximate yield, in OD units, of oligonucleotide per base for each of the common synthesis scales.

Synthesis Scale	OD Units/Base
40 nmol	0.25–0.5
0.2 μmol	1.0–1.5
1 μmol	5
10 μmol	40

Problem 7.3 A 22-mer is synthesized on a 0.2 μmol-scale column. Assuming that synthesis is optimized, how many OD units can be expected?

Solution 7.3

Using the OD units/base value in Table 7.1 for the 0.2 μmol-scale column yields the following equations:

$$22\text{ bases} \times \frac{1\text{ OD unit}}{\text{base}} = 22\text{ OD units}$$

$$22\,\text{bases} \times \frac{1.5\ \text{OD units}}{\text{base}} = 33\,\text{OD units}$$

Therefore, this synthesis should produce between 22 and 33 OD units of oligonucleotide.

7.2 MEASURING STEPWISE AND OVERALL YIELD BY THE DIMETHOXYTRITYL (DMT) CATION ASSAY

Overall yield is the amount of full-length product synthesized. In contrast, **stepwise yield** is a measure of the percentage of the nucleic acid molecules on the synthesis column that couple at each base addition. If the reagents and the DNA synthesizer are performing optimally, 98% stepwise yield can be expected. For such a synthesis, the 2% of molecules that do not participate in a coupling reaction will be capped and will be unavailable for reaction in any subsequent cycle. Since, at each cycle, 2% fewer molecules can react with phosphoramidite, the longer the oligo being synthesized, the lower the overall yield.

The trityl group protects the 5′ hydroxyl on the ribose sugar of the most recently added base, preventing it from participating in unwanted side reactions. Prior to addition of the next base, the trityl group is removed by treatment with TCA. Detritylation renders the 5′ hydroxyl available for reaction with the next phosphoramidite. When removed under acidic conditions, the trityl group has a characteristic bright orange color. A fraction collector can be configured to the instrument to collect each trityl fraction as it is washed from the synthesis column during the detritylation step. The fractions can then be diluted with *p*-toluene sulfonic acid monohydrate in acetonitrile and assayed for absorbance at 498 nm. (For accuracy, further dilutions may be necessary to keep readings between 0.1 and 1.0 – within the range of linear absorbance.)

For the following calculations, the first trityl fraction, that generated from the column 3′ base, should not be used since it can give a variable reading depending on the integrity of the trityl group on the support base. Spontaneous detritylation of the column base can occur prior to synthesis, resulting in a low absorbance for the first fraction. Including this fraction in the calculation for overall yield would result in an overestimation of product yield.

7.2.1 Overall yield

Overall yield, the measure of the total product synthesized, is given by the formula

$$\text{overall yield} = \frac{\text{trityl absorbance of last base added}}{\text{trityl absorbance of first base added}} \times 100$$

Problem 7.4 An 18-mer is synthesized on a 0.2 μmol column. Assay of the trityl fractions gave the following absorbance values at 498 nm. What is the overall yield?

Number	Base	Absorbance
1	Column C	0.524
2	G	0.558
3	T	0.541
4	A	0.538
5	A	0.527
6	C	0.515
7	T	0.505
8	G	0.493
9	G	0.485
10	C	0.474
11	A	0.468
12	T	0.461
13	T	0.452
14	T	0.439
15	C	0.427
16	G	0.412
17	A	0.392
18	A	0.379

Solution 7.4

Using the equation for overall yield gives

$$\text{overall yield} = \frac{0.379}{0.558} \times 100 = 68\%$$

Therefore, 68% of the starting column base is made into full-length product; the overall yield is 68%.

7.2.2 Stepwise yield

Stepwise yield, the percent of molecules coupled at each single base addition, is calculated by the formula

$$\text{stepwise yield} = (\text{overall yield})^{1/\text{couplings}}$$

The number of couplings is represented by the number of trityl fractions spanned to calculate the overall yield minus one. (One is subtracted because the trityl fraction from the first base coupled to the column base can be an unreliable indicator of actual coupling efficiency at that step.) An efficient synthesis will have a stepwise yield of 98 ± 0.5%.

A low stepwise yield can indicate reagent problems, either in their quality or in their delivery. Even if a high stepwise yield is obtained, however, the product may still not be of desired quality because some synthesis failures are not detectable by the trityl assay.

Problem 7.5 What is the stepwise yield for the synthesis described in Problem 7.4?

Solution 7.5

In Problem 7.4, an overall yield of 68% is obtained. A total of 17 trityl fractions span the absorbance values used to calculate the overall yield. Using the equation for stepwise yield gives

$$\text{stepwise yield} = 0.68^{1/17-1} = 0.68^{1/16} = 0.68^{0.06} = 0.98$$

($0.68^{0.06}$ is obtained on a calculator by entering ., **6**, **8**, $\boldsymbol{x^y}$, ., **0**, **6**, and =.) Stepwise yield is usually expressed as a percent value:

$$0.98 \times 100 = 98\%$$

Therefore, the stepwise yield is 98%.

Assuming a stepwise yield of 98%, it is possible to calculate the overall yield for any length of oligonucleotide by using the formula

$$\text{overall yield} = 0.98^{(\text{number of bases}-1)} \times 100$$

Problem 7.6 Fresh reagents are used to synthesize a 25-mer. Assuming 98% stepwise yield, what overall yield can be expected?

Solution 7.6

The relationship just shown above yields the following equation:

$$\text{overall yield} = 0.98^{(25-1)} \times 100 = 0.98^{24} \times 100 = 0.62 \times 100 = 62\%$$

(0.98^{24} is obtained on a calculator by entering ., **9**, **8**, $\boldsymbol{x^y}$, **2**, **4**, and =.)

Therefore, an overall yield of 62% can be expected.

Problem 7.7 Assuming a stepwise yield of 95%, what overall yield can be expected for the synthesis of a 25-mer?

Solution 7.7

Using the same relationship gives

$$\text{overall yield} = 0.95^{(25-1)} \times 100 = 0.95^{24} \times 100 = 0.29 \times 100 = 29\%$$

Therefore, an overall yield of 29% can be expected for a 25-mer when the stepwise yield is 95%.

(Notice that a drop in stepwise yield from 98% (Problem 7.6) to 95% (Problem 7.7) results in a dramatic drop in overall yield from 62% to 29%. An instrument and reagents of highest quality are essential to obtaining optimal synthesis of product.)

7.3 CALCULATING MICROMOLES OF NUCLEOSIDE ADDED AT EACH BASE ADDITION STEP

Trityl fraction assays can be used not only to estimate stepwise and overall yields but also to determine the number of micromoles of nucleoside added to the oligonucleotide chain at any coupling step. The number of micromoles of DMT cation released and the number of micromoles of nucleoside added are equivalent. Micromoles of DMT can be calculated by dividing a trityl fraction's absorbance at 498 by the extinction coefficient of DMT according to the formula

$$\mu\text{mol DMT} = \frac{A_{498} \times \text{volume} \times \text{dilution factor}}{e}$$

where A_{498} is the absorbance of the trityl fraction at 498 nm, volume is the final volume of the diluted DMT used for the A_{498} reading, dilution factor is the inverse of the dilution (if one is used) of the trityl fraction into a second tube of 0.1 *M* toluene sulfonic acid in acetonitrile, and $e = 70\,\text{mL}/\mu\text{mol}$, the extinction coefficient of the DMT cation at 498 nm.

Problem 7.8 A trityl fraction from a 0.2 μmol synthesis is brought up to 10 mL with 0.1 *M* toluene sulfonic acid monohydrate in acetonitrile, and the absorbance is read at 498 nm. A reading of 0.45 is obtained. How many micromoles of nucleoside couple at that step?

Solution 7.8

The trityl fraction is brought up to a volume of 10 mL and this volume is used to measure absorbance. No dilution into a second tube is used. Using the earlier formula yields

$$\mu\text{mol DMT} = \frac{0.45 \times 10\,\text{mL}}{70\,\text{mL}/\mu\text{mol}} = 0.06$$

Therefore, since micromoles of DMT are equivalent to micromoles of nucleoside added, 0.06 μmol of nucleoside are added at the coupling step represented by this trityl fraction.

CHAPTER SUMMARY

Oligonucleotides (oligos) are synthetic, short, ssDNA molecules used as probes, primers, or as the building blocks for longer DNA strands. The most popular type of chemistry for making oligonucleotides uses the phosphoramidite method, in which building of the nt chain starts on a column carrying the first 3′ base. Assuming 100% synthesis efficiency, the amount of oligo synthesized on a column is related to the amount of 3′ base originally loaded on the column and the length of the oligo being synthesized. It is given by the expression

$$\mu\text{mol oligo} = \text{base additions} \times \frac{\text{column size (in } \mu\text{mol)}}{\text{base addition}}$$

The maximum theoretical yield of oligonucleotide (in OD units), assuming 100% coupling efficiency at each base, is given by the following expression:

$$\text{OD units} = \text{base additions} \times \frac{\mu\text{mol column size}}{\text{base addition}} \times \frac{10 \text{ OD units}}{\mu\text{mol base}}$$

Overall synthesis yield is calculated as

$$\text{overall yield} = \frac{\text{trityl absorbance of last base added}}{\text{trityl absorbance of first base added}} \times 100$$

Stepwise yield, the percent of molecules coupled at each base addition, is calculated as

$$\text{stepwise yield} = (\text{overall yield})^{1/\text{couplings}}$$

When the stepwise yield is less than 100%, the overall yield can be calculated as

$$\text{overall yield} = \text{stepwise yield}^{(\text{oligo length} - 1)} \times 100$$

The number of μM of nt added at each addition step is calculated as

$$\mu\text{mol DMT} = \frac{A_{498} \times \text{volume} \times \text{dilution factor}}{e}$$

where A_{498} is the absorbance of the trityl fraction at 498 nm, volume is the final volume of the diluted DMT used for the A_{498} reading, dilution factor is the inverse of the dilution (if one is used) of the trityl fraction into a second tube of 0.1 M toluene sulfonic acid in acetonitrile, and $e = 70$ mL/μmol, the extinction coefficient of the DMT cation at 498 nm.

Chapter 8

The polymerase chain reaction (PCR)

INTRODUCTION

The **polymerase chain reaction (PCR)** is a method for the amplification of a specific segment of nucleic acid[a]. A typical PCR contains a template nucleic acid (DNA or RNA), a thermally stable DNA polymerase, the four deoxynucleoside triphosphates (dNTPs), magnesium, oligonucleotide primers, and a buffer. Three steps are involved in the amplification process: (1) In a **denaturation** step, the template nucleic acid is made single-stranded by exposure to an elevated temperature (92–98°C). (2) The temperature of the reaction is then brought to between 65°C and 72°C in an **annealing** step, by which the two primers attach to opposite strands of the denatured template such that their 3′ ends are directed toward each other. (3) In the final step, called **extension**, a thermally stable DNA polymerase adds dNTPs onto the 3′ ends of the two annealed primers such that strand replication occurs across the area between and including the primer annealing sites. These steps make up one **cycle** and are repeated over and over again for, depending on the specific application, from 25 to 40 times.

A very powerful technique, PCR has found use in a wide range of applications. To cite only a few, it is used to examine biological evidence in forensic cases, to identify contaminating microorganisms in food, to diagnose genetic diseases, and to map genes to specific chromosome segments.

8.1 TEMPLATE AND AMPLIFICATION

One of the reasons the PCR technique has found such wide use is that only very little starting template DNA is required for amplification. In forensic applications, for example, a result can be obtained from the DNA recovered from a single hair or from residual saliva on a licked envelope. Typical PCRs performed in forensic analysis use from only 0.5 to 10 ng of DNA.

[a]The PCR process is covered by patents owned by Roche Molecular Systems, Inc. and F. Hoffman-La Roche Ltd.

Calculations for Molecular Biology and Biotechnology. DOI: 10.1016/B978-0-12-375690-9.00008-5
© 2010 Elsevier Inc. All rights reserved.

Problem 8.1 Twenty nanograms of human genomic DNA are needed for a PCR. The template stock has a concentration of 0.2 mg DNA/mL. How many microliters of the stock solution should be used?

Solution 8.1

Conversion factors are used to bring mg and mL quantities to ng and μL quantities, respectively. The equation can be written as follows:

$$\frac{0.2\,\text{mg DNA}}{\text{mL}} \times \frac{1\,\text{mL}}{1 \times 10^3\,\mu\text{L}} \times \frac{1 \times 10^6\,\text{ng}}{\text{mg}} \times x\,\mu\text{L} = 20\,\text{ng DNA}$$

By writing the equation in this manner, terms cancel so that the equation reduces to ng = ng. Multiplying and solving for x yields the following relationship:

$$\frac{(2 \times 10^5\,\text{ng DNA})\,(x)}{1 \times 10^3} = 20\,\text{ng DNA}$$

Numerator values are multiplied together, as are denominator values.

$$(2 \times 10^5\,\text{ng DNA})\,(x) = 2 \times 10^4\,\text{ng DNA}$$

Multiply both sides of the equation by 1×10^3.

$$x = \frac{2 \times 10^4\,\text{ng DNA}}{2 \times 10^5\,\text{ng DNA}} = 0.1$$

Divide both sides of the equation by 2×10^5 ng =DNA.

Therefore, 0.1 μL of the 0.2 mg/mL human genomic DNA stock will contain 20 ng of DNA. It is often the case that a calculation is performed, such as this one, that derives an amount that cannot be measured accurately with standard laboratory equipment. Most laboratory pipettors cannot deliver 0.1 μL with acceptable accuracy. A dilution of the stock, therefore, should be performed to accommodate the accuracy limits of the available pipettors. Most laboratories have pipettors that can deliver 2 μL accurately. We can ask the question, 'into what volume can 2 μL of the stock solution be diluted such that 2 μL of the new dilution will deliver 20 ng of DNA?' The equation for this calculation is then written as follows:

$$\frac{0.2\text{ mg DNA}}{\text{mL}} \times \frac{1\text{ mL}}{1 \times 10^3\,\mu\text{L}} \times \frac{1 \times 10^6\text{ ng}}{\text{mg}} \times \frac{2\,\mu\text{L}}{x\,\mu\text{L}} \times 2\,\mu\text{L} = 20\text{ ng DNA}$$

$$\frac{8\times10^5\ \text{ng DNA}}{(1\times10^3)(x)} = 20\ \text{ng DNA}$$

Cancel terms and multiply numerator values and denominator values.

$$\frac{8\times10^5\ \text{ng DNA}}{1\times10^3} = (20\ \text{ng DNA})x$$

Multiply both sides of the equation by x and solve for x.

$$800\ \text{ng DNA} = (20\ \text{ng DNA})x$$

Simplify the equation.

$$\frac{800\ \text{ng DNA}}{20\ \text{ng DNA}} = x = 40$$

Divide each side of the equation by 20 ng DNA.

Therefore, if 2 μL of the DNA stock is diluted into a total volume of 40 μL (2 μL DNA stock + 38 μL water), then 2 μL of that dilution will contain 20 ng of template DNA.

8.2 EXPONENTIAL AMPLIFICATION

During the PCR process, the product of one cycle serves as template in the next cycle. This characteristic is responsible for PCR's ability to amplify a single molecule into millions. PCR is an exponential process. Amplification of template progresses at a rate of 2^n, where n is equal to the number of cycles. The products of PCR are termed **amplicons**. The DNA segment amplified is called the **target**.

Problem 8.2 A single molecule of DNA is amplified by PCR for 25 cycles. Theoretically, how many molecules of amplicon will be produced?

Solution 8.2

Using the relationship of overall amplification as equal to 2^n, where n is, in this case, 25, gives the following result:

$$\text{overall amplification} = 2^{25} = 33554432$$

Therefore, over 33 million amplicons will be produced.

Problem 8.3 Beginning with 600 template DNA molecules, after 25 cycles of PCR, how many amplicons will be produced?

Solution 8.3

Since each template can enter into the PCR amplification process, 2^{25} should be multiplied by 600:

$$\text{overall amplification} = 600 \times 2^{25} = 2 \times 10^{10} \text{ molecules}$$

Therefore, the amplification of 600 target molecules in a PCR run for 25 cycles should theoretically produce 2×10^{10} molecules of amplified product.

Problem 8.4 If the PCR cited in Problem 8.2 is performed as a 100 μL reaction, how many molecules of amplified product will be present in 0.001 μL?

Solution 8.4

A ratio can be set up that is read '2×10^{10} amplicons is to 100 μL as x amplicons is to 0.001 μL.'

$$\frac{2 \times 10^{10} \text{ amplicons}}{100\ \mu\text{L}} = \frac{x \text{ amplicons}}{0.001\ \mu\text{L}}$$

$$\frac{(2 \times 10^{10} \text{ amplicons})(0.001\ \mu\text{L})}{100\ \mu\text{L}} = x \text{ amplicons}$$

$$x = 2 \times 10^{5} \text{ amplicons}$$

Therefore, 2×10^{5} amplicons are present in 0.001 μL.

A note on contamination control: As shown in Problem 8.4, very small quantities of a PCR can contain large amounts of product. Laboratories engaged in the use of PCR, therefore, must take special precautions to avoid contamination of a new reaction with product from a previous one. Even quantities of PCR product present in an aerosol formed during the opening of a reaction tube are enough to overwhelm a new reaction. Contamination of a workspace by amplified product is of particular concern in facilities that amplify the same locus (or loci) over and over again, as is the case in forensic and paternity testing laboratories.

Problem 8.5 One hundred nanograms of a 242 bp fragment is produced in a PCR. Two nanograms of human genomic DNA are used as starting template. What amount of amplification occurs?

Solution 8.5

The first step in solving this problem is to determine how many copies are represented by a particular amount of DNA (in this case, 2 ng of human genomic DNA). Assume that the target sequence is present in only one copy in the genome. Although there would be two copies of the target sequence in all human cells (except egg, sperm, or red blood cells), there would be one copy in the haploid human genome. The haploid human genome consists of approximately 3×10^9 bp. Therefore, there is one copy of the target sequence per 3×10^9 bp. The molecular weight of a bp (660 g/*M*) and Avogadro's number can be used to convert this to an amount of DNA per gene copy. Since this amount will be quite small, for convenience the answer can be expressed in picograms:

$$\frac{3 \times 10^9 \text{ bp}}{\text{copy}} \times \frac{660 \text{ g}}{\text{mol bp}} \times \frac{1 \text{ mol bp}}{6.023 \times 10^{23} \text{ bp}} \times \frac{1 \times 10^{12} \text{ pg}}{\text{g}} = \frac{3.3 \text{ pg}}{\text{copy}}$$

Therefore, 3.3 pg of human genomic DNA contains one copy of the target sequence.

The number of copies of the target sequence contained in 2 ng of human genomic DNA can now be determined:

$$2 \text{ ng} \times \frac{1 \times 10^3 \text{ pg}}{\text{ng}} \times \frac{1 \text{ copy}}{3.3 \text{ pg}} = 600 \text{ copies}$$

Therefore, 600 copies of the target sequence are contained in 2 ng of human genomic DNA template.

By a similar method, the weight of one copy of the 242 bp product can be determined as follows:

$$\frac{242 \text{ bp}}{\text{copy}} \times \frac{660 \text{ g}}{\text{mol bp}} \times \frac{1 \text{ mol bp}}{6.023 \times 10^{23} \text{ bp}} \times \frac{1 \times 10^{12} \text{ pg}}{\text{g}} = \frac{2.7 \times 10^{-7} \text{ pg}}{\text{copy}}$$

Therefore, a 242 bp fragment (representing one copy of PCR product) weighs 2.7×10^{-7} pg.

The number of copies of a 242 bp fragment represented by 100 ng of that fragment can now be determined:

$$100\,\text{ng} \times \frac{1 \times 10^{3}\,\text{pg}}{\text{ng}} \times \frac{\text{copy}}{2.7 \times 10^{-7}\,\text{pg}} = 3.7 \times 10^{11}\ \text{copies}$$

Therefore, there are 3.7×10^{11} copies of the 242 bp PCR product in 100 ng.

The amount of amplification can now be calculated by dividing the number of copies of the target sequence at the end of the PCR by the number of copies of target present in the input template DNA:

$$\frac{3.7 \times 10^{11}\ \text{copies}}{600\ \text{copies}} = 6.2 \times 10^{8}$$

Therefore, the PCR creates a (6.2×10^{8})-fold increase in the amount of amplified DNA segment.

8.3 POLYMERASE CHAIN REACTION (PCR) EFFICIENCY

The formula 2^n used to calculate PCR amplification can be employed to describe any system that proceeds by a series of doubling events, whereby 1 gives rise to 2, 2 gives rise to 4, 4 to 8, 8 to 16, 16 to 32, etc. This relationship describes the PCR if the reaction is 100% efficient. No reaction performed in a test tube, however, will be 100% efficient. To more accurately represent reality, the relationship must be modified to account for actual reaction efficiency. The equation then becomes

$$Y = (1 + E)^n$$

where Y is the degree of amplification, n is the number of cycles, and E is the mean efficiency of amplification at each cycle; the percent of product from one cycle serving as template in the next.

Loss of amplification efficiency can occur for any number of reasons, including loss of DNA polymerase enzyme activity, depletion of reagents, and incomplete denaturation of template, as occurs when the amount of product increases in later cycles. Amplification efficiency can be determined experimentally by taking a small aliquot out from the PCR at the beginning and end of each cycle (from the third cycle on) and electrophoresing that sample on a gel. In other lanes on that gel are run standards with known numbers of molecules. Intensities of the bands (as determined by staining or densitometry) are compared for quantification. The efficiency, expressed as a decimal, is then the final copy number at the end of a cycle divided by the number of copies at the beginning of that cycle.

To take into account the amount of starting template in a PCR (represented by X), the equation becomes

$$Y = X(1 + E)^n$$

Use of this relationship requires an understanding of logarithms and antilogarithms. A **common logarithm**, as will be used here, is defined as the exponent indicating the power to which 10 must be raised to produce a given number. An **antilogarithm** is found by reversing the process used to find the logarithm of a number; e.g., if $y = \log x$, then $x = \text{antilog } y$. Before this equation can be converted to a more useful form, two laws of logarithms must be stated, the **Product Rule for Logarithms** and the **Power Rule for Logarithms**. (More discussion on logarithms can be found in Chapter 3.)

Product Rule for Logarithms: For any positive numbers M, N, and a (where a is not equal to one), the logarithm of a product is the sum of the logarithms of the factors.

$$\log_a MN = \log_a M + \log_a N$$

Power Rule for Logarithms: For any positive number M and a (where a is not equal to one), the logarithm of a power of M is the exponent times the logarithm of M.

$$\log_a M^p = p\log_a M$$

The equation $Y = X(1 + E)^n$ can now be written in a form that will allow more direct calculation of matters important in PCR. Taking the common logarithm of both sides of the equation and using both the Product and Power Rules for Logarithms yields

$$\log Y = \log X + n\log(1 + E)$$

This is the form of the equation that can be used when calculating product yield or when calculating the number of cycles it takes to generate a certain amount of product from an initial template concentration.

Problem 8.6 An aliquot of template DNA containing 3×10^5 copies of a target gene is placed into a PCR. The reaction has a mean efficiency of 85%. How many cycles are required to produce 2×10^{10} copies?

Solution 8.6

For the equation $Y = X(1 + E)^n$, $Y = 2 \times 10^{10}$ copies, $X = 3 \times 10^5$ copies, and $E = 0.85$. Placing these values into the equation gives the following relationship:

$$2 \times 10^{10} = (3 \times 10^5)(1 + 0.85)^n$$

$\frac{2 \times 10^{10}}{3 \times 10^5} = 1.85^n$ Divide each side of the equation by 3×10^5.

$6.7 \times 10^4 = 1.85^n$ Simplify the left side of the equation.

$\log 6.7 \times 10^4 = \log 1.85^n$ Take the common logarithm of both sides of the equation.

$\log 6.7 \times 10^4 = n \log 1.85$ Use the Power Rule for Logarithms.

$\frac{\log 6.7 \times 10^4}{\log 1.85} = n$ Divide both sides of the equation by log 1.85.

$\frac{4.8}{0.27} = n = 17.8$ Substitute the log values into the denominator and numerator. (To find the log of a number on a calculator, enter that number and then press the ***log*** key.)

Therefore, it will take 18 cycles to produce 2×10^{10} copies of amplicon starting from 3×10^5 copies of template.

Problem 8.7 A PCR amplification proceeds for 25 cycles. The initial target sequence was present in the reaction at 3×10^5 copies. At the end of the 25 cycles, 4×10^{12} copies have been produced. What is the efficiency (E) of the reaction?

Solution 8.7

Using the equation $Y = X(1 + E)^n$, $Y = 4 \times 10^{12}$ copies, $X = 3 \times 10^5$ copies, and $n = 25$, the equation then becomes

$$4 \times 10^{12} = (3 \times 10^5)(1 + E)^{25}$$

$$\frac{4 \times 10^{12}}{3 \times 10^5} = (1 + E)^{25}$$

$1.3 \times 10^7 = (1 + E)^{25}$	Divide each side of the equation by 3×10^5 and simplify.
$\log 1.3 \times 10^7 = \log(1 + E)^{25}$	Take the logarithm of both sides of the equation.
$\log 1.3 \times 10^7 = 25 \log(1 + E)$	Use the Power Rule for Logarithms.
$7.1 = 25 \log(1 + E)$	Substitute the log of 1.3×10^7 into the equation. (The log of 1.3×10^7 is found on a calculator by entering **1**, **.**, **3**, **EXP**, **7**, **log**.)
$\frac{7.1}{25} = \log(1 + E)$	Divide each side of the equation by 25 and simplify the left side of the equation.
$0.28 = \log(1 + E)$	
$\text{antilog } 0.28 = \text{antilog}[\log(1 + E)]$	Take the antilog of each side of the equation.
$\text{antilog } 0.28 = 1 + E$	
$1.91 = 1 + E$	To find the antilog of 0.28 on the calculator, enter **10**, x^y, **.28**, and =.
$1.91 - 1 = E$	Subtract 1 from both sides of the equation.
$0.91 = E$	

Therefore, the PCR is 91% efficient ($0.91 \times 100 = 91\%$).

8.4 CALCULATING THE T_m OF THE TARGET SEQUENCE

The T_m (**melting temperature**) of dsDNA is the temperature at which half of the molecule is in double-stranded conformation and half is in single-stranded form. It does not matter where along the length of the molecule the single-stranded or double-stranded regions occur so long as half of the entire molecule is in single-stranded form. It is desirable to know the melting temperature of the amplicon to help ensure that an adequate denaturation temperature is chosen for thermal cycling, one that will provide an optimal amplification reaction.

DNA is called a **polyanion** because of the multiple negative charges along the sugar–phosphate backbone. The two strands of DNA, since they carry like charges, have a natural tendency to repel each other. In a water environment free of any ions, DNA will denature easily. Salt ions such as sodium (Na^+) and magnesium (Mg^{+2}), however, can complex to the negative charges on the phosphate groups of DNA and will counteract the tendency of the two strands to repel each other. In fact, under conditions of high salt, two strands of DNA can be made to anneal to each other even if they contain a number of mismatched bases between them.

Most PCRs contain salt. Potassium chloride (KCl) is frequently added to a PCR to enhance the activity of the polymerase. Magnesium is a cofactor of the DNA polymerase enzyme. It is provided in a PCR by the salt magnesium chloride ($MgCl_2$). The presence of potassium (K^+) and magnesium ions will affect the melting temperature of duplex nucleic acid and should be a consideration when making a best calculation of T_m.

Other factors affect the T_m of duplex DNA. Longer molecules, having more bps holding them together, require more energy to denature them. Likewise, T_m is affected by the G + C content of the DNA. Since Gs and Cs form three hydrogen bonds between them (as compared with two hydrogen bonds formed between As and Ts), DNAs having higher G + C content will require more energy (a higher temperature) to denature them.

These considerations are taken into account in the formula given by Wetmur and Sninsky (1995) for calculating the melting temperature of long duplex DNAs:

$$T_m = 81.5 + 16.6\log\left[\frac{[\text{SALT}]}{1.0 + 0.7[\text{SALT}]}\right] + 0.41(\%\text{G} + \text{C}) - \frac{500}{L}$$

where L = the length of the amplicon in bp. %G + C is calculated using the formulae

$$\%\text{G} + \text{C} = \frac{\text{G} + \text{C bases in the amplicon}}{\text{total number of bases in the amplicon}} \times 100$$

and

$$[\text{SALT}] = [\text{K}^+] + 4[\text{Mg}^{2+}]^{0.5}$$

In this last equation, any other monovalent cation, such as Na^+, if in the PCR, can be substituted for K^+.

For the thermal cycling program, a denaturation temperature should be chosen 3–4°C higher than the calculated T_m to ensure adequate separation of the two amplicon strands.

Problem 8.8 An 800 bp PCR product having a G + C content of 55% is synthesized in a reaction containing 50 m*M* KCl and 2.5 m*M* $MgCl_2$. What is the amplicon's T_m?

Solution 8.8

First, calculate the salt concentration ([SALT]). The formula given for this requires that the concentrations of K^+ and Mg^{+2} be expressed in *M* rather than m*M* concentration. For KCl this is calculated as follows:

$$50\text{ m}M\text{ KCl} \times \frac{1\,M}{1000\text{ m}M} = 0.05\,M\text{ KCl}$$

Therefore, the KCl concentration is 0.05 *M*.

The *M* concentration of $MgCl_2$ is calculated as follows:

$$2.5\text{ m}M\text{ MgCl}_2 \times \frac{1\,M}{1000\text{ m}M} = 0.0025\,M\text{ MgCl}_2$$

Therefore, the $MgCl_2$ concentration is 0.0025 *M*.

Placing these values into the equation for calculating salt concentration gives the following relationship:

$$[\text{SALT}] = [0.05] + 4[0.0025]^{0.5}$$

$[0.0025]^{0.5}$ on the calculator is obtained by entering **0.0025**, pressing the x^y key, entering **0.5**, and pressing the = key.

$$[\text{SALT}] = 0.05 + 4(0.05)$$

Solving this equation gives the following result:

$$[\text{SALT}] = 0.05 + 0.2 = 0.25$$

Placing this value into the equation for T_m yields the following equation:

$$T_m = 81.5 + 16.6\log\left[\frac{[0.25]}{1.0 + 0.7[0.25]}\right] + 0.41(55) - \frac{500}{800}$$

The equation is simplified to give the following relationship:

$$T_m = 81.5 + 16.6\log[0.213] + 0.41(55) - 0.63$$

Calculating the log of 0.213 (enter **0.213** on the calculator and press the ***log*** key) and further simplifying the equation gives the following result:

$$T_m = 81.5 + 16.6(-0.672) + 22.55 - 0.63$$

$$T_m = 81.5 + (-11.155) + 22.55 - 0.63 = 92.3$$

Therefore, the melting temperature of the 800 bp amplicon is 92.3°C. Programming the thermal cycler for a denaturation temperature of 96°C should ensure proper denaturation at each cycle.

8.5 PRIMERS

Polymerase chain reaction primers are used in concentrations ranging from 0.05 to 2 μM.

Problem 8.9 A primer stock consisting of the two primers needed for a PCR has a concentration of 25 μ*M*. What volume of the primer stock should be added to a 100 μL reaction so that the primer concentration in the reaction is 0.4 μM?

Solution 8.9

The calculation is set up in the following manner:

$$25\,\mu M \text{ primer} \times \frac{x\,\mu\text{L}}{100\,\mu\text{L}} = 0.4\,\mu M \text{ primer}$$

The equation is then solved for *x*.

$$\frac{(25\,\mu M)x}{100} = 0.4\,\mu M$$

$$(25\,\mu M)x = 40\,\mu M$$

$$x = \frac{40\,\mu M}{25\,\mu M} = 1.6$$

Therefore, 1.6 μL of the 25 μ*M* primer stock should be added to a 100 μL PCR to give a primer concentration of 0.4 μ*M*.

Problem 8.10 A 100 μL PCR contains primers at a concentration of 0.4 μ*M*. How many picomoles of primer are in the reaction?

Solution 8.10

Remember, 0.4 μ*M* is equivalent to 0.4 μmol/L. A series of conversion factors is used to bring a molar concentration into a mole quantity:

$$\frac{0.4\ \mu\text{mol primer}}{\text{L}} \times \frac{1\text{L}}{1 \times 10^6\ \mu\text{L}} \times \frac{1 \times 10^6\ \text{pmol}}{\mu\text{mol}} \times 100\ \mu\text{L} = 40\ \text{pmol primer}$$

Therefore, there are 40 pmol of primer in the reaction.

Problem 8.11 A 100 μL PCR is designed to produce a 500 bp fragment. The reaction contains 0.2 μ*M* primer. Assuming a 100% efficient reaction and that the other reaction components are not limiting, what is the theoretical maximum amount of product (in μg) that can be made?

Solution 8.11

The first step is to calculate how many micromoles of primer are in the reaction. (0.2 μ*M* is equivalent to 0.2 μmol/L.)

$$\frac{0.2\ \mu\text{mol primer}}{\text{L}} \times \frac{1\text{L}}{1 \times 10^6\ \mu\text{L}} \times 100\ \mu\text{L} = 2 \times 10^{-5}\ \mu\text{mol primer}$$

Therefore, the reaction contains 2×10^{-5} μmol of primer. Assuming a 100% efficient reaction, each primer will be extended into a complete DNA strand. Therefore, 2×10^{-5} μmol of primer will make 2×10^{-5} μmol of extended products. Since in this case the product is 500 bp in length, its molecular weight must be part of the equation used to calculate the amount of total product made:

$$500\ \text{bp} \times \frac{660\ \text{g/mol}}{\text{bp}} \times \frac{1\text{mol}}{1 \times 10^6\ \mu\text{mol}} \times \frac{1 \times 10^6\ \mu\text{g}}{\text{g}} \times 2 \times 10^{-5}\ \mu\text{mol} = 6.6\ \mu\text{g}$$

Therefore, in this reaction, 6.6 μg of 500 bp PCR product can be synthesized.

Problem 8.12 Twenty pg of an 8000 bp recombinant plasmid are used in a 50 μL PCR in which, after 25 cycles, 1 μg of a 400 bp PCR fragment is produced. The reaction contains 0.2 μ*M* primer.

a) How much of the primer is consumed during the synthesis of the product?

b) How much primer remains after 25 cycles?

c) How many cycles would be required to completely consume the entire reaction's supply of primer?

Solution 8.12(a)

Since each PCR fragment contains primer incorporated into each strand in the ratio of one primer to one DNA strand, the picomolar concentration of PCR fragment will be equivalent to the picomolar concentration of primer. The first step is to calculate the molecular weight of the 400 bp PCR fragment. This is accomplished by multiplying the size of the fragment by the molecular weight of a single bp:

$$400\text{ bp} \times \frac{660\text{ g/mol}}{\text{bp}} = \frac{2.64 \times 10^5\text{ g}}{\text{mol}}$$

Therefore, the 400 bp PCR fragment has a molecular weight of 2.64×10^5 g/mol. This value can then be used in an equation to convert the amount of 400 bp fragment produced into picomoles:

$$1\,\mu\text{g} \times \frac{1\text{ g}}{1 \times 10^6\,\mu\text{g}} \times \frac{1\text{ mol}}{2.64 \times 10^5\text{ g}} \times \frac{1 \times 10^{12}\text{ pmol}}{\text{mol}} = 3.8\text{ pmol}$$

Therefore, 3.8 pmol of 400-bp fragment were synthesized in the PCR. If 3.8 pmol of fragment is produced, then 3.8 pmol of primer is consumed by incorporation into product.

Solution 8.12(b)

To calculate how much primer remains after 25 cycles, we must first calculate how much primer we started with. Primer was in the reaction at a concentration of 0.2 μ*M*. We can calculate how many picomoles of primer are represented by 0.2 μ*M* primer in the following manner (remember, 0.2 μ*M* primer is equivalent to 0.2 μmol primer/L):

$$\frac{0.2\,\mu\text{mol primer}}{\text{L}} \times \frac{1 \times 10^6\text{ pmol}}{\mu\text{mol}} \times \frac{1\text{ L}}{1 \times 10^6\,\mu\text{L}} \times 50\,\mu\text{L} = 10\text{ pmol primer}$$

Therefore, the reaction began with 10 pmol of primer. And 3.8 pmol of primer are consumed in the production of PCR fragment (Solution 8.12a). This amount is subtracted from the 10 pmol of primer initially contained in the reaction:

$$10\text{ pmol} - 3.8\text{ pmol} = 6.2\text{ pmol}$$

Therefore, 6.2 pmol of primer remain unincorporated after 25 cycles.

Solution 8.12(c)

The solution to this problem requires the use of the general equation $Y = X(1 + E)^n$. We will solve for n, the exponent. As for the other components, from the information given we need to determine Y, the ending number of copies of fragment; X, the initial number of target molecules; and E, the overall efficiency of the reaction. The starting number of target sequence is calculated by first determining the weight of one copy of the 8000 bp recombinant plasmid:

$$\frac{8000\text{ bp}}{\text{copy}} \times \frac{660\text{ g}}{\text{mol bp}} \times \frac{1\text{ mol bp}}{6.023 \times 10^{23}\text{ bp}} = \frac{8.8 \times 10^{-18}\text{ g}}{\text{copy}}$$

Therefore, each copy of the 8000 bp recombinant plasmid weighs 8.8×10^{-18} g. This value can then be used as a conversion factor to determine how many copies of the recombinant plasmid are represented by 20 pg, the starting amount of template:

$$20\text{ pg} \times \frac{\text{g}}{1 \times 10^{12}\text{ pg}} \times \frac{\text{copy}}{8.8 \times 10^{-18}\text{ g}} = 2.3 \times 10^{6}\text{ copies}$$

Therefore, there are 2.3×10^6 copies of starting template.

After 25 cycles, 1 μg of 400 bp product is made. To determine how many copies this represents, the weight of one copy must be determined:

$$\frac{400\text{ bp}}{\text{copy}} \times \frac{660\text{ g}}{\text{mol bp}} \times \frac{1\text{ mol bp}}{6.023 \times 10^{23}\text{ bp}} = \frac{4.4 \times 10^{-19}\text{ g}}{\text{copy}}$$

Therefore, each 400 bp PCR fragment weighs 4.4×10^{-19} g. Using this value as a conversion factor, 1 μg of product can be converted to number of copies in the following way:

$$1\,\mu\text{g} \times \frac{\text{g}}{1 \times 10^{6}\,\mu\text{g}} \times \frac{\text{copy}}{4.4 \times 10^{-19}\text{ g}} = 2.3 \times 10^{12}\text{ copies}$$

Therefore, after 25 cycles, 2.3×10^{12} copies of the 400 bp PCR fragment have been made.

Using the starting number of copies of template and the number of copies of PCR fragment made after 25 cycles, the efficiency (*E*) of the reaction can now be calculated.

$$2.3 \times 10^{12} \text{ copies} = 2.3 \times 10^{6} \text{ copies}(1 + E)^{25}$$

$$\frac{2.3 \times 10^{12}}{2.3 \times 10^{6}} = (1 + E)^{25}$$

$1 \times 10^{6} = (1 + E)^{25}$	Divide each side of the equation by 2.3×10^{6}.
$\log 1 \times 10^{6} = 25 \log(1 + E)$	Take the logarithm of both sides of the equation and use the Power Rule for Logarithms.
$\frac{6}{25} = \log(1 + E)$ $0.24 = \log(1 + E)$	Take the log of 1×10^{6} (on the calculator, enter **1**, press the **EXP** key, enter **6**, then press the ***log*** key) and divide each side of the equation by 25.
$1.74 = 1 + E$	Take the antilog of both sides of the equation. To find the antilog of 0.24 on the calculator, enter **0.24** then the **10^x** key.
$1.74 - 1 = E$	Subtract 1 from both sides of the equation to find *E*.
$0.74 = E$	

Therefore, *E* has a value of 0.74.

The reaction contains 10 pmol of primer. Ten picomoles of primer will make 10 pmol of product. The number of copies this represents is calculated as follows:

$$10 \text{ pmol} \times \frac{1 \text{mol}}{1 \times 10^{12} \text{ pmol}} \times \frac{6.023 \times 10^{23} \text{ copies}}{\text{mol}} = 6.023 \times 10^{12} \text{ copies}$$

Therefore, 10 pmol of primer can make 6.023×10^{12} copies of PCR fragment.

We now have all the information we need to calculate the number of cycles required to make 6.023×10^{12} copies (10 pmol) of product. For the following equation, *Y*, the final number of copies of product, is 6.023×10^{12} copies; *X*, the initial number of copies of target sequence, is

2.3×10^6 copies; and E, the efficiency of the PCR, is 0.74. We will solve for the exponent n, the number of PCR cycles, as follows:

$$6.023 \times 10^{12} = (2.3 \times 10^6)(1 + 0.74)^n$$

$\frac{6.023 \times 10^{12}}{2.3 \times 10^6} = (1.74)^n$ Divide each side of the equation by 2.3×10^6 and simplify the term in parentheses.

$2.62 \times 10^6 = (1.74)^n$

$\log 2.62 \times 10^6 = n \log 1.74$ Take the log of both sides of the equation and use the Power Rule for Logarithms.

$\frac{\log 2.62 \times 10^6}{\log 1.74} = n$ Divide each side of the equation by log 1.74.

$\frac{6.42}{0.24} = n = 26.75$ Take the log of the values on the left side of the equation and divide these values to give n.

Therefore, after 27 cycles, the 10 pmol of primer will have been converted into 10 pmol of PCR fragment. Many protocols use primer quantities as small as 10 pmol. Using a relatively small amount of primer can increase a reaction's specificity. For the reaction in this example, using more than 27 cycles would provide no added benefit.

8.6 PRIMER T_m

For an optimal PCR, the two primers used in a PCR should have similar melting temperatures. The optimal annealing temperature for a PCR experiment may actually be several degrees above or below that of the T_m of the least stable primer. The best annealing temperature should be derived empirically. The highest annealing temperature giving the highest yield of desired product and the least amount of background amplification should be used for routine amplification of a particular target. The formulas described next can be used to give the experimenter a good idea of where to start.

A number of different methods for calculating primer T_m can be found in the literature, and several are commonly used. A very quick method for estimating the T_m of an oligonucleotide primer is to add 2°C for each A or T nt present in the primer and 4°C for each G or C nt. Expressed as an equation, this relationship is

$$T_m = 2°(A + T) + 4°(G + C)$$

This method of calculating T_m, however, is valid only for primers between 14 and 20 nts in length.

Problem 8.13 A primer with the sequence ATCGGTAACGATTACATTC is to be used in a PCR. Using the formula above, what is the T_m of this oligonucleotide?

Solution 8.13

The oligo contains 12 As and Ts and 7 Gs and Cs.

$$2° \times 12 = 24° \quad \text{and} \quad 4° \times 7 = 28°$$

Adding these two values together gives

$$24° + 28° = 52°$$

Therefore, by this method of calculating melting temperature, the primer has a T_m of 52°C.

8.6.1 Calculating T_m based on salt concentration, G/C content, and DNA length

The equation introduced earlier to calculate the T_m of a target sequence can also be used to determine the T_m of a primer:

$$T_m = 81.5 + 16.6\log\left[\frac{[\text{SALT}]}{1.0 + 0.7[\text{SALT}]}\right] + 0.41(\%\text{G} + \text{C}) - \frac{500}{L}$$

where [SALT] is the concentration of the monovalent cation in the PCR (typically KCl) and the concentration of magnesium expressed in moles per liter and determined as

$$[\text{SALT}] = [\text{K}^+] + 4[\text{Mg}^{2+}]^{0.5}$$

L is equal to the number of nucleotides in the primer. As pointed out earlier, the %G + C term in this equation should *not* be expressed as a decimal (50% G + C, not 0.50 G + C). This formula takes into account the salt concentration, the length of the primer, and the G + C content.

Problem 8.14 A primer with the sequence ATCGGTAACGATTACATTC is to be used in a PCR. The PCR contains 0.05 *M* KCl and 2.5 m*M* magnesium. Using the formula just given, what is the T_m of this oligonucleotide?

Solution 8.14

A 0.05 *M* KCl solution is equivalent to 0.05 mol KCl/L. The primer, a 19-mer, has 7 Gs and Cs. It has a %G + C content of 37%.

$$\frac{7 \text{ nucleotides}}{19 \text{ nucleotides}} \times 100 = 37\%$$

The salt concentration is

$$[\text{SALT}] = [0.05] + 4[0.0025]^{0.5} = 0.25$$

The T_m of the primer is calculated as follows:

$$T_m = 81.5° + 16.6\log\left[\frac{[0.25]}{1.0 + 0.7[0.25]}\right] + 0.41(37) - \frac{500}{19}$$

$$T_m = 81.5° + 16.6\log[0.213] + 15.2 - 26.3$$

$$T_m = 81.5° + (-11.2) + 15.2 - 26.3 = 59.2°\text{C}$$

Therefore, by this method of calculation, the primer has a T_m of 59.2°C. (Compare this with 52°C calculated in Problem 8.13 for the same primer sequence.)

8.6.2 Calculating T_m based on nearest-neighbor interactions

Wetmur and Sninsky (1995) describe a formula for calculating primer T_m that uses nearest-neighbor interactions between adjacent nts within the oligonucleotide. Experiments by Breslauer et al. (1986) demonstrated that it is base sequence rather than base composition that dictates the stability and melting characteristics of a particular DNA molecule. For example, a DNA molecule having the structure

$$^{5'}\text{AT}^{3'}$$
$$_{3'}\text{TA}_{5'}$$

will have different melting and stability properties than a DNA molecule with the structure

$$^{5'}\text{TA}^{3'}$$
$$_{3'}\text{AT}_{5'}$$

even though they have the same base composition. The stability of a DNA molecule is determined by its free energy ($\Delta G°$). The melting behavior of a DNA molecule is dictated by its enthalpy ($\Delta H°$).

Free energy is a measure of the tendency of a chemical system to react or change. It is expressed in kilocalories per mole (kcal/mol). (A **calorie** is the amount of heat at a pressure of one atmosphere required to raise the temperature of one gram of water by 1°C.) In a chemical reaction, the energy released or absorbed in that reaction is the difference (Δ) between the energy of the reaction products and the energy of the reactants. Under conditions of constant temperature and pressure, the energy difference is defined as ΔG (the **Gibbs free-energy change**). The **standard free-energy change**, $\Delta G°$, is the change in free energy when one mole of a compound is formed at 25°C at one atmosphere pressure.

Enthalpy is a measure of the heat content of a chemical or physical system. The **enthalpy change,** $\Delta H°$, is the quantity of heat released or absorbed at constant temperature, pressure, and volume. It is equal to the enthalpy of the reaction products minus the enthalpy of the reactants. It is expressed in units of kcal/mol.

Tables 8.1 and 8.2 show nearest-neighbor interactions for enthalpy and free energy.

Table 8.1 Enthalpy ($\Delta H°$) values for nearest-neighbor nucleotides (in kcal/mol). Reproduced from Quartin and Wetmur (1989).

First nucleotide	Second nucleotide			
	dA	dC	dG	dT
dA	−9.1	−6.5	−7.8	−8.6
dC	−5.8	−11.0	−11.9	−7.8
dG	−5.6	−11.1	−11.0	−6.5
dT	−6.0	−5.6	−5.8	−9.1

Table 8.2 Free-energy ($\Delta G°$) values for nearest-neighbor nucleotides (in kcal/mol). Reproduced from Quartin and Wetmur (1989).

First nucleotide	Second nucleotide			
	dA	dC	dG	dT
dA	−1.55	−1.40	−1.45	−1.25
dC	−1.15	−2.30	−3.05	−1.45
dG	−1.15	−2.70	−2.30	−1.40
dT	−0.85	−1.15	−1.15	−1.55

Free energy and enthalpy of nearest-neighbor interactions are incorporated into the following equation for calculating primer melting temperature (Wetmur & Sninsky, 1995):

$$T_m = \frac{T^\circ \Delta H^\circ}{\Delta H^\circ - \Delta G^\circ + RT^\circ \ln(C)} + 16.6 \log\left[\frac{(\text{SALT})}{1.0 + 0.7(\text{SALT})}\right] - 269.3$$

where

$$(\text{SALT}) = (\text{K}^+) + 4(\text{Mg}^{+2})^{0.5}$$

$T°$ is the temperature in degrees Kelvin (K). Standard state (25°C) is assumed for this calculation. Actual degree increments are equivalent between K and °C. Zero°C is equivalent to 273.2K. $T°$ in K, therefore, is 25° + 273.2° = 298.2K. R, the molar gas constant, is 1.99cal/mol K. C is the molar concentration of primer. $\Delta H°$ is calculated as the sum of all nearest-neighbor $\Delta H°$ values for the primer plus the enthalpy contributed by 'dangling end' interactions (template sequence extending past the bases complementary to the 5′ and 3′ ends of the primer). A primer will have two dangling ends when annealing to the template. The enthalpy contribution ($\Delta H°_e$) from a single dangling end is −5000cal/mol. $\Delta H°$ is calculated by the following formula:

$$\Delta H^\circ = \sum\nolimits_{nn} (\text{each } \Delta H^\circ{}_{nn}) + \Delta H^\circ{}_e$$

(The subscript '*nn*' designates nearest-neighbor interaction. The subscript '*e*' designates the dangling-end interaction.)

$\Delta G°$, the free energy, is calculated as the sum of all nearest-neighbor $\Delta G°$ values for the primer plus free energy contributed by the initiation of base-pair formation (+2200cal/mol) plus the $\Delta G°$ contributed by a single dangling end (−1000cal/mol). It is calculated by the following formula:

$$\Delta G^\circ = \sum\nolimits_{nn} (\text{each } \Delta G^\circ{}_{nn}) + \Delta G^\circ{}_e + \Delta G^\circ{}_i$$

(The subscript '*nn*' designates nearest-neighbor interaction. The subscripts '*e*' and '*i*' designate dangling-end interactions and initiation of base-pair formation, respectively.)

Problem 8.15 A primer with the sequence ATCGGTAACGATTACATTC is to be used in a PCR at a concentration of 0.4 μ*M*. The PCR contains 0.05 *M* KCl and 2.5 m*M* Mg^{2+}. Assume two dangling ends. Using the formula given earlier, what is the oligonucleotide's T_m?

Solution 8.15

The salt concentrations (K^+ and Mg^{2+}) must both be expressed as molar concentrations. 2.5 m*M* Mg^{2+} is equivalent to 0.0025 *M*, as shown here:

$$2.5\,\text{m}M\,\text{Mg}^{2+} \times \frac{1\,M}{1000\,\text{m}M} = 0.0025\,M\,\text{Mg}^{2+}$$

The total salt concentration is therefore

$$(\text{SALT}) = (0.05) + 4(0.0025)^{0.5} = (0.05) + (0.2) = 0.25$$

As shown in the following table, the sum of the $\Delta H°$ nearest-neighbor interactions is −142 kcal/mol.

Nearest-neighbor bases	$\Delta H°$
AT	−8.6
TC	−5.6
CG	−11.9
GG	−11.0
GT	−6.5
TA	−6.0
AA	−9.1
AC	−6.5
CG	−11.9
GA	−5.6
AT	−8.6
TT	−9.1
TA	−6.0
AC	−6.5
CA	−5.8
AT	−8.6
TT	−9.1
TC	−5.6
Total $\Delta H°$	**−142.0**

The ΔH°_e value is calculated as follows:

$$\frac{-5000\ \text{cal/mol}}{\text{dangling end}} \times 2\ \text{dangling ends} = -10\,000\ \text{cal/mol}$$

This is equivalent to −10 kcal/mol:

$$-10\,000\ \text{cal/mol} \times \frac{\text{kcal/mol}}{1000\ \text{cal/mol}} = -10\ \text{kcal/mol}$$

The total ΔH° is calculated as follows:

$$\Delta H^\circ = -142\ \text{kcal/mol} + (-10\ \text{kcal/mol}) = -152\ \text{kcal/mol}$$

Therefore, the ΔH° for this primer is −152 kcal/mol. This is equivalent to −152 000 cal/mol:

$$-152\ \text{kcal/mol} \times \frac{1000\ \text{cal/mol}}{\text{kcal/mol}} = -152\,000\ \text{cal/mol}$$

As shown in the following table, the sum of the ΔG° nearest-neighbor interactions is −27.3 kcal/mol.

Nearest-neighbor bases	ΔG°_{nn}
AT	−1.25
TC	−1.15
CG	−3.05
GG	−2.30
GT	−1.40
TA	−0.85
AA	−1.55
AC	−1.40
CG	−3.05
GA	−1.15
AT	−1.25
TT	−1.55
TA	−0.85
AC	−1.40
CA	−1.15
AT	−1.25
TT	−1.55
TC	−1.15
Total ΔG°	**−27.3**

The $\Delta G°$ for the contribution from the dangling ends is determined as follows:

$$\Delta G°_e = 2 \text{ dangling ends} \times \frac{-1000 \text{ cal/mol}}{\text{dangling end}} \times \frac{\text{kcal/mol}}{1000 \text{ cal/mol}} = -2 \text{ kcal/mol}$$

$\Delta G°$ is then calculated by adding the values determined earlier to the $\Delta G°$ for initiation of base-pair formation ($\Delta G°_i = +2200$ cal/mol $= 2.2$ kcal/mol).

$$\Delta G° = -27.3 \text{ kcal/mol} + (-2 \text{ kcal/mol}) + 2.2 \text{ kcal/mol} = -27.1 \text{kcal/mol}$$

Therefore, $\Delta G°$ for this primer is -27.1 kcal/mol, which is equivalent to $-27\,100$ cal/mol:

$$-27.1 \text{kcal/mol} \times \frac{1000 \text{ cal/mol}}{\text{kcal/mol}} = -27\,100 \text{ cal/mol}$$

The primer is at a concentration of 0.4 μM. This is converted to a molar (M) concentration as follows:

$$0.4\ \mu M \times \frac{1 M}{1 \times 10^6\ \mu M} = 4 \times 10^{-7}\ M$$

Placing these values into the Wetmur and Sninsky equation gives the following relationship:

$$T_m = \frac{T° \Delta H°}{\Delta H° - \Delta G° + RT° \ln(C)} + 16.6 \log\left[\frac{(\text{SALT})}{1.0 + 0.7(\text{SALT})}\right] - 269.3$$

$$T_m = \frac{298.2 \times (-152\,000)}{-152\,000 - (-27\,100) + 1.99(298.2) \ln(4 \times 10^{-7})} + 16.6 \log\left[\frac{(0.25)}{1.0 + 0.7(0.25)}\right] - 269.3$$

$$T_m = \frac{-45\,326\,400}{-152\,000 + 27\,100 + (593.4) \ln(4 \times 10^{-7})} + 16.6 \log[0.21] - 269.3$$

$$T_m = \frac{-45\,326\,400}{-124\,900 + (593.4)(-14.73)} + (16.6)(-0.68) - 269.3$$

$$T_m = \frac{-45\,326\,400}{-124\,900 + (-8{,}741)} + (-11.3) - 269.3$$

$$T_m = 339.2 - 11.3 - 269.3 = 58.6°$$

Therefore, by this method of calculation, the primer has a T_m of 58.6°C.

8.7 DEOXYNUCLEOSIDE TRIPHOSPHATES (dNTPs)

Deoxynucleoside triphosphates (dATP, dCTP, dGTP, and dTTP) are the building blocks of DNA. They should be added to a PCR in equimolar amounts and, depending on the specific application, are used in concentrations ranging from 20 to 200 μ*M* each. In preparing a series of PCRs, the four dNTPs can be prepared as a master mix, an aliquot of which is added to each PCR in the series.

Problem 8.16 A dNTP master mix is prepared by combining 40 μL of each 10 m*M* dNTP stock. Four μL of this dNTP master mix is added to a PCR having a final volume of 50 μL.

a) What is the concentration of each dNTP in the master mix?

b) What is the concentration of total dNTP in the PCR? Express the concentration in micromolarity.

Solution 8.16(a)

The dNTP master mix contains 40 μL of each of the four dNTPs. The total volume of the dNTP master mix, therefore, is 4 × 40 μL = 160 μL. Diluting each dNTP stock into this final volume gives

$$10\ \text{m}M\ \text{each dNTP} \times \frac{40\ \mu\text{L}}{160\ \mu\text{L}} = 2.5\ \text{m}M\ \text{each dNTP}$$

Therefore, the master mix contains 2.5 m*M* of each dNTP.

Solution 8.16(b)

The total concentration of dNTP in the master mix is 10 m*M* (2.5 m*M* dATP + 2.5 m*M* dCTP + 2.5 m*M* dGTP + 2.5 m*M* dTTP = 10 m*M* total dNTP). Dilution of 4 μL of the dNTP master mix into a 50 μL PCR yields the following concentration:

$$10\ \text{m}M\ \text{total dNTP} \times \frac{4\ \mu\text{L}}{50\ \mu\text{L}} \times \frac{1 \times 10^3\ \mu M}{\text{m}M} = 800\ \mu M\ \text{total dNTP}$$

Therefore, the 50 μL PCR contains a total dNTP concentration of 800 μ*M*.

Problem 8.17 A 50 μL PCR is designed to produce a 300 bp amplified product. The reaction contains 800 μ*M* total dNTP. The primers used for the amplification are an 18-mer and a 22-mer. Assume that neither the primers nor any other reagent in the reaction are limiting and that the reaction is 100% efficient.

a) With a dNTP concentration of 800 μ*M* dNTP, how much DNA can be produced in this reaction?

b) How many copies of the 300 bp fragment could theoretically be produced?

c) How much primer is required to synthesize the number of calculated copies of the 300 bp fragment?

Solution 8.17(a)

The first step is to calculate the number of micromoles of dNTP in the reaction. (800 μ*M* dNTP is equivalent to 800 μmol dNTP/L.)

$$\frac{800\ \mu\text{mol dNTP}}{\text{L}} \times \frac{1\text{L}}{1 \times 10^6\ \mu\text{L}} \times 50\ \mu\text{L} = 0.04\ \mu\text{mol dNTP}$$

Therefore, the reaction contains 0.04 μmol of dNTP.

A dNTP is a triphosphate. When added onto an extending DNA strand by the action of a DNA polymerase, a diphosphate is released. The nt, as incorporated into a strand of DNA, has a molecular weight of 330 g/mol. Therefore, the number of micromoles of dNTP is equal to the number of micromoles of nt of the form incorporated into DNA. The amount of DNA that can be made from 0.04 μmol of nucleotide is calculated by multiplying this amount by the molecular weight of a single nt. Multiplication by a conversion factor allows expression of the result in micrograms.

$$0.04\ \mu\text{mol nt} \times \frac{330\ \text{g/mol}}{\text{nt}} \times \frac{1\text{mol}}{1 \times 10^6\ \mu\text{mol}} \times \frac{1 \times 10^6\ \mu\text{g}}{\text{g}} = 13.2\ \mu\text{g}$$

Therefore, 0.04 μmol of dNTP can theoretically be converted into 13.2 μg of DNA.

Solution 8.17(b)

A 300 bp PCR fragment, since it is double-stranded, consists of 600 nts. Each PCR fragment is made by extension of annealed primers. The primers in this reaction are 18 and 22 nts in length. Forty nts (18 nt + 22 nt = 40 nt) of each PCR fragment, therefore, are made up from primer. Another 560 nts (600 nt − 40 nt from the primer) of each PCR fragment are made from the dNTP pool. The 560 nts represent one copy of the target sequence. Their weight is calculated as follows:

$$\frac{560\ \text{nt}}{\text{copy}} \times \frac{330\ \text{g}}{\text{mol nt}} \times \frac{1\text{mol nt}}{6.023 \times 10^{23}\ \text{nt}} \times \frac{1 \times 10^6\ \mu\text{g}}{\text{g}} = \frac{3.1 \times 10^{-13}\ \mu\text{g}}{\text{copy}}$$

Therefore, each copy of PCR fragment contains $3.1 \times 10^{-13}\,\mu g$ of nt contributed from the dNTP pool. Multiplying this value by the amount of DNA that can be made from the dNTPs (13.2 μg, as calculated in Solution 8.17a) gives the following result:

$$\frac{1\text{ copy}}{3.1 \times 10^{-13}\,\mu g} \times 13.2\,\mu g = 4.3 \times 10^{13}\text{ copies}$$

Therefore, the dNTPs in this reaction can synthesize 4.3×10^{13} copies of the 300 bp PCR fragment.

Solution 8.17(c)

Since the primers are incorporated into each copy of the PCR product, the number of micromoles of primers needed to synthesize 4.3×10^{13} copies is equivalent to the number of micromoles of product. To solve this problem, then, the number of micromoles of 300 bp product must first be determined:

$$4.3 \times 10^{13}\text{ copies} \times \frac{1\text{ mol}}{6.023 \times 10^{23}\text{ copies}} \times \frac{1 \times 10^{6}\,\mu\text{mol}}{\text{mol}}$$
$$= 7.1 \times 10^{-5}\,\mu\text{mol 300 bp product}$$

Therefore, $7.1 \times 10^{-5}\,\mu$mol of primer in the 50 μL reaction are needed. To convert this value to a micromolar concentration, the quantity is expressed as an amount per liter:

$$\frac{7.1 \times 10^{-5}\,\mu\text{mol primer}}{50\,\mu L} \times \frac{1 \times 10^{6}\,\mu L}{L} = \frac{1.4\,\mu\text{mol primer}}{L} = 1.4\,\mu M\text{ primer}$$

Therefore, each primer must be present in the reaction at a concentration of 1.4 μ*M* to support the synthesis of 4.3×10^{13} copies of 300 bp PCR product when the total dNTP concentration is 800 μ*M*.

■

8.8 DNA POLYMERASE

A number of DNA polymerases isolated from different thermophilic bacteria have been successfully used for PCR amplification. Although they may have varying characteristics as to exonuclease activity, processivity, fidelity, and specific activity, they have been chosen as suitable for PCR because they are thermally stable; they retain their activity even after repeated exposure to the high temperatures used to denature the template nucleic acid.

Problem 8.18 Two units of *Taq* DNA polymerase are added to a 100 μL PCR. The enzyme has a specific activity of 292 000 units/mg and a molecular weight of 94 kilodaltons (kDa). How many molecules of the enzyme are in the reaction?

Solution 8.18

Solving this problem requires the use of Avogadro's number (6.023×10^{23} molecules/mol). A molecular weight of 94 kDa is equivalent to 94 000 g/mol. These values are incorporated into the following equation to give the number of enzyme molecules:

$$2 \text{ units enzyme} \times \frac{\text{mg}}{292\,000 \text{ units enzyme}} \times \frac{\text{g}}{1000 \text{ mg}} \times \frac{\text{mol}}{94\,000 \text{ g}} \times \frac{6.023 \times 10^{23} \text{ molecules}}{\text{mol}} = x \text{ molecules}$$

Simplifying the equation gives the following result:

$$\frac{1.2 \times 10^{24} \text{ molecules}}{2.7 \times 10^{13}} = 4.4 \times 10^{10} \text{ molecules}$$

Therefore, the reaction contains 4.4×10^{10} molecules of *Taq* DNA polymerase.

8.8.1 Calculating DNA polymerase's error rate

All DNA polymerases, including those used for PCR, display a certain misincorporation rate by which a base other than that complementary to the template is added to the 3′ end of the strand being extended. Most DNA polymerases have a 3′ to 5′ exonuclease 'proofreading' function that removes incorrect bases from the 3′ end of the extending strand. Once the incorrect base has been removed, the polymerase activity of the enzyme adds the correct base and extension continues.

Hayashi (1994) describes a method for calculating the fraction of product fragments having the correct sequence after a PCR. It is given by the following equation, yielding the percent of PCR fragments containing no errors from misincorporation:

$$F(n) = \frac{[1 + E(P)]^n}{(E + 1)^n} \times 100$$

where $F(n)$ is the fraction of strands with correct sequence after n cycles; E is the efficiency of amplification; n is the number of cycles; and P is the probability of producing error-free PCR fragments in one cycle of amplification in a no-hit Poisson distribution, where λ in the Poisson equation, $P = e^{-\lambda}\lambda^r/r!$, is equal to the DNA polymerase error rate multiplied by the length of the amplicon in nts.

Calculating probability by Poisson distribution was described in Chapter 3 for analysis of the fluctuation test. It utilizes the general formula

$$P = \frac{e^{-\lambda}\lambda^r}{r!}$$

where e is the base of the natural logarithm system and r designates the number of 'hits.' $r!$ (r factorial) designates the product of each positive integer from r down to and including 1. For the zero ('no hit' or no errors incorporated) case, $r = 0$ and the Poisson equation becomes

$$P = \frac{e^{-\lambda}\lambda^0}{0!}$$

When solving P for the no-hit case, remember first that 0! is equal to 1 and second that any number raised to the exponent 0 is equal to 1. The no-hit case is then

$$P = \frac{e^{-\lambda}(1)}{1} = e^{-\lambda}$$

Problem 8.19 Under a defined set of reaction conditions, *Taq* DNA polymerase is found to have an error rate of 2×10^{-5} errors per nt polymerized. An amplicon of 500 bp is synthesized in the reaction. The reaction has an overall amplification efficiency of 0.85. After 25 cycles, what percent of the fragments will have the same sequence as that of the starting template (i.e., will not contain errors introduced by the DNA polymerase)?

Solution 8.19

The Poisson distribution probability for the no-hit case is given by the following equation:

$$P = \frac{e^{-\lambda}\lambda^0}{0!}$$

For this problem, λ is equal to the error rate multiplied by the amplicon size, given by the following equation:

$$\lambda = (2 \times 10^{-5})(500) = 0.01$$

Substituting this λ value into the Poisson equation gives the following relationship:

$$P = e^{-0.01} = 0.99$$

(To find $e^{-0.01}$ on a calculator, enter 0.01, change the sign by pressing the **+/−** key, then press the ***e*^x^** key (on some calculators, it may be necessary first to press the **SHIFT** key to gain access to the e^x function.))

Bringing the value for P into the equation for calculating the percentage of PCR fragments with no errors gives the following equation:

$$F(n) = \frac{[1 + E(P)]^n}{(E + 1)^n} \times 100$$

$$F(n) = \frac{[1 + 0.85\,(0.99)]^{25}}{(0.85 + 1)^{25}} \times 100$$

Simplifying the equation gives the following relationship:

$$F(n) = \frac{[1.84]^{25}}{(1.85)^{25}} \times 100$$

Both the numerator and the denominator must be raised to the 25th power. To determine $[1.84]^{25}$ on a calculator, enter 1.84, press the $\boldsymbol{x^y}$ key, enter 25, then press the = key.

The equation then gives the following result:

$$F(n) = \frac{4.17 \times 10^6}{4.78 \times 10^6} \times 100 = 87.2$$

Therefore, after 25 cycles, 87.2% of the amplicons will have the correct sequence.

8.9 QUANTITATIVE POLYMERASE CHAIN REACTION (PCR)

The ability to measure minute amounts of a specific nucleic acid can be crucial to obtaining a thorough understanding of the processes involved in gene expression and infection. Because of the PCR's unique sensitivity and because of its capacity to amplify as few as only several molecules to greater than a million-fold, it has been exploited for use in nucleic acid quantitation. No other technique can be used to detect so small an amount of target so efficiently or so quickly. Although a number of different schemes have been devised to employ the PCR as a quantitative tool, the technique called **competitive PCR** has emerged as the most reliable, non-instrument-based approach.

In a competitive PCR assay, a series of replicate reaction tubes is prepared in which two templates are coamplified. One template, the target sequence being quantified (its quantity is unknown), is added as an extract in identical volumes to each tube. Depending on the application, this target template might be RNA transcribed from an induced gene of interest or an RNA present in biological tissue as a consequence of viral infection. The other template is a competitor added to each PCR as a dilution series in an increasing and known amount. The two templates, the target and the competitor, to ensure that they will be amplified with the same efficiency, have similar but distinguishable lengths, similar sequence, and the same primer-annealing sequences. If RNA is the nucleic acid to be quantified, then the competitor template should also be RNA. The RNAs must first be reverse transcribed into cDNA prior to PCR amplification. (The technique used to prepare cDNA from RNA followed by PCR amplification is termed **RT-PCR**.) During reverse transcription and PCR, the competitor template competes with the target sequence for the primers and reaction components.

Following amplification, the reaction products are separated by gel electrophoresis. The amount of product in each band can be determined by several different methods. If a radioactively labeled dNTP is included in the PCR, autoradiography or liquid scintillation spectrometry can be used to detect the relative amounts of product. Staining with ethidium bromide and gel photography can be used in conjunction with densitometric scanning to quantitate products. If fluorescent dye-labeled primers are used, the products can be quantitated on an automated DNA sequencer.

If quantitation is done either by assessing radioisotope incorporation or by fluorescent/densitometric scanning of an ethidium bromide-stained gel, the quantities obtained for the smaller fragments must be corrected to reflect actual molar amounts. For example, smaller DNA fragments will incorporate less ethidium bromide than larger fragments, even though the actual number of molecules between them may be equivalent. Smaller fragments, therefore, will fluoresce less intensely. To be able to compare their quantities directly, the amount of the smaller fragment should be multiplied by the ratio of bp of the larger fragment to bp of the smaller fragment.

In any PCR, the amount of PCR fragment produced is proportional to the amount of starting template. The ratio of PCR products in the competitive PCR assay reflects the amount of each template initially present in the reaction. However, the total amount of PCR product cannot exceed some maximum value as limited by the amounts of dNTP or active *Taq* polymerase. As the amount of competitor template is increased in the series of reactions, the amount of product from the target template being quantitated will decrease as the competitor effectively competes for the reaction components. In that reaction, where the initial amounts of target and competitor template are equivalent, equal amounts of their products will accumulate. Therefore, at the point where the quantity of the product from the target and the quantity of the product from the competitor template are equal (the **equivalence point**), the starting concentration of target prior to RT-PCR is equal to the known amount of starting competitor template in that reaction.

It is unlikely that, by chance, any one reaction in the series will produce exactly equivalent amounts of product. The equivalence point, therefore, must be determined by constructing a plot relating the logarithm of the ratio of the amount of competitor PCR product/amount of target PCR product to the logarithm of the amount of initial competitor template. Although such a plot may suggest a straight line, regression analysis will need to be conducted to estimate the line of best fit through the data points. The equivalence point on the regression line is equal to 0 on the *y* axis. (At the equivalence point, the ratio of competitor/target is equal to 1. The logarithm of 1 is 0.) Regression will be described more fully in the following problem.

Problem 8.20 Ten reaction tubes are prepared for the quantitation of an RNA present in an extract from viral-infected cells. Ten μL of extract are placed in each of the 10 tubes. Into the reaction tubes are placed a competitor RNA having the same primer-annealing sites as the target RNA and a similar base sequence. The competitor template is added to the 10 tubes in the following amounts: 0, 2, 10, 20, 50, 100, 250, 500, 1000, and 5000 copies. Each reaction is treated with reverse transcriptase to make cDNA. The reactions are then amplified by PCR. RT-PCR of the target RNA yields a fragment 250 bp in length, while that of the competitor template produces a 200 bp fragment. Following amplification, the products are run on a 2% agarose gel in the presence of 0.5 μg/mL ethidium bromide. A photograph of the gel is taken under UV light, and the fluorescence intensity of each band is quantitated by densitometric scanning. The results shown in Figure 8.1 are obtained. Densitometric scanning gives the values presented in Table 8.3.

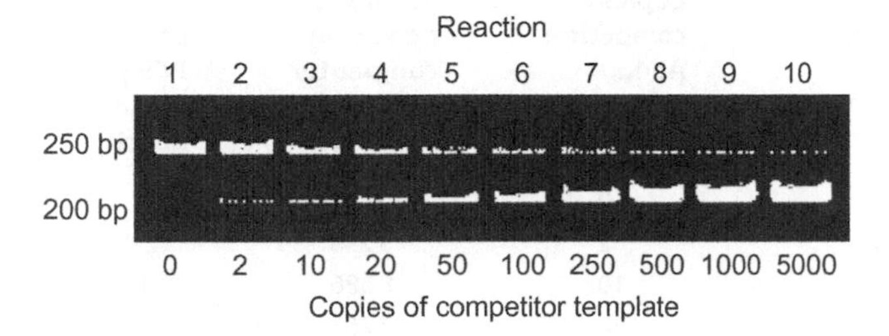

FIGURE 8.1 Agarose gel of competitive polymerase chain reaction (PCR) assay. The target yields a 250 bp product. The competitive template yields a 200 bp product.

Table 8.3 Scanning data for the gel photograph in Figure 8.1.

Reaction tube	Copies of competitor added	Scanning density of competitor RT-PCR product	Scanning density of target RT-PCR product
1	0	0	24822
2	2	3298	28437
3	10	8686	18218
4	20	7761	16872
5	50	14590	13520
6	100	16477	11212
7	250	21994	7115
8	500	25050	5173
9	1000	26598	5091
10	5000	31141	1951

Solution 8.20

The PCR product generated from the competitor RNA is smaller than that generated by the target RNA. When run on an agarose gel and stained with ethidium bromide, smaller fragments will stain with less intensity than larger fragments, even though their molar amounts may be equivalent, because the shorter molecules will bind less ethidium bromide. To be

Table 8.4 Corrected scanning density values for the 200 bp competitor RNA polymerase chain reaction (PCR) product for Problem 8.20. Each value is multiplied by 250/200 (bp size of target PCR product/bp size of competitor RNA PCR product). For example, for reaction two, $3298 \times 250/200 = 4123$.

Reaction tube	Copies of competitor Added	Scanning density of competitor RT-PCR product	Corrected competitor PCR product scanning density
1	0	0	0
2	2	3298	4123
3	10	8686	10858
4	20	7761	9701
5	50	14590	18238
6	100	16477	20596
7	250	21994	27492
8	500	25050	31312
9	1000	26598	33247
10	5000	31141	38926

able to directly compare the amount of competitor PCR product to that of the target, the scanning density values of the competitor must be corrected by multiplying them by the ratio of the size of the larger fragment over the smaller fragment. For this experiment, the product generated by the target RNA is 250 bp. That generated by the competitor RNA is 200 bp. Therefore, the scanning density values for each competitor product should be multiplied by 250/200. These values are entered in Table 8.4.

The *y* axis for our plot will be the log of the corrected competitor PCR product scanning density/target PCR product scanning density. Table 8.5 presents these values. The *x* axis is the log of the number of copies of competitor RNA added to each PCR in the assay. Table 8.6 provides these values. A plot can now be generated using log copies of competitor RNA on the *x* axis and log competitor/target on the *y* axis (Figure 8.2).

The points on the graph in Figure 8.2 suggest a straight-line relationship, but not all points fall exactly on a straight line. Such a graph is called a **scatter plot** because the points are scattered across the grid. The object then becomes to draw a line through the scatter plot that will best fit the data. Such a best-fit line can be calculated by a technique called **regression analysis** (also called **linear regression** or the **method of least**

Table 8.5 Competitor polymerase chain reaction (PCR) product scanning density divided by the target PCR product scanning density is given in the fourth column. The log of these values (*y* axis values) are shown in the fifth column.

Reaction tube	Corrected competitor PCR product scanning density	Scanning density of target RT-PCR product	Competitor divided by target	Log of competitor divided by target
1	0	24822	–	–
2	4123	28437	0.145	–0.839
3	10858	18218	0.596	–0.225
4	9701	16872	0.575	–0.240
5	18238	13520	1.349	0.130
6	20596	11212	1.837	0.264
7	27492	7115	3.864	0.587
8	31312	5173	6.053	0.782
9	33247	5091	6.531	0.815
10	38926	1951	19.953	1.300

Table 8.6 Log values for the number of copies of competitor RNA placed in each tube of the quantitative PCR assay for Problem 8.20. For example, for reaction two, in which two copies of the competitive template RNA are placed in the reaction, the log of 2 is 0.3.

Reaction tube	Copies of competitor added	Log of competitor copies
1	0	–
2	2	0.3
3	10	1.0
4	20	1.3
5	50	1.7
6	100	2.0
7	250	2.4
8	500	2.7
9	1000	3.0
10	5000	3.7

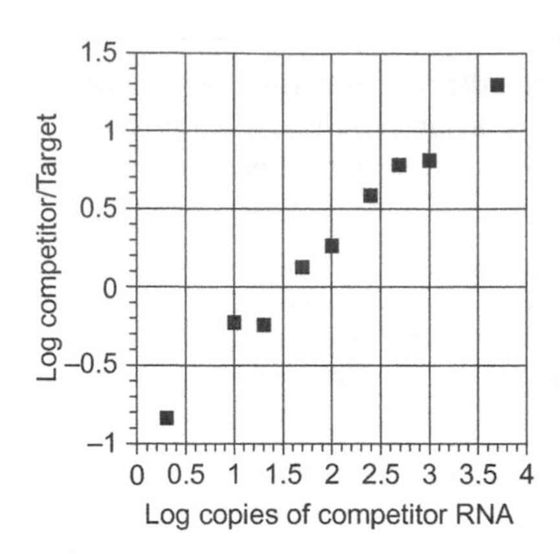

■ **FIGURE 8.2** Scatter plot of log competitor/target vs. log copies of competitor RNA.

squares). The values on the *x* axis are called the **independent** or **predictor variables**. They are chosen by the experimenter and are therefore not random. (In this example, the log of the number of copies of competitor RNA placed in the series of competitive PCRs is the independent variable.) The values on the *y* axis are the **dependent** or **response variables**. (In this example, the log of the corrected fluorescence of the competitor PCR product gel band divided by the fluorescence of the target PCR product band is the dependent variable.) Regression analysis creates a best-fit line that minimizes the distance of all the data points from the line. The regression line drawn to best fit the scatter plot has the formula

$$y = mx + b$$

This equation, in fact, describes any straight line in two-dimensional space. The *y* term is the variable in the vertical axis, *m* is the slope of the line, *x* is the variable on the horizontal axis, and *b* is the value of *y* where the line crosses the vertical axis (called the **intercept**).

For the best-fit regression line,

$$b = \overline{y} - m\overline{x}$$

where *y* (overstrike) is the mean *y* value, *x* (overstrike) is the mean *x* value, and

$$m = \frac{\sum_{i=1}^{n}(x_i - \overline{x})(y_i - \overline{y})}{\sum^{n}(x_i - \overline{x})^2}$$

Table 8.7 Calculation of the sum of all x and y values.

Reaction	X_i (log copies of competitor added (see Table 8.6))	Y_i (log of competitor divided by target (see Table 8.5))
2	0.3	−0.839
3	1.0	−0.225
4	1.3	−0.240
5	1.7	0.130
6	2.0	0.264
7	2.4	0.587
8	2.7	0.782
9	3.0	0.815
10	3.7	1.300
Total	**18.1**	**2.574**

where n designates the number of reactions (the number of data points), i refers to a specific PCR in the series of competitive reactions ($i = 1$ to n), and Σ is used as summation notation to indicate the sum of all terms in the series for each value of i.

We can now use our data to determine the regression line. The first step is to calculate the mean values for x_i and y_i. This is done by adding all the values for x_i and y_i (Table 8.7) and then dividing each sum by the total number of samples used to obtain that sum. (The values for reaction one, in which no competitor template was added, will not be used for these calculations because this reaction provides no useful information other than to give us peace of mind that a 200 bp product would not be produced in the absence of competitor template.)

The mean value for x is calculated as follows:

$$\bar{x} = \frac{18.1}{9} = 2.0$$

The mean value for y is calculated as follows:

$$\bar{y} = \frac{2.574}{9} = 0.286$$

The values for m and b in the equation for the regression line can now be calculated. Table 8.8 gives the values necessary for these calculations.

Table 8.8 Calculation of values necessary to determine m and b for the regression line of best fit.

x_i	y_i	$(x_i -\ )$	$(y_i -\ )$	$(x_i -\ )^2$	$(y_i -\ )^2$	$(x_i -\ ) \times (y_i -\ )$
0.3	−0.839	−1.7	−1.125	2.89	1.266	1.913
1.0	−0.225	−1.0	−0.511	1.00	0.261	0.511
1.3	−0.240	−0.7	−0.526	0.49	0.277	0.368
1.7	0.130	−0.3	−0.156	0.09	0.024	0.047
2.0	0.264	0.0	−0.022	0.00	0.000	0.000
2.4	0.587	0.4	0.301	0.16	0.091	0.120
2.7	0.782	0.7	0.496	0.49	0.246	0.347
3.0	0.815	1.0	0.529	1.00	0.280	0.529
3.7	1.300	1.7	1.014	2.89	1.028	1.724
18.1	**2.574**			**9.01**	**3.473**	**5.559**

From Table 8.8, we have the following values:

$$\sum^{9}(x_i - \bar{x})^2 = 9.01$$

and

$$\sum_{i=1}^{9}(x_i - \bar{x})(y_i - \bar{y}) = 5.559$$

With these values, we can calculate m:

$$m = \frac{\sum_{i=1}^{9}(x_i - \bar{x})(y_i - \bar{y})}{\sum^{9}(x_i - \bar{x})^2} = \frac{5.559}{9.01} = 0.617$$

Knowing the value for m, the value for b can be calculated:

$$\begin{aligned} b &= \bar{y} - m\bar{x} \\ &= 0.286 - 0.617(2.0) \\ &= 0.286 - 1.234 = -0.948 \end{aligned}$$

Therefore, substituting the m and b values into the equation for a straight line gives

$$y = 0.617x + (-0.948)$$

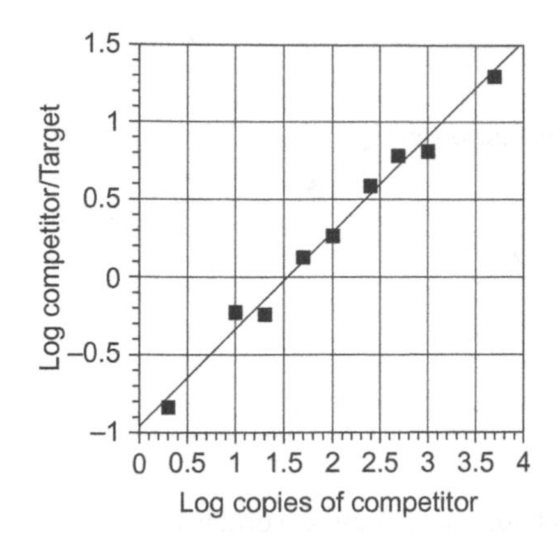

FIGURE 8.3 Line of best fit using the regression equation $y = 0.617x + (-0.948)$.

A regression line can now be plotted. Since a line is defined by two points, we need to determine two values for the equation $y = mx + b$. The regression line always passes through the coordinate (*x* (overstrike), *y* (overstrike)), in this example (2.0, 0.286). For a second point, let us set *x* equal to 3.5. (This choice is arbitrary; any *x* value within the experimental range will do.) Solving for *y* using our regression line equation gives

$$y = 0.617(3.5) + (-0.948)$$
$$y = 2.160 - 0.948$$
$$y = 1.212$$

Plotting these two points, (2.0, 0.286) and (3.5, 1.212), and drawing a straight line between them gives the line of best fit (Figure 8.3).

The equivalence point can be calculated using our equation for the regression line and setting *y* equal to 0, as shown in the following equation:

$$y = 0.617x + (-0.948)$$
$$0 = 0.617x - 0.948$$
$$0.948 = 0.617x$$
$$\frac{0.948}{0.617} = x$$
$$x = 1.536$$

Since this number represents the $\log_{10}$ value of the copies of the competitor, we must find its antilog. On a calculator, this is done by entering 1.536 and then pressing the $\mathbf{10^x}$ key.

$$\text{antilog}_{10} 1.536 = 34$$

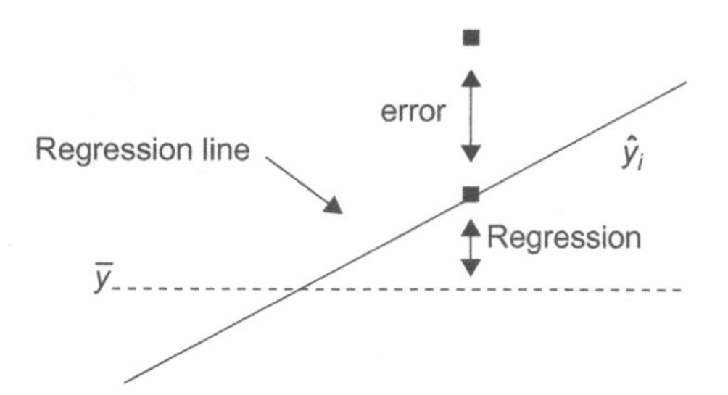

■ **FIGURE 8.4** Error is the amount y_i deviates from $\hat{y}_i$.

Therefore, the aliquot of cell extract contains 34 copies of mRNA template.

How closely the regression line fits the data (or how closely the data approaches our best-fit regression line) can be determined by calculating the **sum of the squared errors**, a measure of the total spread of the *y* values from the regression line. To do this, we will define $\hat{y}_i$ as the value for y_i that lies on the best-fit regression line for each value of x_i, such that

$$\hat{y}_i = a + bx_i$$

Error is defined as the amount y_i deviates from $\hat{y}_i$ and is represented by the vertical distance from each data point to the regression line (Figure 8.4). **Regression** measures the difference between the $\hat{y}_i$ value and the mean for *y* ($\bar{y}$).

The measure of how well the regression line fits the plot is given by the **squared correlation, R^2**. It is the proportion of the total $(y - \bar{y})^2$ values accounted for by the regression, or a measure of how many y_i values fall directly on the best-fit regression line. It is given by the following equation, using the square of the regression values for each point:

$$R^2 = \frac{\sum_{i=1}^{n} (\hat{y}_i - \bar{y})^2}{\sum_{i=1}^{n} (y_i - \bar{y})^2}$$

R^2 can also be calculated using the error values for each point. It is found using the following formula:

$$R^2 = 1 - \frac{\sum_{i=1}^{n} (y_i - \hat{y})^2}{\sum_{i=1}^{n} (y_i - \bar{y})^2}$$

Table 8.9 Values for calculating R^2.

x_i	y_i		Regression		Error	
0.3	−0.839	−0.763	−1.049	1.100	−0.076	0.006
1.0	−0.225	−0.331	−0.617	0.381	0.106	0.011
1.3	−0.240	−0.146	−0.432	0.187	−0.094	0.009
1.7	0.130	0.101	−0.185	0.034	0.029	0.001
2.0	0.264	0.286	0.000	0.000	−0.022	0.000
2.4	0.587	0.533	0.247	0.061	0.054	0.003
2.7	0.782	0.718	0.432	0.187	0.064	0.004
3.0	0.815	0.903	0.617	0.381	−0.088	0.008
3.7	1.300	1.335	1.049	1.100	−0.035	0.001
Totals				**3.431**		**0.043**

Table 8.9 provides the values necessary to calculate R^2 by either method. Previously, it was calculated that the mean for x is 2.0 and the mean for y is 0.286. Values for $\hat{y}_i$ are calculated using the equation

$$\hat{y}_i = mx_i + b$$

where b is equal to −0.948 and m is equal to 0.617 (as calculated previously).

From Table 8.9, the following values are derived:

$$\sum_{i=1}^{9} (\hat{y}_i - \bar{y})^2 = 3.431$$

$$\sum_{i=1}^{9} (y_i - \hat{y}_i)^2 = 0.043$$

The sum of $(y_i - \bar{y})^2$ was calculated in Table 8.8 and was determined to be 3.473. Using the first equation for calculating R^2 gives the following result:

$$R^2 = \frac{\sum_{i=1}^{n} (\hat{y}_i - \bar{y})^2}{\sum_{i=1}^{n} (y_i - \bar{y})^2}$$

$$R^2 = \frac{3.431}{3.473} = 0.988$$

The same result is obtained using the alternate method:

$$R^2 = 1 - \frac{\sum_{i=1}^{n}(y_i - \hat{y})^2}{\sum_{i=1}^{n}(y_i - \bar{y})^2}$$

$$R^2 = 1 - \frac{0.043}{3.473}$$

$$= 1 - 0.012 = 0.988$$

Therefore, the plotted regression line has an R^2 value of 0.988.

The closer the R^2 value is to 1.0, the better the fit of the regression line. An R^2 value of 1.0 corresponds to a perfect fit; all values for y lie on the regression line. In real-world experimentation, R^2 will be less than 1.0. R^2 will equal zero if x_i and y_i are completely independent; that is, in knowing a value for x_i, you will in no way be able to predict a corresponding value for y_i. An R^2 value of 0 might be expected if the scatter plot is completely random; that is, if there are data points all over the graph, with no relationship between x and y.

An alternate measure of how well the regression line fits the plot of experimental data is given by the **correlation coefficient, *r***, represented by the equation

$$r = (\text{sign of } m)\sqrt{R^2}$$

where r is a positive value if the regression line goes up to the right (has a positive slope) and negative if the line goes down to the right (has a negative slope). If a relationship exists such that if one value (x) changes then the other (y) does so in a related manner, the two values are said to be correlated. The value of r ranges from -1 to 1, with 1 indicating a perfect linear relationship between x and y, and a positive slope, -1, indicating a perfect linear relationship with a negative slope. Zero indicates no linear relationship.

For this example, m is a positive value ($+0.617$) and the equation for r is

$$r = \sqrt{R^2}$$

$$= \sqrt{0.988} = 0.994$$

Therefore, the regression line has a correlation coefficient, r, of 0.994.

CHAPTER SUMMARY

The polymerase chain reaction (PCR) is used to amplify a specific segment of DNA and has applications in gene cloning, DNA sequencing, mutation detection, and forensics. The PCR process generates replicated DNA strands (assuming perfect efficiency) at a rate of 2^n, where n is equal to the number of cycles of denaturation, annealing, and extension. As PCRs are not 100% efficient, amplification is more accurately expressed as

$$Y = (1 + E)^n$$

where Y is the degree of amplification, n is the number of PCR cycles, and E is the mean amplification efficiency. The number of copies of starting template, X, can be incorporated into the amplification equation as follows:

$$Y = X(1 + E)^n$$

The target stand's melting temperature (T_m) can be calculated as

$$T_m = 81.5 + 16.6 \log\left[\frac{[\text{SALT}]}{1.0 + 0.7[\text{SALT}]}\right] + 0.41(\%\text{G} + \text{C}) - \frac{500}{L}$$

where L = the length of the amplicon in bp. %G + C is calculated using the formulae

$$\%\text{G} + \text{C} = \frac{\text{G} + \text{C bases in the amplicon}}{\text{total number of bases in the amplicon}} \times 100$$

and

$$[\text{SALT}] = [\text{K}^+] + 4[\text{Mg}^{2+}]^{0.5}$$

In this last equation, any other monovalent cation, such as Na^+, if in the PCR, can be substituted for K^+.

There are several methods for calculating the T_m of the PCR primers. For short primers (between 14 and 20 nts in length), a quick calculation of primer T_m can be performed using the simple equation

$$T_m = 2^\circ(\text{A} + \text{T}) + 4^\circ(\text{G} + \text{C})$$

which takes into account the number of times each nt residue appears within the oligonucleotide. Another formula frequently used to calculate T_m is the following:

$$T_m = 81.5^\circ + 16.6(\log[M^+]) + 0.41(\%\text{G} + \text{C}) - (500/N)$$

where $[M^+]$ is the concentration of monovalent cation in the PCR (for most PCRs, this would be KCl) expressed in moles per liter and N is equal to the number of nts in the primer. The %G + C term in this equation should *not* be expressed as a decimal (50% G + C, not 0.50 G + C). This formula takes into account the salt concentration, the length of the primer, and the G + C content. It can be used for primers that are 14–70 nts in length. The most rigorous method for calculating primer T_m is by using nearest-neighbor interactions:

$$T_m = \frac{T^\circ \Delta H^\circ}{\Delta H^\circ - \Delta G^\circ + RT^\circ \ln(C)} + 16.6 \log\left[\frac{(\text{SALT})}{1.0 + 0.7(\text{SALT})}\right] - 269.3$$

where

$$(\text{SALT}) = (\text{K}^+) + 4(\text{Mg}^{+2})^{0.5}$$

and T° is temperature (in degrees Kelvin), ΔH° is the enthalpy change, ΔG° is the standard free-energy change, R is the molar gas constant, and C is the molar primer concentration.

The concentrations of primer, dNTPs, magnesium, template, DNA polymerase and all other components can be determined by standard dimensional analysis.

A DNA polymerase's error rate during PCR amplification can be calculated as

$$F(n) = \frac{[1 + E(P)]^n}{(E + 1)^n} \times 100$$

where $F(n)$ is the fraction of strands with correct sequence after n cycles, E is the efficiency of amplification, n is the number of cycles, and P is the probability of producing error-free PCR fragments in one cycle of amplification in a no-hit Poisson distribution. Calculating probability by Poisson distribution utilizes the general formula

$$P = \frac{e^{-\lambda}\lambda^r}{r!}$$

where e is the base of the natural logarithm system and r designates the number of 'hits.' $r!$ (r factorial) designates the product of each positive integer from r down to and including 1. Here, λ in the Poisson equation $P = e^{-\lambda}\lambda^r/r!$ is equal to the DNA polymerase error rate multiplied by the length of the amplicon in nts. For the zero ('no hit' or no errors incorporated) case, $r = 0$ and the Poisson equation becomes

$$P = \frac{e^{-\lambda}\lambda^0}{0!}$$

The polymerase chain reaction can be used to quantitate nucleic acids in a competitive reaction between an unknown template and a dilution series of known, calibrated template. Linear regression is used to determine the equation for the line of best fit that can then be used to calculate the equivalence point, where the starting concentration of target prior to PCR is equal to the known amount of starting competitor template in the competitive PCR. At the equivalence point, y in the equation for a straight line ($y = mx + b$) is set to zero. Calculating m and b from the regression line allows us to determine x at y equal to zero. The antilog of x yields the starting number of template copies.

REFERENCES

Breslauer, K.J., R. Frank, H. Blöcker, and L.A. Marky (1986). Predicting DNA duplex stability from the base sequence. *Proc. Natl. Acad. Sci. USA* 83:3746–3750.

Wetmur, J.G., and J.J. Sninsky (1995). Nucleic acid hybridization and unconventional bases. In *PCR Strategies* (M.A. Innis, D.H. Gelfand, and J.J. Sninsky, eds.). Academic Press, San Diego, CA, pp. 69–83.

FURTHER READING

Gilliland, G., S. Perrin, K. Blanchard, and H.F. Bunn (1990). Analysis of cytokine mRNA and DNA: detection and quantitation by competitive polymerase chain reaction. *Proc. Natl. Acad. Sci. USA* 87:2725–2729.

Hayashi, K. (1994). Manipulation of DNA by PCR. In *The Polymerase Chain Reaction* (K. Mullis, F. Ferré, and R.A. Gibbs, eds.). Birkhäuser, Boston, pp. 3–13.

Piatak, M. Jr., K.-C. Luk, B. Williams, and J.D. Lifson (1993). Quantitative competitive polymerase chain reaction for accurate quantitation of HIV DNA and RNA species. *BioTechniques* 14:70–80.

Quartin, R.S., and J.G. Wetmur (1989). Effect of ionic strength on the hybridization of oligodeoxynucleotides with reduced charge due to methylphosphonate linkages to unmodified oligodeoxynucleotides containing the complementary sequence. *Biochemistry* 28:1040–1047.

Chapter 9

The real-time polymerase chain reaction (RT-PCR)

INTRODUCTION

Since its introduction in 1983, PCR has found extensive use in a wide range of applications including gene cloning, gene mapping, mutation detection, DNA sequencing, and human identification. As discussed in the preceding chapter, PCR is also a valuable tool for measuring gene expression, and an example of the technique known as competitive PCR is shown as one of the methods that can be used to calculate the amount of mRNA made during the transcription of a specific gene. Since the PCR products synthesized in that protocol are separated and quantified by gel electrophoresis, the method is an end point assay. That is, the products are analyzed after the reaction that makes them has completed. Innovations in both instrumentation and in fluorescent dye chemistry have spurred the rapid development of methods that can detect PCR products *as* they are being made, in real time.

A number of real-time PCR methods have been described but two have emerged as the most popular. One uses a DNA-binding molecule called **SYBR® green**, a dye that binds to dsDNA but not to ssDNA and, when so bound, fluoresces. During the cycling reaction, the sample will produce an increasing amount of fluorescent signal as more and more double-stranded product is generated to which the SYBR green dye can attach. The amount of fluorescence in the reaction at any particular time, therefore, is directly related to the number of dsDNA molecules present in the reaction. The downside of SYBR green, however, is that it will bind and fluoresce all double-stranded products in the reaction whether they are specific products, nonspecific products, primer dimers, or other amplification artifacts.

The other real-time PCR method is known as the **TaqMan®** or **5′ nuclease assay**. It uses a **dye-labeled probe** that anneals to one of the template strands close to and downstream from one of the two PCR primers. A fluorescent dye, referred to as the **reporter**, is attached to the probe's 5′ end. On the probe's 3′ end is another molecule, called a **quencher**, which absorbs the energy from the light source used to excite the reporter dye. When the reporter and the quencher are connected to each other through the intervening probe, the quencher reduces the fluorescent signal of the

Calculations for Molecular Biology and Biotechnology. DOI: 10.1016/B978-0-12-375690-9.00009-7
© 2010 Elsevier Inc. All rights reserved.

reporter dye. However, during PCR, Taq DNA polymerase, extending the primer on the probe's target strand, displaces and degrades the annealed probe through the action of its 5′ to 3′ exonuclease function. The reporter dye is thereby released from its molecular attachment to the quencher and it fluoresces. The more PCR product generated, the more probe that can bind to that product. The more probe bound, the more reporter dye released during the amplification process and the more signal generated. Fluorescent signal, therefore, is directly related to the amount of input template.

Whether using SYBR green or TaqMan® probes, the relationship between signal intensity and the amount of template in a real-time PCR provides a reliable means to both quantitate nucleic acids and to assay for the presence or absence of specific gene sequences.

9.1 THE PHASES OF REAL-TIME PCR

In a PCR, the target DNA sequence is amplified at an exponential rate – one template is replicated into two products, two into four, four into eight, etc. As a first approximation, that statement is true enough. Exponential amplification, however, cannot be sustained forever. Usually by the 35th cycle (but depending on a number of factors), the reaction has slowed its pace. The primers and dNTPs are no longer in abundant excess, the DNA polymerase has lost a degree of its activity, complete denaturation becomes less efficient, and products are broken down by the polymerase's nuclease activity. As these events begin to happen, the reaction enters a linear phase (Figure 9.1), in which the template fails to double completely. Eventually, by the 40th cycle, the reaction has entered a plateau phase, in which amplification has all but ceased.

Real-time PCR relies on the ability of the instrument to detect that fractional cycle of amplification where sufficient PCR product has accumulated to generate a signal above the **background** fluorescent noise – the signal at the limits of the instrument's detection that fluctuates during the early cycles of the amplification. The PCR cycle number at which signal can be discriminated from background noise is referred to as the **cycle threshold** or, by abbreviation, the **C_T value**. It is axiomatic in PCR that, all things being equal (no inhibition or instrument anomalies), the greater the amount of input nucleic acid used as starting template, the greater the amount of product that can be made and the sooner it can be detected. So it is with real-time PCR: the greater the amount of starting template, the sooner the cycle threshold is reached and the lower the C_T value. This forms the basis of DNA quantification. For a 100% efficient real-time PCR, the amount of

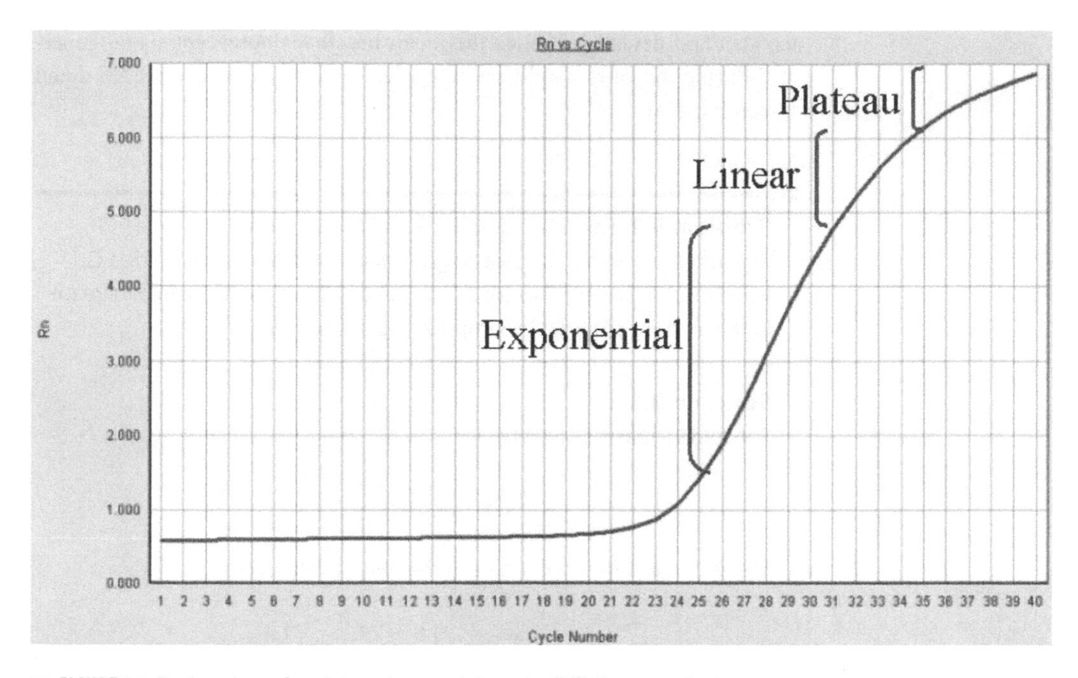

■ **FIGURE 9.1** The three phases of a real-time polymerase chain reaction (PCR). The exponential phase actually begins as early as the third cycle. However, the instrument cannot detect the products generated in the early cycles. The plot shows PCR cycle number on the *x* axis and normalized reporter signal (R_n) on the *y* axis. R_n is the ratio of the fluorescent signal intensity of the reporter dye to the fluorescent signal intensity of a passive reference dye (usually the red dye ROX) present in all reactions.

product doubles each cycle. One makes two, two makes four, four makes eight, eight makes sixteen, and so on. Likewise, a change in a C_T value of one (from a higher to a lower value) represents a doubling of the target molecule. A change of two C_T values represents a four-fold increase in the amount of target. A change of three C_T values represents an eight-fold increase in amplified target, and so on. Mathematically, a fold change in the C_T value (ΔC_T) is equal to $2^{-\Delta CT}$ ($\Delta C_T = 2^{-\Delta CT}$ fold difference).

The C_T value is automatically determined by the real-time PCR instrument's software but it can also be set manually, in which case the operator determines a **threshold** line. The threshold line, whether determined automatically or manually, defines the C_T value by where it intersects the amplification curve. The threshold cycle should be designated within that exponential area of the amplification curve where reaction components are in an abundant, nonlimiting supply. On the real-time PCR instruments manufactured by Applied Biosystems, the threshold, by default, is set at ten standard deviations above the mean baseline fluorescent signal generated during the early amplification cycles (typically between the third and fifteenth cycle).

Problem 9.1 Two DNA samples, *A* and *B*, are run in a real-time PCR experiment in which the *β-actin* gene is assayed. Sample *A* yields a C_T value of 21.8. Sample *B* yields a C_T value of 23.2. What is the fold increase in the amount of target in Sample *A* over Sample *B*?

Solution 9.1

The fold increase is given by

$$\text{Fold increase} = 2^{-\Delta C_T}$$

The ΔC_T, the difference between the two C_T values, is

$$\Delta C_T = 21.8 - 23.2 = -1.4$$

The fold increase is then

$$2^{-\Delta C_T} = 2^{-(-1.4)} = 2^{1.4} = 2.64$$

Therefore, there is a 2.64-fold difference in the amount of amplified product in Sample *A* compared to Sample *B*.

If the ΔC_T is a positive number – if we had subtracted 21.8 from 23.2 to give us 1.4 – the exponent would be negative and this would give us a fraction:

$$2^{-\Delta C_T} = 2^{-1.4} = 0.379$$

Because there is an inverse relationship between the C_T value and template amount – the lower the C_T value, the greater the amount of amplifiable template present – to determine the fold increase between the two samples, we would need to take the reciprocal of the calculated value (divide 1 by the value) to give the fold increase:

$$1/0.379 = 2.64$$

And again, we show a 2.64-fold increase in Sample *A* over Sample *B*.

9.2 CONTROLS

A typical real-time PCR experiment will include several controls. A **no template control** (NTC) is a reaction that carries all the reagents, primers, and probes necessary for a real-time PCR except for the template DNA. The NTC is used to see whether signal can be generated in the absence of a target nucleic acid. It can detect contamination and any type of primer or probe interactions that can produce fluorescent signal that might confound the results.

An **exogenous control** is a well-characterized nucleic acid or a synthesized construct of RNA or DNA spiked into each reaction at a known concentration. It can serve as an **internal positive control** (IPC) to differentiate between PCR inhibition and a true negative reaction. It is also used to account for problems of inconsistency that might occur when preparing the Master Mix. An **endogenous control** is a nucleic acid present in the preparation of the target gene that, with the appropriate primers and probe, can also be detected in the real-time PCRs. It can serve as an active reference to normalize for any differences in the amount of total mRNA target added to the real-time PCR.

Real-time PCRs must also be normalized for signal intensity. **Normalization**, the correction for fluctuations in signal intensity between what should be identical samples, is performed at two levels, one to correct for variation in the concentration of PCR reagents and the other to compensate for differences in the amounts of RNA isolated from different tissue samples (when using real-time PCR to measure gene expression). Many real-time PCR experiments are performed in microtiter plates having anywhere from 48 to 384 wells, each containing a sample and amplification reagents. Pipetting that many components in parallel can introduce sample-to-sample variation – some wells may get more or less reagents than others. Differences in component concentration between parallel samples can result in variability in the calculated levels of the genes under study. This variability can be corrected for by including a **passive reference** within the PCR mix against which the signal generated during amplification is compared. In the mixes supplied by Life Technologies, this passive reference is the dye known as **ROX** (red to the optics of the ABI instruments). It is called a passive reference because it neither participates as a necessary component of the amplification nor does it inhibit it. It merely glows red.

Imagine that you are doing a real-time PCR gene expression experiment and into one well of the microtiter plate you accidentally place 200 ng of total mRNA isolated from liver cells and into another you place only 100 ng of total mRNA isolated from muscle cells. Even if, in actuality, the target gene is expressed at equivalent levels in both cell types, it would appear that the gene in liver cells was expressed two-fold higher than in muscle cells. Normalization, in which the signal intensity of the reference dye is divided by the signal intensity of the passive reference dye, can correct for this discrepancy. This yields a value called $\mathbf{R_n}$, for **normalized reporter** (see Figure 9.1).

9.3 ABSOLUTE QUANTIFICATION BY THE TAQMAN ASSAY

The goal of **absolute quantification** is to determine the actual concentration of a nucleic acid within a sample. That concentration may be expressed by weight or by copy number but, either way, it requires that a standard curve be generated from a serially diluted DNA sample having a concentration determined by independent means. If prepared properly, the diluted standards will yield a straight line when their C_T values are plotted against the log of the DNA concentration for each dilution. For most applications, the target on the standard is a single-copy gene – a gene present only once in a haploid genome. An unknown DNA sample, assayed using the same amplification primers and TaqMan probe as those used for the standard, is quantified based on its C_T value compared to the standard curve. For the most accurate results, the concentration of the unknown sample should fall within the bounds of the concentrations represented by the standard.

9.3.1 Preparing the standards

As the instruments used for data collection are capable of detecting such a large dynamic range of fluorescent signal, the DNA standard can be diluted over several orders of magnitude. The following problem will demonstrate how these dilutions are prepared.

Problem 9.2 The Quantifiler™ Human DNA Quantification Kit marketed by Life Technologies for quantification of forensic DNA samples contains a human DNA standard with a concentration of 200 μng/μL. The protocol requires that you make a series of eight dilutions of the standard into TE buffer. The diluted samples will have DNA concentrations of 50, 16.7, 5.56, 1.85, 0.62, 0.21, 0.068, and 0.023 ng/μL. You prepare the dilutions by aliquoting 10 μL from one tube into the next for each sample.

a) What is the dilution factor (what fold dilution) is needed to make the first tube containing DNA at a concentration of 50 ng/μL?

b) What is the dilution factor of the second tube containing DNA at a concentration of 16.7 ng/μL?

c) Into what volume of TE buffer should 10 μL of the human DNA stock solution (at 200 ng/μL) be pipetted to make the first standard having a DNA concentration of 50 ng/μL?

d) Into what volume of TE buffer should 10 μL of the 50 ng/μL dilution be pipetted to prepare the second standard dilution tube so that it has a concentration of 16.7 ng/μL?

e) What are the dilutions you will need to make to prepare the entire dilution series?

Solution 9.2(a)

The stock solution of human DNA has a concentration of 200 ng/μL. A 10 μL aliquot of the stock is used to prepare a dilution having a concentration of 50 ng/μL. To determine the dilution factor, we calculate a relationship of ratios. We want the number one as the numerator of our dilution factor expressed as a fraction (rather than as a decimal). We prepare the equation so that it can be stated '50 ng/μL is to 200 ng/μL as 1 is to *x*.' We then solve for *x*:

$$\frac{50\,\text{ng}/\mu\text{L}}{200\,\text{ng}/\mu\text{L}} = \frac{1}{x}$$

The ng/μL units cancel. Using algebra to solve for *x* we have:

$$\frac{50x}{200} = 1$$

$$50x = 200$$

$$x = \frac{200}{50} = 4$$

The relationship, then, is

$$\frac{50\,\text{ng}/\mu\text{L}}{200\,\text{ng}/\mu\text{L}} = \frac{1}{4}$$

Therefore, we use a dilution factor of ¼ to make the first standard having a concentration of 50 ng/μL.

The dilution can also be expressed as a 'fold' value. In this case, we divide the initial concentration by the concentration of the diluted standard:

$$\frac{200\,\text{ng}/\mu\text{L}}{50\,\text{ng}/\mu\text{L}} = 4$$

Therefore, we are diluting the stock solution four-fold. We can also say we are making a '4X' dilution of the stock.

Solution 9.2(b)

The dilution factor used to prepare a concentration of 16.7 ng/μL from a 10 μL aliquot taken from the first dilution (at 50 ng/μL) is calculated by solving for *x* in the following equation.

$$\frac{16.7\ \text{ng}/\mu\text{L}}{50\ \text{ng}/\mu\text{L}} = \frac{1}{x}$$

$$\frac{16.7x}{50} = 1$$

$$16.7x = 50$$

$$x = \frac{50}{16.7} = 3$$

We then have

$$\frac{16.7\ \text{ng}/\mu\text{L}}{50\ \text{ng}/\mu\text{L}} = \frac{1}{3}$$

Therefore, we use a dilution factor of ⅓ to make the 16.7 ng/μL standard from an aliquot of the 50 ng/μL standard.

Solution 9.2(c)

We are going to take 10 μL from the stock DNA solution having a concentration of 200 ng/μL and dilute that into a volume of TE buffer to make a standard with a concentration of 50 ng/μL. We calculated in Solution 9.1(a) that we are going to make a ¼ dilution. We can calculate the volume of TE needed in our second dilution tube by preparing a relationship of ratios that states '1 is to 4 as 10 μL is to *x* μL.' We then solve for *x*:

$$\frac{1}{4} = \frac{10\ \mu\text{L}}{x\ \mu\text{L}}$$

$$\frac{x}{4} = 10$$

$$x = 10 \times 4 = 40$$

Therefore, 10 μL of the 200 ng/μL DNA stock solution should go into a *total* volume of 40 μL. Thirty μL of TE should be placed into the first dilution tube and 10 μL of the 200 ng/μL DNA standard should be added to it to make a standard having a concentration of 50 ng/μL.

We could have also solved this problem by using the more general formula

$$200\,\text{ng}/\mu\text{L} \times \frac{10\,\mu\text{L}}{x\,\mu\text{L}} = 50\,\text{ng}/\mu\text{L}$$

and then solve for *x*:

$$\frac{2000\,\text{ng}/\mu\text{L}}{x} = 50\,\text{ng}/\mu\text{L}$$

$$2000\,\text{ng}/\mu\text{L} = (50\,\text{ng}/\mu\text{L})x$$

$$\frac{2000}{50} = x = 40$$

Both methods tell us that the first dilution of our standard should have a final volume of 40 μL.

Solution 9.2(d)

We are going to take 10 μL of the first dilution having a concentration of 50 ng/μL into a volume of TE such that the DNA concentration in this second standard tube has a concentration of 16.7 ng/μL. We have already determined in Solution 9.1(b) that we need to dilute the first standard by ⅓. To determine what volume of TE should be in the second dilution tube, we can set up a relationship of ratios as we did in Solution 9.1(c):

$$\frac{1}{3} = \frac{10\,\mu\text{L}}{x\,\mu\text{L}}$$

$$\frac{x}{3} = 10$$

$$x = 10 \times 3 = 30$$

Therefore, the final volume of the second dilution having a DNA concentration of 16.7 ng/μL should have a final volume of 30 μL. We will need to put 20 μL of TE into this second tube and we will add 10 μL of the first standard to it to make a final volume of 30 μL.

Solution 9.2(e)

We need to prepare a set of standards having DNA concentrations of 50, 16.7, 5.56, 1.85, 0.62, 0.21, 0.068, and 0.023 ng/μL. We have determined in the steps described above that the first two standards in the series are made by taking 10 μL of the DNA stock solution (at 200 ng DNA/μL) into 30 μL TE and then aliquoting 10 μL of the first standard into 20 μL of TE to make the second. Dividing the concentration of each standard by the concentration of the next higher standard, we see that each standard following the second requires a 3X dilution:

$$\frac{16.7}{5.56} = 3 \quad \frac{5.56}{1.85} = 3 \quad \frac{1.85}{0.62} = 3 \quad \frac{0.62}{0.21} = 3 \quad \frac{0.21}{0.068} = 3 \quad \frac{0.068}{0.023} = 3$$

Therefore, since each standard including and after the second one requires a ⅓ dilution and since we are transferring 10 μL between each one, our dilution series will be:

$$200\ \text{ng}/\mu\text{L} \times \frac{10\ \mu\text{L}}{40\ \mu\text{L}} \times \frac{10\ \mu\text{L}}{30\ \mu\text{L}} \times \frac{10\ \mu\text{L}}{30\ \mu\text{L}} \times \frac{10\ \mu\text{L}}{30\ \mu\text{L}} \times \frac{10\ \mu\text{L}}{30\ \mu\text{L}} \times \frac{10\ \mu\text{L}}{30\ \mu\text{L}} \times \frac{10\ \mu\text{L}}{30\ \mu\text{L}} \times \frac{10\ \mu\text{L}}{30\ \mu\text{L}}$$

ng/μL: 50 16.7 5.56 1.85 0.62 0.21 0.068 0.023

9.3.2 Preparing a standard curve for quantitative polymerase chain reaction (qPCR) based on copy number

Some quantitative real-time PCR experiments are performed to determine how many copies of a specific gene are present in an unknown sample. This type of quantification requires the creation of a standard curve based on gene copy number. The gene copy number for the unknown is derived from its C_T value and extrapolation from a standard curve of C_T vs. number of gene copies. The gene used for quantification in the DNA sample used as the standard should be present only once (in a single copy) per haploid genome. For human DNA, any number of single-copy genes can be used for this purpose including *RNaseP*, Factor IX, plasminogen activator inhibitor type a, DNA topoisomerase I, and dihydrofolate reductase, among others. Creating a standard curve requires that we know the size, in bp, of the genome we are using as the standard. (Genome sizes can be found on the website cbs.dtu.dk/databases/DOGS/index.html, maintained by the Center for Biological Sequence Analysis.)

Problem 9.3 We will generate a standard curve of human DNA in which samples contain 500 000, 50 000, 5000, 500, 50, and 5 copies of *RNaseP*, a single-copy gene. What amount of human DNA, by weight, should be aliquoted into a series of separate tubes such that each contains the desired number of copies of the single gene?

Solution 9.3

Since the DNA of the single-copy gene is a part of the human genome and since the easiest way to measure out a genome is by its weight, we will need to determine how much the human genome weighs. First, we will calculate the weight of a single bp using Avogadro's number and the molecular weight of a bp, which we will take to be 660 g/mole:

$$1\,bp \times \frac{660\,\text{g/mole}}{\text{bp}} \times \frac{1\,\text{mole}}{6.023 \times 10^{23}\,\text{bp}} = \frac{1.096 \times 10^{-21}\,\text{g}}{\text{bp}}$$

Therefore, one bp weighs 1.096×10^{-21} g. We can now calculate the weight of the human genome, which we will take to contain 3×10^9 bp:

$$3 \times 10^9\,\text{bp} \times \frac{1.096 \times 10^{-21}\,\text{g}}{\text{bp}} = 3.3 \times 10^{-12}\,\text{g}$$

Therefore, the human genome, 3×10^9 bps in length, weighs 3.3×10^{-12} grams (g). We can convert this to pg (1×10^{-12} g), a more convenient unit to work with for such a small amount, as follows:

$$3.3 \times 10^{-12}\,\text{g} \times \frac{1 \times 10^{12}\,\text{pg}}{\text{g}} = 3.3\,\text{pg}$$

Therefore, the human genome weighs 3.3 pg. This represents the weight of the haploid genome – the genome composed of 23 chromosomes as found in the sperm or egg. A single-copy gene would be present only once in a haploid genome but twice in a diploid genome, as found in all the other human body cells having a nucleus. For the standard curve, therefore, 3.3 pg of human DNA is equivalent to one copy of a single-copy gene.

We can now calculate how many pg of human DNA we will need for each sample making up our standard curve, as follows:

$$500\,000\ \text{gene copies} \times \frac{3.3\,\text{pg DNA}}{\text{gene copy}} = 1.65 \times 10^6\,\text{pg DNA}$$

$$50\,000\ \text{gene copies} \times \frac{3.3\,\text{pg DNA}}{\text{gene copy}} = 1.65 \times 10^5\,\text{pg DNA}$$

$$5000\ \text{gene copies} \times \frac{3.3\,\text{pg DNA}}{\text{gene copy}} = 1.65 \times 10^4\,\text{pg DNA}$$

$$500 \text{ gene copies} \times \frac{3.3\,\text{pg DNA}}{\text{gene copy}} = 1650\,\text{pg DNA}$$

$$50 \text{ gene copies} \times \frac{3.3\,\text{pg DNA}}{\text{gene copy}} = 165\,\text{pg DNA}$$

$$5 \text{ gene copies} \times \frac{3.3\,\text{pg DNA}}{\text{gene copy}} = 16.5\,\text{pg DNA}$$

Therefore, the tube for 500 000 copies of *RNaseP* should receive 1.65×10^6 pg of human DNA, the tube for 50 000 copies of *RNaseP* should receive 1.65×10^5 pg of human DNA, and so on, as shown in the calculations above.

Problem 9.4 Continuing with the experiment described in Problem 9.3, we will place 10 μL of each diluted standard into a real-time PCR.

a) What will be the concentrations of human DNA in the dilution series such that we will be pipetting the desired number of copies of *RNaseP* into each reaction?

b) The stock solution of human DNA has a concentration of 2 μg/μL. How should the stock solution be diluted such that each dilution, when it is prepared, has a final volume of 200 μL?

Solution 9.4(a)

In Solution 9.3, we calculated the weight of human DNA representing each copy number standard from 500 000 down to 5 copies of *RNaseP*. We will create a dilution series spanning these amounts and, from those tubes, we will aliquot 10 μL from each into 50 μL real-time PCRs. To calculate the concentration of each dilution tube such that 10 μL contains the desired number of copies, we divide each weight calculated in Solution 9.3 by 10:

For 500 000 copies, we have

$$\frac{1.65 \times 10^6\,\text{pg}}{10\,\mu\text{L}} = 165\,000\,\text{pg}/\mu\text{L}$$

For 50 000 copies, we have

$$\frac{1.65 \times 10^5\,\text{pg}}{10\,\mu\text{L}} = 16\,500\,\text{pg}/\mu\text{L}$$

For 5000 copies, we have

$$\frac{1.65 \times 10^4\ \text{pg}}{10\ \mu\text{L}} = 1650\ \text{pg}/\mu\text{L}$$

For 500 copies, we have

$$\frac{1650\ \text{pg}}{10\ \mu\text{L}} = 165\ \text{pg}/\mu\text{L}$$

For 50 copies, we have

$$\frac{165\ \text{pg}}{10\ \mu\text{L}} = 16.5\ \text{pg}/\mu\text{L}$$

For 5 copies, we have

$$\frac{16.5\ \text{pg}}{10\ \mu\text{L}} = 1.65\ \text{pg}/\mu\text{L}$$

Therefore, the dilution series should have the concentrations shown above so that, when 10 μL is pipetted from each one, it will contain the desired number of gene copies, from 500 000 to 5.

Solution 9.4(b)

The stock solution of human DNA has a concentration of 2 μg/μL. We first need to convert the units of concentration to pg/μL to match with the units we have calculated for our dilution series:

$$\frac{2\ \mu\text{g}}{\mu\text{L}} \times \frac{1 \times 10^6\ \text{pg}}{\mu\text{g}} = \frac{2 \times 10^6\ \text{pg}}{\mu\text{L}}$$

Therefore, 2 μg/μL is equivalent to 2×10^6 pg/μL.

Since each dilution, from 165 000 to 16 500 pg/μL, from 16 500 to 1650 pg/μL, and continuing down stepwise to 1.65 pg/μL, represents a 10X dilution (a 1/10 dilution factor), we should be placing 20 μL of the next highest dilution into a total volume of 200 μL water to make those dilutions (as shown by a relationship of ratios below):

$$\frac{1}{10} = \frac{x}{200}$$

$$x = \frac{200}{10} = 20$$

Therefore, each dilution (except for the first) is made by aliquoting 20 μL from the previous dilution into a total volume of 200 μL of water (or buffer) to make the next dilution.

We now need to calculate how much of the DNA stock solution at 2 μg/μL we should aliquot into a total volume of 200 μL to make a DNA concentration of 165 000 pg/μL for the first dilution.

$$\frac{2 \times 10^6\,\text{pg}}{\mu\text{L}} \times \frac{x\,\mu\text{L}}{200\,\mu\text{L}} = \frac{1.65 \times 10^5\,\text{pg}}{\mu\text{L}}$$

Solving for *x* gives us the aliquot to be diluted into a final volume of 200 μL:

$$\frac{(2 \times 10^6\,\text{pg})(x)}{\mu\text{L}} = \frac{(1.65 \times 10^5\,\text{pg})(200)}{\mu\text{L}}$$

$$x = \frac{3.3 \times 10^7}{2 \times 10^6} = 16.5$$

Therefore, to make the first dilution, we aliquot 16.5 μL from the DNA stock solution having a concentration of 2 μg/μL into 183.5 μL of water to give a final volume of 200 μL and a concentration of 1.65×10^5 pg/μL. We make 10-fold dilutions for all the remaining standards as follows (the concentrations in pg/μL appear below each dilution):

$$2\,\mu\text{g}/\mu\text{L} \times \frac{16.5\,\mu\text{L}}{200\,\mu\text{L}} \times \frac{20\,\mu\text{L}}{200\,\mu\text{L}} \times \frac{20\,\mu\text{L}}{200\,\mu\text{L}} \times \frac{20\,\mu\text{L}}{200\,\mu\text{L}} \times \frac{20\,\mu\text{L}}{200\,\mu\text{L}} \times \frac{20\,\mu\text{L}}{200\,\mu\text{L}}$$

stock 1.65×10^5 1.65×10^4 1650 165 16.5 1.65

Taking 10 μL from each dilution into a real-time PCR will deliver the desired number of copies calculated above. Shown below is the proof of that statement for a 10 μL aliquot taken from the first dilution:

$$\frac{1.65 \times 10^5\,\text{pg}}{\mu\text{L}} \times \frac{1\,\text{gene copy}}{3.3\,\text{pg}} \times 10\,\mu\text{L} = \frac{1.65 \times 10^6\,\text{gene copies}}{3.3}$$

$$= 500\,000\,\text{gene copies}$$

9.3.3 The standard curve

The real-time PCR instrument's software can present the analysis of the standard curve in several different formats. Figure 9.2 shows the amplification curves for the human DNA dilution series used in the Life

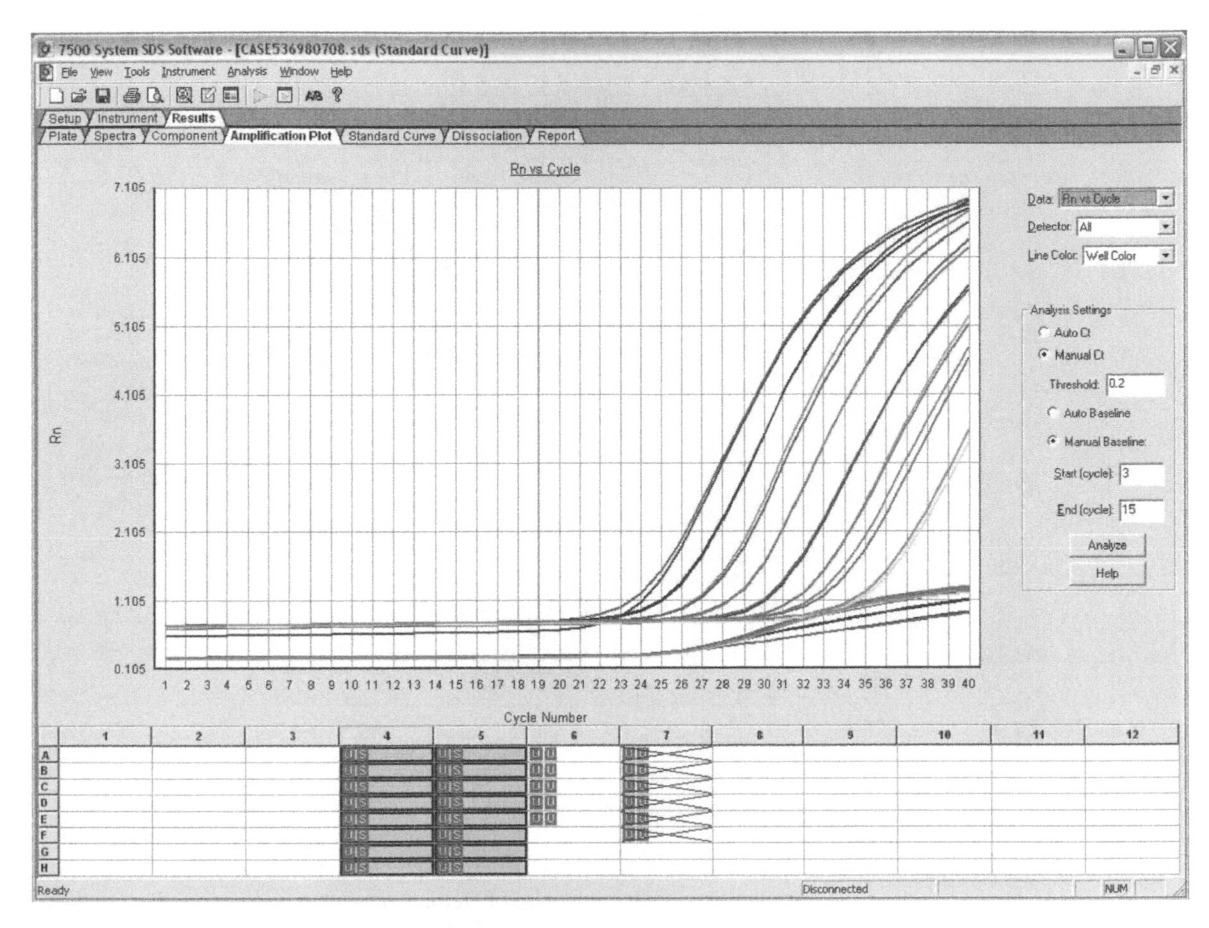

■ **FIGURE 9.2** An amplification plot of R_n (*y* axis) vs. cycle number (*x* axis) for the Quantifiler™ Human DNA Quantification Kit showing the relationship between DNA amount and fluorescence; higher DNA concentrations show fluorescent signal earlier. The group of shallow curves in the lower right corner of the plot is contributed by an internal positive control (IPC) assayed using its own unique set of primers and probe.

Technologies' Quantifiler™ Human DNA Quantification Kit. Real-time PCRs were run for each dilution in duplicate. The leftmost curve represents the amplification kinetics of the TaqMan assay for the 50 ng/μL dilution, and the rightmost curve that for 0.023 ng/μL. Curves for 16.7, 5.56, 0.62, 0.21, and 0.068 ng/μL dilutions lie between them. These curves show the relationship between DNA template amount and C_T value: the higher the DNA concentration, the earlier that signal can be detected and the lower the C_T value. The green horizontal line is the threshold value determined by the instrument's software. The C_T value for each dilution is the fractional

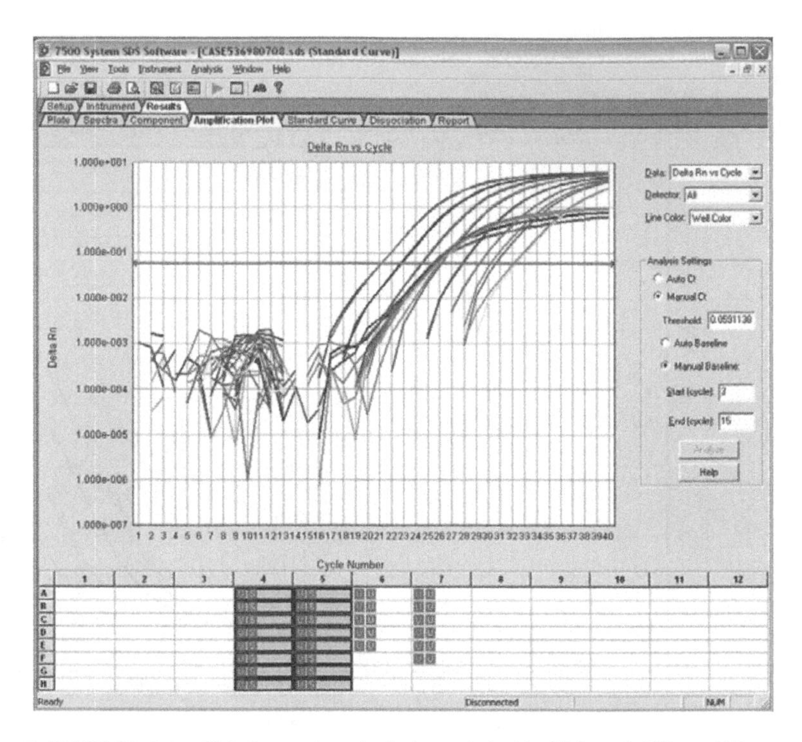

■ **FIGURE 9.3** A plot of Delta R_n vs. cycle number for the standards of the ABI Quantifiler™ Human DNA Detection Kit analyzed on an Applied Biosystems 7500 Real-time PCR System. Delta R_n is the change in normalized reporter signal (R_n minus baseline).

cycle number where the plot of Delta R_n (the change in normalized reporter signal) on the *y* axis vs. cycle number on the *x* axis intersects the threshold. The threshold and, therefore, the C_T value, should always fall within the exponential phase of the amplification reaction. A baseline is determined in the early part of the run, usually between the third and fourteenth cycles, where there is little change in fluorescent signal. Figure 9.2 also shows a group of curves of lower magnitude (in the lower right corner of the plot) intersecting several of the curves of the standard. These are contributed by an IPC spiked into each real-time PCR assay for the purpose of detecting reaction inhibition that might be carried into the reaction as a contaminant of the template DNA prep.

Figure 9.3 shows the same data plotted as the delta R_n (the normalized reporter minus baseline fluorescence on the *y* axis) vs. cycle number (on the *x* axis). The green line on this plot represents the threshold automatically set by the software – at a point that is above the baseline but sufficiently low enough on the amplification curve to be within the exponential

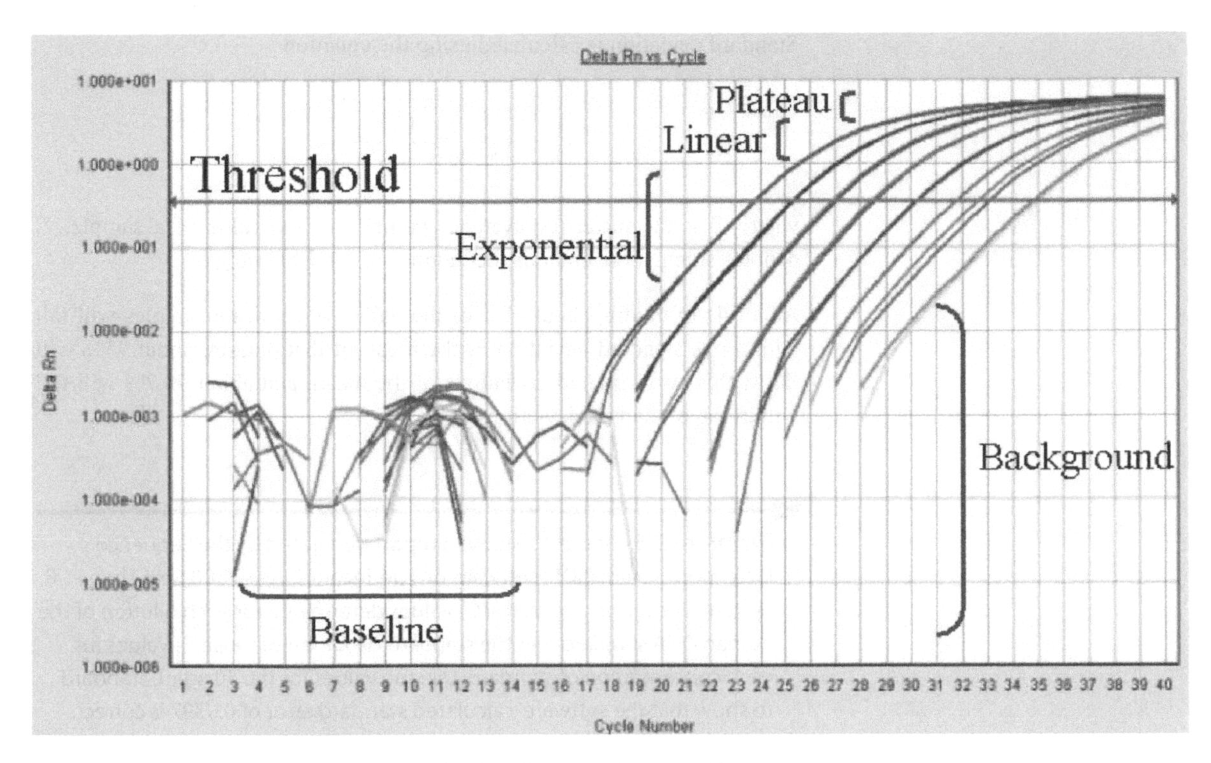

■ **FIGURE 9.4** Amplification plot for the dilution series of the Quantifiler™ Human DNA standard. The polymerase chain reaction (PCR) primers and probe target the telomerase gene. Shown are the phases of the PCR (exponential, linear, and plateau); the background fluorescent signal; the baseline; and the threshold where fluorescent signal, within the exponential phase of amplification, can be distinguished from background noise

phase of amplification. The intersection of the threshold line with the amplification curve defines the C_T. Figure 9.4 shows the features of the amplification plot labeled.

9.3.4 **Standard deviation**

The **standard deviation** (abbreviated as *s*) is the most commonly used indicator of precision – how close experimental values are to one another. For a series of experimental outcomes, it is an indication of the variability of that dataset about the mean (average) value. A small standard deviation indicates that the values are all close to the mean. A large standard deviation indicates that many of the data points are far from the average value. If preparing multiple trials for a single sample, the standard deviation value can give an indication of how careful you are with your experimental technique. A high value can point to pipetting errors between samples.

Standard deviation is calculated using the equation

$$s = \sqrt{\frac{\sum(x-\bar{x})^2}{n-1}}$$

where $\bar{x}$ is the mean (or average) value, x is the value of a sample, Σ denotes summation, and n is the number of replicate samples.

As a rule of thumb, about 68% of the values of replicate samples will fall within one standard deviation of the mean of that dataset, about 95% will fall within two standard deviations of the mean, and about 99.7% will fall within three standard deviations.

Problem 9.5 Figure 9.5 shows a report generated by the Sequence Detection System (SDS) software on Life Technologies 7500 Real-Time PCR System. The report shows the C_T values determined for each dilution of the human DNA standard and the standard deviation of those C_T values for each concentration of standard. Use the values for the 50 ng/μL standard to show that the software-calculated standard error of 0.0397 is correct.

Solution 9.5

The 50 ng/μL standard ('std 1' in Figure 9.5) was run in duplicate. The software determined that those two samples had C_T values of 22.88 and 22.93 cycles. The mean value for these two samples is

$$\begin{aligned}\bar{x} &= \frac{22.8784 + 22.9345}{2}\\ &= \frac{45.8129}{2}\\ &= 22.90645\end{aligned}$$

We will now calculate the sum of the squares of each value minus the mean divided by n − 1 (this value is also known as the **variance**):

$$\begin{aligned}\sum\frac{(x-\bar{x})^2}{n-1} &= \frac{(22.8784 - 22.90645)^2 + (22.9345 - 22.90645)^2}{2-1}\\ &= \frac{(0.02805)^2 + (0.02805)^2}{1}\\ &= 0.0007868 + 0.0007868 = 0.00157361\end{aligned}$$

Placing this value into our equation for the standard deviation yields

$$s = \sqrt{0.00157361} = 0.0397$$

This standard deviation value matches that in the table of Figure 9.5 for 'std 1.' This value has units of cycles.

Setup | Instrument | **Results**

Plate | Spectra | Component | Amplification Plot | Standard Curve | Dissociation | **Report**

Well	Sample Name	Detector	Task	Ct	StdDev Ct	Quantity
A4	std 1	IPC	Unknown	29.4136	0.065	
A4	std 1	Q	Standard	22.8784	0.0397	50
A5	std 1	IPC	Unknown	29.5055	0.065	
A5	std 1	Q	Standard	22.9345	0.0397	50
B4	std 2	IPC	Unknown	28.1869	0.0706	
B4	std 2	Q	Standard	24.4561	0.0584	16.7
B5	std 2	IPC	Unknown	28.2867	0.0706	
B5	std 2	Q	Standard	24.3735	0.0584	16.7
C4	std 3	IPC	Unknown	27.8442	0.0458	
C4	std 3	Q	Standard	26.0598	0.072	5.56
C5	std 3	IPC	Unknown	27.9089	0.0458	
C5	std 3	Q	Standard	26.1616	0.072	5.56
D4	std 4	IPC	Unknown	27.8627	0.0499	
D4	std 4	Q	Standard	27.7225	0.118	1.85
D5	std 4	IPC	Unknown	27.9332	0.0499	
D5	std 4	Q	Standard	27.5557	0.118	1.85
E4	std 5	IPC	Unknown	28.0472	0.0293	
E4	std 5	Q	Standard	29.313	0.0498	0.62
E5	std 5	IPC	Unknown	28.0058	0.0293	
E5	std 5	Q	Standard	29.3834	0.0498	0.62
F4	std 6	IPC	Unknown	28.0784	0.0455	
F4	std 6	Q	Standard	30.9448	0.0205	0.21
F5	std 6	IPC	Unknown	28.1428	0.0455	
F5	std 6	Q	Standard	30.9158	0.0205	0.21
G4	std 7	IPC	Unknown	28.0729	0.0311	
G4	std 7	Q	Standard	31.8097	0.288	0.068
G5	std 7	IPC	Unknown	28.1169	0.0311	
G5	std 7	Q	Standard	32.2168	0.288	0.068
H4	std 8	IPC	Unknown	28.3082	0.0655	
H4	std 8	Q	Standard	34.0747	0.127	0.023
H5	std 8	IPC	Unknown	28.2156	0.0655	
H5	std 8	Q	Standard	33.8948	0.127	0.023

■ **FIGURE 9.5** A C_T report of duplicate dilutions of the human DNA standard from 50 ng/μL down to 0.0023 ng/μL DNA concentrations. The table shows on which well position of a 96-well plate each sample was run, the sample name (std 1 at 50 ng/μL to std 8 at 0.0023 ng/μL), the detector (the sample being assayed, either 'Q' for a quantifiler standard or 'IPC' for an internal positive control), the task denoting the nature of the sample, the C_T value, the standard deviation of the C_T values for each dilution ('StdDev Ct'), and the quantity of DNA in ng/μL.

9.3.5 Linear regression and the standard curve

In Chapter 5, we used a standard curve and linear regression analysis to calculate the concentration of DNA stained with PicoGreen. In the example provided, regression analysis was performed using Microsoft Excel. The real-time PCR software also performs regression analysis when plotting the log of the DNA concentration (on the x axis) vs. the C_T value for each dilution (on the y axis). The line of best fit is determined and written as the general equation $y = mx + b$. The analysis software provides the line's R^2 value as well as its slope (m) and its y intercept (b). Figure 9.6 shows the standard curve generated for a run using the Applied Biosystems' Quantifiler® assay, in which a slope of -3.287829 and a y intercept of 28.518219 are calculated.

The R^2 value, the correlation coefficient, is a measure of how closely the data points fall on the regression line. If the data points are all on the line, the R^2 value will be close to one. When that is the case, the C_T value (on the y axis) accurately predicts the DNA concentration (on the x axis). If the R^2 value is zero, then a value on one axis cannot predict the value on the other. If the R^2 value is greater than 0.99, you can have confidence that the C_T value will reliably predict the DNA concentration.

Note: PCR efficiency should be evaluated using a minimum of three replicates of each concentration of diluted standard. Only two replicates are used in this example for reasons of space.

Problem 9.6 Use the values for the slope and intercept shown in Figure 9.6 to show that the software-calculated DNA concentration of 0.676031 ng/μL for Evidence Item 1 (shown in Figure 9.7) is correct.

Solution 9.6

The general equation for the regression line plotting the standard curve is $y = mx + b$, where y is the C_T value for a sample, m is the slope, x is the log of the DNA concentration, and b is the y intercept. The slope, m, for the standard curve shown in Figure 9.6 is -3.287829. The y intercept is 28.518219 cycles. As shown in Figure 9.7, Evidence Item 1 in this assay has a C_T value of 29.0773 cycles. Placing these values into the equation for the regression line gives

$$y = mx + b$$

$$29.0773 = (-3.287829)x + 28.518219$$

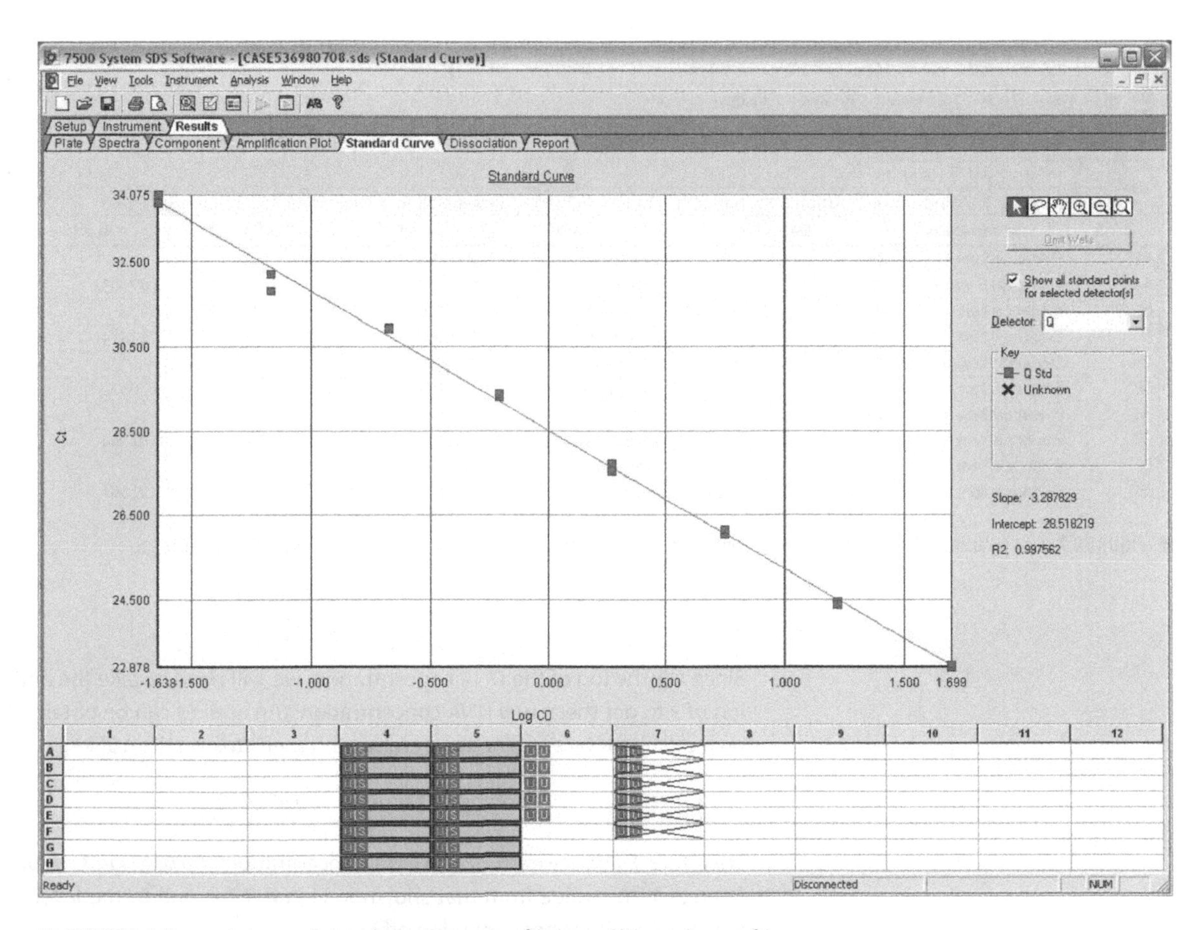

■ **FIGURE 9.6** The standard curve calculated for the dilution series of the human DNA control as part of the Life Technologies Quantifiler® Human DNA Quantification Kit. The Sequence Detection Systems (SDS) software for the 7500 Real-Time PCR System generates a standard curve of the dilution series of the human control DNA and uses linear regression to plot a line of best fit and to provide the *m* (slope) and *b* (y intercept) values for that line when expressed by the general equation $y = mx + b$.

Solving for x yields

$$\frac{29.0773 - 28.518219}{-3.287829} = x$$

$$\frac{0.559081}{-3.287829} = x$$

$$= -0.170046$$

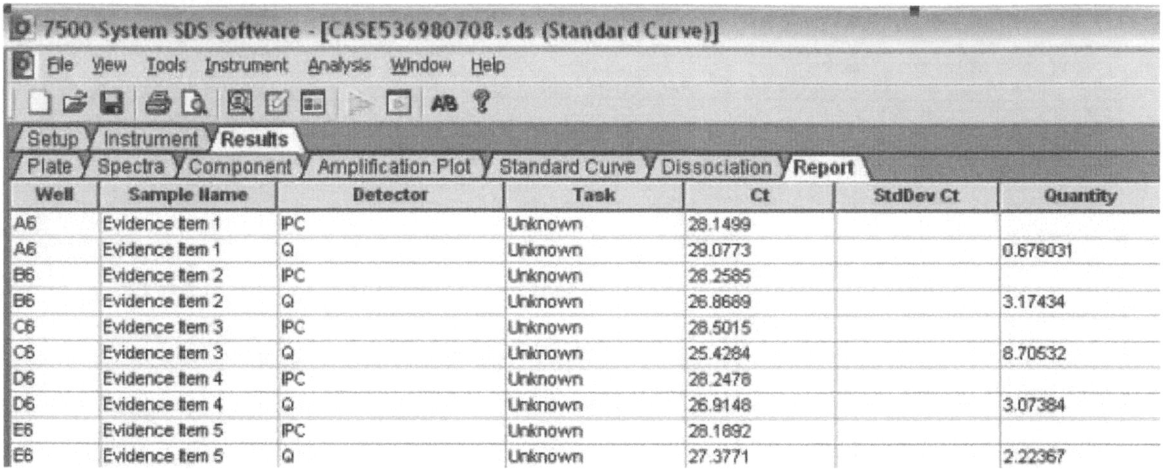

7500 System SDS Software - [CASE536980708.sds (Standard Curve)]

File View Tools Instrument Analysis Window Help

Setup / Instrument / Results

Plate / Spectra / Component / Amplification Plot / Standard Curve / Dissociation / Report

Well	Sample Name	Detector	Task	Ct	StdDev Ct	Quantity
A6	Evidence Item 1	IPC	Unknown	28.1499		
A6	Evidence Item 1	Q	Unknown	29.0773		0.676031
B6	Evidence Item 2	IPC	Unknown	28.2585		
B6	Evidence Item 2	Q	Unknown	26.8689		3.17434
C6	Evidence Item 3	IPC	Unknown	28.5015		
C6	Evidence Item 3	Q	Unknown	25.4284		8.70532
D6	Evidence Item 4	IPC	Unknown	28.2476		
D6	Evidence Item 4	Q	Unknown	26.9148		3.07384
E6	Evidence Item 5	IPC	Unknown	28.1892		
E6	Evidence Item 5	Q	Unknown	27.3771		2.22367

■ **FIGURE 9.7** Concentrations for unknown samples calculated using the values for the regression line plotting the standard curve shown in Figure 9.6.

Since x is the log of the DNA concentration, we will need to take the antilog of x to get the actual DNA concentration. (An antilog can be obtained on a calculator by entering the number and then striking the **10^x^** key.)

$$\text{antilog} - 0.170046 = 0.676012$$

Therefore, Evidence Item 1 has a DNA concentration of 0.676 ng/μL. (A difference in this value from that shown in Figure 9.7 is due to the level to which values have been rounded off.)

9.4 AMPLIFICATION EFFICIENCY

Amplification efficiency in PCR is a measure of how much template is converted to amplified product during each cycle of the exponential phase of the reaction. At 100% efficiency, the amount of product exactly doubles each cycle – each dsDNA molecule produces two dsDNA molecules. It is best measured only during the early exponential phase as it will naturally decline to zero as the reaction enters the stationary phase and components are used up, the primers compete with product for binding to the denatured strands, and the enzyme loses its activity. Even during the exponential phase, however, there are a number of distractions that can prevent the DNA polymerase enzyme from manufacturing duplicate molecules at that kind of pace. The reaction can be made suboptimal by inhibitors, other enzymes carried over from previous preparative reactions, and various contaminants that copurify with the template.

PCR amplification, as we saw in the previous chapter, is described by the equation

$$Y = X(1+E)^n$$

where Y is the amount of PCR product, X is the starting number of template molecules, E is the efficiency of amplification, and n is equal to the number of PCR cycles. In real-time PCR using a TaqMan® probe, Y, the amount of product generated during amplification, is proportional to the amount of fluorescent signal. We can therefore write real-time PCR amplification as proceeding according to the equation

$$R_f = R_i(1+E)^n$$

where R_f is equivalent to the final fluorescence signal of the reporter dye, R_i is equivalent to the reporter dye's initial fluorescent signal, E is the amplification efficiency, and n is the number of cycles.

Amplification efficiency (E) is determined from a standard curve using the following formula:

$$EXPONENTIAL\ AMPLIFICATION = 10^{(-1/slope)}$$

$$EFFICIENCY = (10^{(-1/slope)} - 1) \times 100$$

A 100% efficient reaction will have a slope of −3.32. Put another way, during exponential amplification, a 100% efficient reaction will double every cycle and will produce a 10-fold increase in PCR product every 3.32 cycles. Mathematically, this is stated as

$$\log_2 10 = 3.3219$$

or

$$2^{3.3219} = 10$$

A slope with a more negative value (−3.85, for example) will represent a PCR having efficiency less than 100% and in need of optimization. A slope having a more positive value (−2.8, for example) indicates possible pipetting errors or problems with the template (degradation or inhibition).

Note: Unfortunately, amplification efficiency is defined in a couple of different ways in the literature. In some publications, it is defined as $10^{(-1/\text{slope})}$, where 'slope' refers to the slope of the line generated by a plot of log DNA concentration (on the y axis) vs. C_T values (on the x axis) for a dilution series of a standard template. For a perfectly efficient reaction, the slope is -3.32. By this definition, we have

$$10^{\left(\frac{-1}{-3.32}\right)} = 10^{0.3} = 2$$

Of course, this does not mean that the reaction is 200% efficient ($2 \times 100 = 200\%$). The other interpretation of amplification efficiency uses the equation shown above, where 1 is subtracted from $10^{(-1/\text{slope})}$ and this value is multiplied by 100 to give the more familiar form of a percent. This ambiguity is noted here to remind you to be aware of the definition being used in the protocol you are following.

Problem 9.7 A real-time PCR is found to be 90% efficient. Starting with just one template molecule, what is the amount of product made in this 90% efficient reaction compared with the amount of product made in a 100% efficient reaction after 20 cycles?

Solution 9.7

We will use the following equation to solve this problem:

$$Y = X(1 + E)^n$$

For the 100% efficient reaction, E is equivalent to 1 and we have

$$Y = 1(1 + 1)^{20} = 2^{20} = 1048576 \text{ copies}$$

For a 90% efficient reaction, E is equivalent of 0.9 and we have

$$Y = 1(1 + 0.9)^{20} = 1.9^{20} = 375900 \text{ copies}$$

Dividing the 90% efficient reaction by the 100% efficient reaction gives

$$\frac{375900 \text{ copies}}{1048576 \text{ copies}} = 0.36$$

Multiplying this value by 100 tells us that, after 20 cycles, the 90% efficient reaction makes only 36% of what could be made if the amplification reaction were 100% efficient.

Problem 9.8 If we start out with 8000 molecules of target in an 85% efficient PCR, how many product molecules will we generate after 10 cycles?

Solution 9.8

We will use the following relationship:

$$Y = X(1 + E)^n$$

In this equation, X is the number of starting molecules. We have, therefore,

$$\begin{aligned} Y &= 8000 \times (1 + 0.85)^{10} \\ &= 8000 \times 1.85^{10} \\ &= 8000 \times 470 \\ &= 3760000 \text{ molecules} \end{aligned}$$

Therefore, after 10 cycles, an 85% efficient PCR containing 8000 molecules of template will generate 3 760 000 molecules of product.

Problem 9.9 In Figure 9.6, the standard curve's regression line was determined to have a slope of −3.287 829. What is the amplification efficiency for these reactions?

Solution 9.9

Using the equation above, we have

$$\begin{aligned} E &= (10^{(-1/-3.287829)} - 1) \times 100 \\ &= (10^{0.304152} - 1) \times 100 \\ &= (2.014429 - 1) \times 100 \\ &= 1.014429 \times 100 \\ &= 101.44\% \end{aligned}$$

Therefore, the reactions that generated the standard curve shown in Figure 9.6 have an amplification efficiency of approximately 101%.

During the exponential phase of synthesis, the increase in fluorescence from cycle to cycle is a reflection of the amplification efficiency. Although PCR efficiency affects the C_T value, amplification efficiencies between 90% and 110% are generally considered to be of a quality capable of generating reliable real-time PCR data. Reactions having lower efficiency will also exhibit lower sensitivity.

Problem 9.10 For a 100% efficient PCR, how many C_T values should there be between the mean of two three-fold dilution samples?

Solution 9.10

The slope of a line plotting C_T values (on the *y* axis) vs. the log of DNA concentration (on the *x* axis) for a 100% efficient reaction has a slope of −3.32. The slope of a line is equal to its 'rise over the run;' that is, the change in *y* divided by the change in *x*. We can choose any numbers for the *x* axis to represent two samples that differ by a three-fold dilution. Here we will choose 16.7 ng/μL and 50 ng/μL.

$$\text{slope} = \frac{\text{rise}}{\text{run}} = \frac{\Delta y}{\Delta x} = \frac{\Delta y}{16.7\,ng/\mu L - 50\,ng/\mu L} = -3.32$$

Since the *x* axis scale is the log of DNA concentration, we will need to take the log values of those two numbers. We then have

$$\frac{\Delta y}{1.22 - 1.70} = -3.32$$

Solving for Δy gives

$$\frac{\Delta y}{-0.48} = -3.32$$

$$\Delta y = (-3.32) \times (-0.48) = 1.60$$

Therefore, there are 1.6 C_T values between the mean of two three-fold dilution samples.

9.5 MEASURING GENE EXPRESSION

The ability to accurately measure gene expression is critical to the understanding of how cells descend into disease and how they may respond to therapy. What changes in transcription send a cell spiraling into a cancerous state? What genes are turned off and what genes are turned on in response to a medication being tested to treat an illness or in response to viral infection?

Using real-time PCR for quantifying gene expression has a number of attractive features, not least of which is its sensitivity – it has the ability to detect as little as a single copy of a specific transcript. In addition, real-time PCR has the capacity to produce accurate data over a wide dynamic range (it is able to accurately detect a small amount of transcript and a large amount of transcript), it offers high throughput (many samples can be processed simultaneously), and it can distinguish between transcripts having very similar sequences.

Reactions that measure differences in gene expression between genes, between cell types, and between treatments, inherently carry with them the potential for a high degree of variability. This can be introduced by pipetting error or by varying efficiencies of either the PCR amplification or the reverse transcription reaction used to convert mRNA to cDNA. Variability arises from within the instrument used to measure the changes in light intensity generated from the real-time PCR. It arises from instability of the reagents used in the assay. It comes from many sources. As a consequence, real-time PCR as a means to measure gene expression requires the generous use of controls designed to set the experimenter's mind at ease and to convince them that the effect on expression that they may be privileged to witness is, in fact, real.

One such safeguard against the gremlins that haunt the laboratory, as we have discussed previously, is the endogenous control. It is used as an active reference to normalize for differences in the amount of total RNA added to a real-time PCR. The most frequently used endogenous controls are the **housekeeping genes** – those genes that are always expressed at the more-or-less constant background level that is needed by the cell to carry out its basic processes of living. Examples of the housekeeping genes include β-actin, GAPDH (glyceraldehydes 3-phosphate dehydrogenase), and 18S RNA. An endogenous control gene should be on the same sample of nucleic acid as the target gene of interest. However, it can be amplified in a separate well on the same plate as the target or it can be amplified right along with the target in the same well in a **multiplex reaction** using different-colored dye-labeled probes.

Measuring gene expression by real-time PCR is most often performed as a relative assay. That is, you do not necessarily care exactly how many mRNA molecules are made from a particular gene, you just want to know what level of transcript is made compared (relative) to some other gene. This difference is usually expressed as a fold difference. Relative quantification, for example, can be used to determine the difference in expression of the heat-shock genes in cancerous cells compared to their expression in normal, healthy cells. It is used to compare the expression of a specific gene in one tissue type compared with its expression in a different tissue type, particularly when those tissues have been treated in some way with a drug or other agent. The gene the response of which is being measured is called the **target**. Its expression is compared to a reference gene known as the **calibrator** that often represents the untreated control. Relative quantification experiments also include an endogenous control – a gene having a consistent level of expression in all experimental samples. There are several ways to measure relative quantification.

9.6 RELATIVE QUANTIFICATION – THE $\Delta\Delta C_T$ METHOD

The **ΔΔCT method**, also referred to as the **Comparative C_T method**, is a means of measuring relative quantification and was first described by Livak and Schmittgen in 2001. It determines the relative change in gene expression between a target gene under investigation and that of a calibrator (reference) gene. Although the assay is designed to measure the amount of transcript (RNA) made in a particular cell, what goes into the real-time PCR is actually cDNA made by reverse transcription of total RNA recovered from the cells the gene expression of which is under study. The experimenter may be interested in the target gene's response to a drug, to infection, to disease, or to some other stimulus. Most frequently, the untreated control is used as the calibrator. An endogenous, internal control – a gene already present in the sample – is included for purposes of normalizing for any differences in the amount of cellular RNA placed into the reverse transcriptase reactions when cDNA is generated. The housekeeping genes – those genes expressed at a constant, background level such as GAPDH, ribosomal RNA, DNA topoisomerase, *RNaseP*, or *β-actin* – are usually used as an endogenous control.

The difference between the C_T of the target gene ($C_{T,\ target}$) and the C_T of the endogenous control ($C_{T,\ ec}$) is the ΔC_T of the sample:

$$C_{T,\ target} - C_{T,\ ec} = \Delta C_T$$

Likewise, the ΔC_T of the calibrator is equal to the C_T of the target gene in the untreated (calibrator) sample minus the C_T of the endogenous control in the untreated (calibrator) sample:

$$C_{T,\ calibrator\ target} - C_{T,\ calibrator\ ec} = \Delta C_{T,\ calibrator}$$

The term $\Delta\Delta C_T$ is calculated as the ΔC_T of the target gene in the treated sample minus the ΔC_T of the target in the untreated, calibrator sample:

$$\Delta\Delta C_T = \Delta C_{T,\ target\ in\ treated\ sample} - \Delta C_{T,\ target\ in\ calibrator\ sample}$$

The calibrator, since it is untreated, should have no change in its ΔΔCT value during the course of the experiment. Its change, therefore, is equivalent to zero. Since 2^0 equals one, the calibrator gene's expression is unity.

When the $\Delta\Delta C_T$ method is used to measure gene expression, therefore, the results are expressed as a 'fold' change in the expression level of the target gene normalized to the endogenous control and relative to the calibrator. It is given by the equation:

$$\text{Relative Fold Change} = 2^{-\Delta\Delta C_T}$$

9.6.1 The $2^{-\Delta\Delta C_T}$ method – deciding on an endogenous reference

Using the $2^{-\Delta\Delta C_T}$ method for comparative analysis of gene expression requires that transcription of the internal control gene – the gene used as the endogenous reference – not be affected by the experimental treatment. To test whether or not this is the case, reactions are performed that mimic the conditions under which the comparative study will be run. However, only the reference gene need be assayed by real-time PCR in this initial experiment.

Evaluating whether or not an experimental treatment can affect expression of the reference gene is a form of **hypothesis testing**. In this case, we are testing what is called the **null hypothesis** (designated H_0). The null hypothesis makes the claim that the results we see in this experiment are the consequences of pure chance and that the experimental treatment has no effect on the reference gene's mean C_T value. If the null hypothesis is true, the average C_T value for each group of samples should be the same.

If we were only comparing two means derived from two groups, we could use a statistical test called the ***t*-test**. However, for this experiment, we will be comparing multiple mean sample values and, therefore, it is more appropriate to use a statistical approach called one-way **ANOVA** (for **analysis of variance**). (Another statistical approach called **multiway ANOVA** should be used testing whether or not two different factors, rather than just one, affect an outcome.)

Variance is a measure of variability and refers to the dispersion of values around a mean. It is calculated as the square of the standard deviation (s) which, as you may recall from earlier in this chapter, is the square root of the sum of the difference of each value from the mean divided by one less than the number of samples:

$$s = \sqrt{\frac{\sum (x - \bar{x})^2}{n - 1}}$$

Variance is this value squared:

$$\text{variance} = s^2 = \frac{\sum(x - \bar{x})^2}{n - 1}$$

In the statistical approach of ANOVA, we calculate the variance between groups of samples as well as the variance of samples within each individual group. We then compare those values. The first step is to calculate the sample average (mean value) for each group. Let us say we have groups a, b, and c. The mean for each group is calculated as

$$\bar{a} = \frac{a_1 + a_2 + \ldots + a_n}{n}$$

$$\bar{b} = \frac{b_1 + b_2 + \ldots + b_n}{n}$$

$$\bar{c} = \frac{c_1 + c_2 + \ldots + c_n}{n}$$

where n is the number of samples in each group.

We then calculate the average of all the averages. We will designate this value $\bar{\bar{x}}$. It is calculated as

$$\bar{\bar{x}} = \frac{\bar{a} + \bar{b} + \bar{c} + \ldots}{m}$$

where m is equal to the number of groups (e.g., the number of time points taken during a time course study).

We then calculate the sample variance of the averages. This is also called the **between-groups variance** and we will designate it s^{*2}. It is calculated as

$$s^2 = \frac{(\bar{a} - \bar{\bar{x}})^2 + (\bar{b} - \bar{\bar{x}})^2 + (\bar{c} - \bar{\bar{x}})^2 + \ldots}{m - 1}$$

We also calculate the sample variance (s^2) for each individual group. This is also called the **within-groups variance**. It is calculated for Group a as

$$s_a{}^2 = \frac{(a_1 - \bar{a})^2 + (a_2 - \bar{a})^2 + \ldots + (a_n - \bar{a})^2}{n - 1}$$

which can also be written as

$$s_a{}^2 = \frac{\sum(a - \bar{a})^2}{n - 1}$$

The variance for each group (a, b, and c) is calculated in this same manner.

Using the within-group variances, we calculate the average of all the sample variances as

$$s^2 = \frac{s_a^{\ 2} + s_b^{\ 2} + s_c^{\ 2} + \ldots}{m}$$

We now use these values to calculate the ***F* statistic**, the ratio of the two variances – the between-groups variance to the within-groups variances. It is used under the assumption that the different groups have similar standard deviations (they have similar variability about their mean values). The F statistic will tell us whether the sample means of several groups are equal. If they are, we can accept the null hypothesis that the experimental treatment has no effect on the expression of the reference gene. The F statistic is calculated as

$$F = \frac{n \times s^{*2}}{s^2}$$

where n is the number of measurements in each individual group. If the F value is large, the variability between groups (the numerator) must be large in comparison to the variability within each group (the denominator). We would have to reject the null hypothesis in this case. However, if the numerator (the between-groups variance) is equal to or smaller than the denominator (the within-groups variance), the F statistic will have a lower value and we should be able to accept the null hypothesis and conclude that there is no difference between the groups.

What, however, is the critical value for the F statistic, the cutoff point, at which we can safely make the determination to accept or reject the null hypothesis? That value is provided by the ***F* distribution**, an asymmetric curve describing the probability of encountering any particular F statistic (Figure 9.8). If the F statistic is greater than the critical value, we reject the null hypothesis. If the F statistic is less than the critical value, we accept the null hypothesis.

The critical cutoff value within the F distribution can be found in an ***F* distribution table** (Table 9.1). A number of resources on the internet provide F distribution tables at several **levels of significance** (1, 5, and 10%). We will use the 5% level of significance, meaning that there is a 5% chance that we could be in error in rejecting the null hypothesis when, in fact, it is actually true.

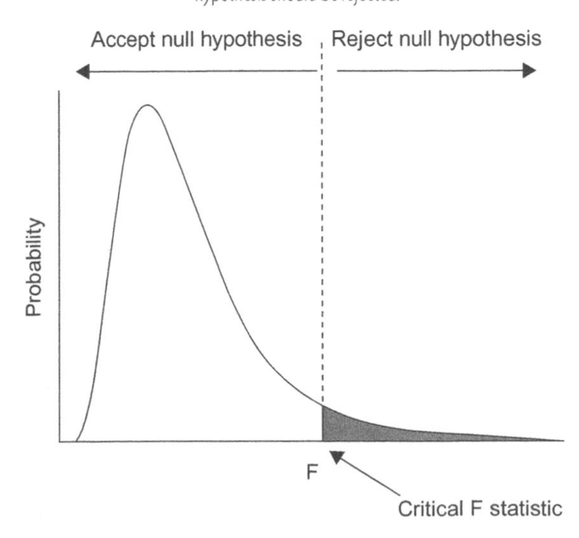

■ **FIGURE 9.8** An *F* distribution curve describes the probability for any particular *F* statistic at a defined level of confidence (typically, 1, 5, or 10%). The portion of the curve to the left of the critical *F* statistic shows those *F* values for which the null hypothesis should be accepted. If the calculated *F* statistic is greater than the critical value, the null hypothesis should be rejected.

Table 9.1 An F distribution table showing the critical values for the F statistic at a 5% level of significance. The table is read by intersecting a column corresponding to the degrees of freedom of the F statistic's numerator (df_n) with the row corresponding to the degrees of freedom of the F statistic's denominator (df_d).

		df_n												
		2	3	4	5	6	7	8	9	10	12	15	20	30
df_d	2	19.00	19.16	19.25	19.30	19.33	19.35	19.37	19.38	19.40	19.41	19.43	19.45	19.46
	3	9.55	9.28	9.12	9.01	8.94	8.89	8.85	8.81	8.79	8.74	8.70	8.66	8.62
	4	6.94	6.59	6.39	6.26	6.16	6.09	6.04	6.00	5.96	5.91	5.86	5.80	5.75
	5	5.79	5.41	5.19	5.05	4.95	4.88	4.82	4.77	4.74	4.68	4.62	4.56	4.50
	6	5.14	4.76	4.53	4.39	4.28	4.21	4.15	4.10	4.06	4.00	3.94	3.87	3.81
	7	4.74	4.35	4.12	3.97	3.87	3.79	3.73	3.68	3.64	3.57	3.51	3.44	3.38
	8	4.46	4.07	3.84	3.69	3.58	3.50	3.44	3.39	3.35	3.28	3.22	3.15	3.08
	9	4.26	3.86	3.63	3.48	3.37	3.29	3.23	3.18	3.14	3.07	3.01	2.94	2.86
	10	4.10	3.71	3.48	3.33	3.22	3.14	3.07	3.02	2.98	2.91	2.85	2.77	2.70
	11	3.98	3.59	3.36	3.20	3.09	3.01	2.95	2.90	2.85	2.79	2.72	2.65	2.57
	12	3.89	3.49	3.26	3.11	3.00	2.91	2.85	2.80	2.75	2.69	2.62	2.54	2.47
	13	3.81	3.41	3.18	3.03	2.92	2.83	2.77	2.71	2.67	2.60	2.53	2.46	2.38
	14	3.74	3.34	3.11	2.96	2.85	2.76	2.70	2.65	2.60	2.53	2.46	2.39	2.31
	15	3.68	3.29	3.06	2.90	2.79	2.71	2.64	2.59	2.54	2.48	2.40	2.33	2.25
	16	3.63	3.24	3.01	2.85	2.74	2.66	2.59	2.54	2.49	2.42	2.35	2.28	2.19
	17	3.59	3.20	2.96	2.81	2.70	2.61	2.55	2.49	2.45	2.38	2.31	2.23	2.15
	18	3.55	3.16	2.93	2.77	2.66	2.58	2.51	2.46	2.41	2.34	2.27	2.19	2.11
	19	3.52	3.13	2.90	2.74	2.63	2.54	2.48	2.42	2.38	2.31	2.23	2.16	2.07
	20	3.49	3.10	2.87	2.71	2.60	2.51	2.45	2.39	2.35	2.28	2.20	2.12	2.04
	21	3.47	3.07	2.84	2.68	2.57	2.49	2.42	2.37	2.32	2.25	2.18	2.10	2.01

Locating the cutoff value for the F statistic in an F distribution table, however, requires that we know the **degrees of freedom** corresponding to the particular dataset used in the calculation. Degrees of freedom describe, in simple terms, the number of values in the final calculation of a statistic that are free to vary. The higher the number of independent pieces of data going into the final estimation of a statistical score, the greater the number of degrees of freedom. In this form of ANOVA, the final statistic is F, in which we are looking at variability of two population groups, variability between groups (the numerator value), and variability within groups (the denominator); we are calculating two statistical estimates to arrive at the F value. The F distribution, therefore, utilizes two values for the degrees of freedom, one for the numerator value and one for the denominator value, with each combination having its own characteristic F distribution. The degrees of freedom for the numerator (df_n) is calculated as $(m - 1)$. Its values are represented by the columns of an F distribution table.

The degrees of freedom for the denominator (df_d) is calculated as $m(n - 1)$. Its values are represented by the rows of an F distribution table.

If the calculated value of the F statistic is less than the critical value taken from the F distribution table, you should accept the null hypothesis and conclude, at that level of significance, that there is no difference between the samples. If the F statistic is greater than the critical value, the null hypothesis should be rejected and you can conclude that there *is* a significant difference between the samples. The experimental treatment, therefore, is the reason for the difference you are observing.

We can even go a step further in bolstering our conviction for the veracity of the null hypothesis by making a judgment as to whether or not a 'statistically significant' relationship exists between the experimental treatment and the average C_T values. That conviction is provided by the ***p*-value**. The p-value is an expression of probability – how probable is it that you could get a result (an F statistic) more extreme than the one you got if the null hypothesis were true. Or, in other words, if you calculated a certain F statistic, what is the probability of obtaining that F statistic? In general terms, the smaller the p-value, the more fervently the null hypothesis can be rejected. A larger p-value suggests there is less evidence against the null hypothesis. Most biologists are content with designating a p-value less than 0.05 (5%) as the indication of significance. At this level of significance, the null hypothesis can be rejected if the p-value is less than or equal to 0.05.

P-values for a particular F statistic can be located in probability tables or by using software freely available on the internet. On such sites (an example is shown in Figure 9.9), you merely enter your F statistic value and the degrees of freedom and the site calculates both the F distribution cutoff value and the p-value.

df numerator =	
df denominator =	
F =	
p =	

Compute

FIGURE 9.9 A p-value calculator for any particular F statistic. The degrees of freedom for the calculated F statistic are entered in the dialogue boxes and the p-value is determined automatically when the **Compute** button is clicked. This calculator is found at davidmlane.com/hyperstat/F_table.html but calculators can also be found at www.danielsoper.com/statcalc/calc07.aspx and stattrek.com/Tables/F.aspx

Problem 9.11 An experiment is conducted to see which housekeeping gene, *HSKG1* or *HSKG2*, will better serve as a reference in a comparative gene expression study. This experiment will mimic that study. Transcription of a heat-shock gene will be monitored following a shift in the incubation temperature from 37° to 43°C. The housekeeping gene chosen as a reference should not be affected by the temperature shift. Thirty cultures of liver cells (grown in six-well plates) are subjected to the heat-shock treatment. Total RNA is extracted from five parallel cultures of heat-shocked liver cells at time zero (just prior to heat shock) and at 1 and 2 hr post temperature shift. Two hundred nanograms of total RNA from each parallel culture are converted to cDNA by reverse transcriptase and placed into real-time PCRs for the assay of either the *HSKG1* or the *HSKG2* gene expression. The results shown in Tables 9.2 and 9.3 are obtained. This

Table 9.2 Results of real-time polymerase chain reaction (PCR) gene expression study for *HSKG1*, as described in Problem 9.11.

Time (hours post heat shock)	Well number	C_T
0	A1	27.3
	B1	27.7
	C1	27.4
	D1	27.2
	E1	27.5
1	F1	27.6
	G1	27.8
	H1	27.5
	A2	27.4
	B2	27.5
2	C2	27.2
	D2	27.1
	E2	27.6
	F2	27.4
	G2	27.3

Table 9.3 Results for real-time polymerase chain reaction (PCR) gene expression study of *HSKG2*, as described in Problem 9.11.

Time (hours post heat shock)	Well number	C_T
0	H2	29.3
	A3	28.7
	B3	29.5
	C3	29.2
	D3	29.7
1	E3	27.1
	F3	27.3
	G3	27.2
	H3	26.8
	A4	27.4
2	B4	26.5
	C4	26.9
	D4	25.8
	E4	26.3
	F4	26.1

experiment yields three groups of C_T values (since there are three time points) with five samples from each group. For the null hypothesis to be true, that there is no effect of the experimental treatment on expression of the housekeeping gene, the mean C_T values for the three groups should be equal. Which gene, *HSKG1* or *HSKG2*, would make a better reference gene?

Solution 9.11

We will use one-way ANOVA with the *F* statistic to evaluate which gene is least affected by the heat-shock treatment and use that gene as a reference in our subsequent comparative gene expression experiment. Using this statistical test, we will see whether we can accept or reject the null hypothesis. It reads:

H_0: There is no difference in the mean C_T values for samples taken at 0, 1, and 2 hr post temperature shift. Any difference we do see is the result of mere chance.

We will choose that gene as a reference for which we can accept the null hypothesis.

Since raw C_T values are calculated from a log-linear plot of fluorescent signal vs. cycle number, C_T is an exponential function, not a linear one. If we were to calculate variance based on the raw C_T numbers, we would not be accurately representing the true variation. However, raw C_T values can be converted to linear values by using the term 2^{-CT}. We will use the converted values when calculating the variance for the two housekeeping genes, as shown in the following tables (Tables 9.4 and 9.5).

We will first calculate the average mean ($\bar{\bar{x}}$) for each gene over the three time points as

$$\bar{\bar{x}} = \frac{\bar{a} + \bar{b} + \bar{c}}{m}$$

where *m* is the number of groups (numbers are 10^{-9}).

For the *HSKG1* gene,

$$\bar{\bar{x}} = \frac{5.61 + 5.09 + 6.01}{3} = \frac{16.71}{3} = 5.57$$

For the *HSKG2* gene,

$$\bar{\bar{x}} = \frac{1.58 + 6.74 + 12.33}{3} = \frac{20.65}{3} = 6.88$$

We now calculate the between-groups variance (s^{*2}) for each gene as

$$s^{*2} = \frac{(\bar{a} - \bar{\bar{x}})^2 + (\bar{b} - \bar{\bar{x}})^2 + (\bar{c} - \bar{\bar{x}})^2}{m - 1}$$

Table 9.4 Variance calculations for the three time points from the temperature shift experiment for *HSKG1*, as described in Problem 9.11.

Hr	Well	C_T	2^{-C_T}	$\bar{x}$ ($\times 10^{-9}$)	$(x-\bar{x})$ ($\times 10^{-9}$)	$(x-\bar{x})^2$ ($\times 10^{-9}$)	$\sum(x-\bar{x})^2$ ($\times 10^{-9}$)	$s^2 \frac{\sum(x-\bar{x})^2}{n-1}$ ($\times 10^{-9}$)
0	A1	27.3	6.05×10^{-9}	5.61	0.44	0.1936	2.1124	0.5281
	B1	27.7	4.59×10^{-9}		−1.02	1.0404		
	C1	27.4	5.65×0^{-9}		0.04	0.0016		
	D1	27.2	6.49×10^{-9}		0.88	0.7744		
	E1	27.5	5.29×10^{-9}		−0.32	0.1024		
1	F1	27.6	4.92×10^{-9}	5.09	−0.17	0.0289	1.0786	0.2697
	G1	27.8	4.28×10^{-9}		−0.81	0.6561		
	H1	27.5	5.29×10^{-9}		0.20	0.0400		
	A2	27.4	5.65×10^{-9}		0.56	0.3136		
	B2	27.5	5.29×10^{-9}		0.20	0.0400		
2	C2	27.2	6.49×10^{-9}	6.01	0.48	0.2304	2.4333	0.6083
	D2	27.1	6.95×10^{-9}		0.94	0.8836		
	E2	27.6	4.92×10^{-9}		−1.09	1.1881		
	F2	27.4	5.65×10^{-9}		−0.36	0.1296		
	G2	27.3	6.05×10^{-9}		0.04	0.0016		

Table 9.5 Variance calculations for the three time points from the temperature shift experiment for *HSKG2*, as described in Problem 9.11.

Hr	Well	C_T	2^{-C_T}	$\bar{x}$ ($\times 10^{-9}$)	$(x-\bar{x})$ ($\times 10^{-9}$)	$(x-\bar{x})^2$ ($\times 10^{-9}$)	$\sum(x-\bar{x})^2$ ($\times 10^{-9}$)	$s^2 \frac{\sum(x-\bar{x})^2}{n-1}$ ($\times 10^{-9}$)
0	H2	29.3	1.51×10^{-9}	1.58	−0.07	0.0049	0.7631	0.1908
	A3	28.7	2.29×10^{-9}		0.71	0.5041		
	B3	29.5	1.32×10^{-9}		−0.26	0.0676		
	C3	29.2	1.62×10^{-9}		0.04	0.0016		
	D3	29.7	1.15×10^{-9}		−0.43	0.1849		
1	E3	27.1	6.95×10^{-9}	6.74	0.21	0.0441	5.0832	1.2708
	F3	27.3	6.05×10^{-9}		−0.69	0.4761		
	G3	27.2	6.49×10^{-9}		−0.25	0.0625		
	H3	26.8	8.56×10^{-9}		1.82	3.3124		
	A4	27.4	5.65×10^{-9}		−1.09	1.1881		
2	B4	26.5	10.53×10^{-9}	12.33	−1.80	3.2400	47.5375	11.8844
	C4	26.9	7.99×10^{-9}		−4.34	18.8356		
	D4	25.8	17.12×10^{-9}		4.79	22.9441		
	E4	26.3	12.10×10^{-9}		−0.23	0.0529		
	F4	26.1	13.90×10^{-9}		1.57	2.4649		

For the *HSKG1* data, we have (values are at 10^{-9})

$$
\begin{aligned}
s^{*2} &= \frac{(5.61 - 5.57)^2 + (5.09 - 5.57)^2 + (6.01 - 5.57)^2}{3 - 1} \\
&= \frac{(0.04)^2 + (-0.48)^2 + (0.44)^2}{2} \\
&= \frac{0.0016 + 0.2304 + 0.1936}{2} \\
&= \frac{0.4256}{2} \\
&= 0.2128
\end{aligned}
$$

For the *HSKG2* data, we have (values are at 10^{-9})

$$
\begin{aligned}
s^{*2} &= \frac{(1.58 - 6.88)^2 + (6.74 - 6.88)^2 + (12.33 - 6.88)^2}{3 - 1} \\
&= \frac{(-5.30)^2 + (-0.14)^2 + (5.45)^2}{2} \\
&= \frac{28.0900 + 0.0196 + 29.7025}{2} \\
&= \frac{57.8121}{2} \\
&= 28.9061
\end{aligned}
$$

We then calculate the average of the sample variances for each gene as

$$s^2 = \frac{s_a{}^2 + s_b{}^2 + s_c{}^2 + \ldots}{m}$$

where m is the number of groups.

For the *HSKG1* gene, we have (values are at 10^{-9})

$$s^2 = \frac{0.5281 + 0.2697 + 0.6083}{3} = \frac{1.4061}{3} = 0.4687$$

For the *HSKG2* gene, we have (values are at 10^{-9})

$$s^2 = \frac{0.1908 + 1.2708 + 11.8844}{3} = \frac{13.346}{3} = 4.4487$$

We are now ready to calculate the F statistic as

$$F = \frac{ns^{*2}}{s^2}$$

where n is the number of samples within each group (values are at 10^{-9}).

For the *HSKG1* gene,

$$F = \frac{5(0.2128)}{0.4687} = \frac{1.0640}{0.4687} = 2.2701$$

For the *HSKG2* gene,

$$F = \frac{5(28.9061)}{4.4487} = \frac{144.5305}{4.4487} = 32.4883$$

To find our cutoff F value in an F distribution table, we need to know the two degrees of freedom for our F statistic. For the numerator of the F statistic (the between-group variance), the degrees of freedom (df_n) is calculated as $m - 1$ where m is the number of groups. We have three groups and therefore two degrees of freedom. We use this value to tell us in which column of an F distribution table to find our cutoff value. For the denominator of the F statistic (the within-groups variance), the degrees of freedom (df_d) is calculated as $m(n - 1)$ where n is the number of members of each group. Since we have five samples for each of three groups, we have $3(5 - 1) = 12$ degrees of freedom. We use this value to tell us in which row of an F distribution table to find our cutoff value.

For 2 and 12 degrees of freedom, the F distribution for a 5% level of confidence shows a critical value of 3.89 (Figure 9.10). Therefore, at these degrees of freedom, the F statistic has a 95% chance of being less than 3.89 if the null hypothesis is true. Or, stated another way, only 5% of the time would the F statistic produce a value this extreme if the null hypothesis were true. Therefore, if the F statistic we calculated has a value lower than 3.89, we accept the null hypothesis and we can conclude that the experimental treatment has no effect on the gene's C_T value. If the F statistic is larger than this value, then we should reject the null hypothesis and conclude that the experimental treatment (the shift to 43°C) has an affect on the gene's expression. The use of such a gene as an endogenous control would not be a wise choice.

The F statistic for the *HSKG1* gene dataset is 2.2701 while that for the *HSKG2* gene dataset is 32.4883. Since, for *HSKG1*, an F statistic of 2.2701 is less than the critical value of 3.89, we can accept the null hypothesis that shifting the temperature to 43°C does not affect the expression of the *HSKG1* gene. However, since the F statistic of 32.4883 from the *HSKG2* gene dataset is larger than the 3.89 cutoff value, the null hypothesis does not apply to *HSKG2*. Use of *HSKG1*, therefore, is a better choice as a reference gene.

We can also find the p-value for each F statistic by making use of a p-value calculator found on the internet, as shown in Figure 9.11 below:

		df_n												
		2	**3**	**4**	**5**	**6**	**7**	**8**	**9**	**10**	**12**	**15**	**20**	**30**
df_d	**2**	19.00	19.16	19.25	19.30	19.33	19.35	19.37	19.38	19.40	19.41	19.43	19.45	19.46
	3	9.55	9.28	9.12	9.01	8.94	8.89	8.85	8.81	8.79	8.74	8.70	8.66	8.62
	4	6.94	6.59	6.39	6.26	6.16	6.09	6.04	6.00	5.96	5.91	5.86	5.80	5.75
	5	5.79	5.41	5.19	5.05	4.95	4.88	4.82	4.77	4.74	4.68	4.62	4.56	4.50
	6	5.14	4.76	4.53	4.39	4.28	4.21	4.15	4.10	4.06	4.00	3.94	3.87	3.81
	7	4.74	4.35	4.12	3.97	3.87	3.79	3.73	3.68	3.64	3.57	3.51	3.44	3.38
	8	4.46	4.07	3.84	3.69	3.58	3.50	3.44	3.39	3.35	3.28	3.22	3.15	3.08
	9	4.26	3.86	3.63	3.48	3.37	3.29	3.23	3.18	3.14	3.07	3.01	2.94	2.86
	10	4.10	3.71	3.48	3.33	3.22	3.14	3.07	3.02	2.98	2.91	2.85	2.77	2.70
	11	3.98	3.59	3.36	3.20	3.09	3.01	2.95	2.90	2.85	2.79	2.72	2.65	2.57
	12	3.89	3.49	3.26	3.11	3.00	2.91	2.85	2.80	2.75	2.69	2.62	2.54	2.47
	13	3.81	3.41	3.18	3.03	2.92	2.83	2.77	2.71	2.67	2.60	2.53	2.46	2.38
	14	3.74	3.34	3.11	2.96	2.85	2.76	2.70	2.65	2.60	2.53	2.46	2.39	2.31
	15	3.68	3.29	3.06	2.90	2.79	2.71	2.64	2.59	2.54	2.48	2.40	2.33	2.25
	16	3.63	3.24	3.01	2.85	2.74	2.66	2.59	2.54	2.49	2.42	2.35	2.28	2.19
	17	3.59	3.20	2.96	2.81	2.70	2.61	2.55	2.49	2.45	2.38	2.31	2.23	2.15
	18	3.55	3.16	2.93	2.77	2.66	2.58	2.51	2.46	2.41	2.34	2.27	2.19	2.11
	19	3.52	3.13	2.90	2.74	2.63	2.54	2.48	2.42	2.38	2.31	2.23	2.16	2.07
	20	3.49	3.10	2.87	2.71	2.60	2.51	2.45	2.39	2.35	2.28	2.20	2.12	2.04
	21	3.47	3.07	2.84	2.68	2.57	2.49	2.42	2.37	2.32	2.25	2.18	2.10	2.01

■ **FIGURE 9.10** The *F* distribution table for a 5% level of confidence with two degrees of freedom for the *F* statistic's numerator (df_n) and 12 degrees of freedom for the *F* statistic's denominator (df_d) shows a value of 3.89 (circled).

df numerator = 2 df denominator = 12 F = 2.2701 p = Compute	A number of sites on the Internet can be used to calculate *p*-values. Merely enter the degrees of freedom and the F statistic (left), click the Compute button, and the *p*-value will be calculated (right).	df numerator = 2 df denominator = 12 F = 2.2701 p = 0.14583 Compute

■ **FIGURE 9.11** A *p*-value calculator shows that the *F* statistic for the *HSKG1* gene has a *p*-value of 0.1483.

The *p*-value will tell us what the probability is that random sampling would lead to a difference between sample means as large (or larger) than that we observed. Since the *F* statistic for the *HSKG1* dataset has a *p*-value of 0.145 83 (which is greater than the 0.05 level of significance) (Figure 9.11), we can conclude that the experimental treatment has no statistically significant effect on the average C_T values we see. However, the *F* statistic for the *HSKG2* dataset has a *p*-value of 0.000 01. In other words, there is only a 0.001% chance that such an experiment would give an *F* statistic having a value of 32.4883 or greater if there were really no effect of the temperature shift on the mean C_T values for each group. Therefore, there is a statistically significant relationship between the experimental treatment and expression of *HSKG2*. Again, from the analysis of *p*-values, we should conclude that using *HSKG1* as an endogenous control is a far better idea than using *HSKG2*.

9.6.2 The $2^{-\Delta\Delta C_T}$ method – amplification efficiency

The gene you choose as the endogenous control in an experiment using the $\Delta\Delta C_T$ method should meet at least two requirements: (1) it should not be affected by the treatment the response of which is being measured and (2) it must amplify with an efficiency equal to that of the target gene. If the efficiency of the target and the endogenous control are approximately equal, the ΔC_T is proportional to the ratio of the initial copy number of the two genes.

Amplification efficiencies between genes can be assessed for similarity by calculating the differences in their average ΔC_T ($C_{T\ target} - C_{T\ reference}$) values across dilutions of template cDNA. Any differences in the ΔC_T values for varying amounts of input RNA can be graphed using Microsoft Excel a plot of ΔC_T vs. cDNA dilution. If the line of best fit (the regression line) to this plot has a slope close to zero, then it can be concluded that the two genes have similar amplification efficiencies and the $\Delta\Delta C_T$ method is a valid approach to measuring changes in gene expression of the target gene relative to the expression of the reference gene. As a general criterion, if the slope of the regression line is less than 0.1, the efficiencies are close enough to each other that use of the $\Delta\Delta C_T$ method is valid. Looked at another way, it does not matter if the two genes have drastically different C_T values. The key point is whether, if each gene's regression line were plotted on the same graph, they would run parallel to each other. If they do, they have similar amplification efficiencies.

Problem 9.12 *GeneA* (the target) is suspected of having function in the immune system's response to viral infection. In this experiment, its expression level will be measured following infection of cultured liver cells by the hepatitis C virus and that expression level will be compared to that of β-actin (the calibrator for this experiment) by the $2^{-\Delta\Delta C_T}$ (Comparative C_T) method. Since it is necessary that *GeneA* and β-actin have similar amplification efficiencies, an initial experiment is conducted to see if this is true. Two target genes can be shown to have the same amplification efficiency by assessing how their ΔC_T values vary with template dilution. RNA is isolated from cultured liver cells and amounts ranging from 0.01 to 1 ng are converted to cDNA by reverse transcriptase. Triplicate samples for *GeneA* and β-actin cDNAs are amplified by real-time PCR from each input RNA amount. The results in Table 9.6 are obtained. In the experiment described here, are *GeneA* and β-actin amplified with similar efficiencies?

Table 9.6 Dataset of C_T values for the experiment described in Problem 9.12.

Total Input RNA (ng)	*GeneA* C_T	Average *GeneA* C_T	β-actin C_T	Average β-actin C_T
1.000	28.71	28.71	26.60	26.59
1.000	28.72		26.55	
1.000	28.70		26.62	
0.500	29.62	29.63	27.55	27.57
0.500	29.64		27.57	
0.500	29.63		27.60	
0.250	30.65	30.62	28.57	28.60
0.250	30.63		28.63	
0.250	30.57		28.60	
0.100	31.82	31.82	29.75	29.76
0.100	31.85		29.80	
0.100	31.80		29.73	
0.050	32.82	32.82	30.80	30.78
0.050	32.84		30.78	
0.050	32.79		30.75	
0.025	33.75	33.76	31.78	31.75
0.025	33.78		31.75	
0.025	33.74		31.72	
0.010	35.02	35.01	33.05	33.01
0.010	35.05		32.89	
0.010	34.96		33.10	

Solution 9.12

It should not be unexpected that a series of liver cell cultures would give, for any defined set of conditions, a range of different C_T values when assayed for their levels of gene expression. When averages are reported for any dataset, no matter in what field: biology, chemistry, or physics, you will almost always see the reported average followed by a ' $\pm$ ' sign attached to some numerical amount. That sign indicates the **margin of error** for that result. It is an indication of the degree to which the data used to calculate the average value vary in range around that average.

How confident can you be, however, that a reported margin of error encompasses all of the variation possible within a dataset? That assurance is given by the **standard error** – a measure derived from the **standard deviation**, which, similarly to the margin of error, represents the average distance of the data from the mean. We are going to get to the margin or error via the standard deviation.

The standard deviation (s), as we have seen, is given by the formula

$$s = \sqrt{\frac{\Sigma(x - \overline{x})^2}{n - 1}}$$

where the Σ symbol represents the sum, x is a sample, $\overline{x}$ is the average (mean), and n is the number of measurements in the dataset.

The standard deviation is used to calculate the standard error by the following formula:

$$\text{standard error} = \frac{s}{\sqrt{n}}$$

where s is the standard deviation and n is the sample size.

Most datasets encountered in a statistical analysis of experiments dealing with real cause and effect huddle symmetrically on either side of the mean. With a large enough sample size, the data points should distribute themselves in the form of a bell-shaped curve where the mean value bisects that curve right down the middle. Many data points will have values close to the average. As you go further away from the mean in both the plus and minus directions, you will find fewer and fewer data points. Eighty percent of the values should fall within 1.28 standard errors of the mean. Ninety-five percent of the data should fall within 1.96 standard errors of the mean, and 99% of the data should fall with 2.58 standard errors of the mean. The percent values shown here are referred to as **confidence levels**. At the 99% confidence level, for example, you are

Table 9.7 Confidence levels and their corresponding Z-values.

Confidence level	Z-value
80%	1.28
90%	1.64
95%	1.96
98%	2.33
99%	2.58

confident, so to speak, that you have accounted for 99% of all possible outcomes within your sampling. The number of standard errors to add and subtract from the mean to give you the desired confidence level is called the **Z-value**. Confidence levels and their corresponding Z-values are shown in Table 9.7.

We can now calculate the margin of error of the sample average with a defined degree of confidence. The general formula is

$$\text{margin of error} = Z \times \frac{s}{\sqrt{n}}$$

where Z (the Z-value) is the number of standard errors associated with a particular confidence level, s is the standard deviation, and n is the number of samples in the dataset. If we want to have confidence that our margin of error contains 99% of all possible results, the equation becomes

$$\text{margin of error} = 2.58 \times \frac{s}{\sqrt{n}}$$

The standard deviations (s) of the C_T values of *Gene A* and β-actin cDNAs for each input RNA amount are calculated in Tables 9.8 and 9.9 below. These values will be used to determine the margin of error for each input RNA amount.

We can now calculate the margin of error for each input RNA amount for *Gene A* (Table 9.10) and β-actin (Table 9.11).

Now, we can determine the degree to which the average C_T values for *Gene A* and β-actin differ for each input RNA amount by subtracting one from the other (giving us the ΔC_T). However, each average C_T value comes with a standard error. How do we account for these standard error values

Table 9.8 Standard deviation calculations for *GeneA* C_T values for different amounts of input RNA converted to cDNA and amplified and detected by real-time polymerase chain reaction (PCR). Σ denotes sum. The sample size (n) for each input RNA amount is three. The mean values used for the third column were calculated in Table 9.6.

Input RNA (ng)	*GeneA* C_T (x)	$(x - \bar{x})$	$(x - \bar{x})^2$	$\sum(x - \bar{x})^2$	$\frac{\sum(x - \bar{x})^2}{n-1}$	$\sqrt{\frac{\sum(x - \bar{x})^2}{n-1}}$
1.000	28.71	0.00	0.0000	0.0002	0.0001	0.01
	28.72	0.01	0.0001			
	28.70	−0.01	0.0001			
0.500	29.62	−0.01	0.0001	0.0002	0.0001	0.01
	29.64	0.01	0.0001			
	29.63	0.00	0.0000			
0.250	30.65	0.03	0.0009	0.0035	0.0018	0.04
	30.63	0.01	0.0001			
	30.57	−0.05	0.0025			
0.100	31.82	0.00	0.0000	0.0013	0.0007	0.03
	31.85	0.03	0.0009			
	31.80	−0.02	0.0004			
0.050	32.82	0.00	0.0000	0.0013	0.0007	0.03
	32.84	0.02	0.0004			
	32.79	0.03	0.0009			
0.025	33.75	−0.01	0.0001	0.0009	0.0005	0.02
	33.78	0.02	0.0004			
	33.74	−0.02	0.0004			
0.010	35.02	0.01	0.0001	0.0042	0.0021	0.05
	35.05	0.04	0.0016			
	34.96	−0.05	0.0025			

Table 9.9 Standard deviation calculations for β-actin C_T values for different amounts of input RNA converted to cDNA and amplified and detected by real-time polymerase chain reaction (PCR). Σ denotes sum. The sample size (*n*) for each input RNA amount is three. The mean values used for the third column were calculated in Table 9.6.

Input RNA (ng)	β-actin C_T (*x*)	$(x - \bar{x})$	$(x - \bar{x})^2$	$\sum(x - \bar{x})^2$	$\frac{\sum(x - \bar{x})^2}{n-1}$	$\sqrt{\frac{\sum(x - \bar{x})^2}{n-1}}$
1.000	26.60	0.01	0.0001	0.0026	0.0013	0.04
	26.55	−0.04	0.0016			
	26.62	0.03	0.0009			
0.500	27.55	−0.02	0.0004	0.0013	0.0007	0.03
	27.57	0.00	0.0000			
	27.60	0.03	0.0009			
0.250	28.57	−0.03	0.0009	0.0018	0.0009	0.03
	28.63	0.03	0.0009			
	28.60	0.00	0.0000			
0.100	29.75	−0.01	0.0001	0.0026	0.0013	0.04
	29.80	0.04	0.0016			
	29.73	−0.03	0.0009			
0.050	30.80	0.02	0.0004	0.0013	0.0007	0.03
	30.78	0.00	0.0000			
	30.75	−0.03	0.0009			
0.025	31.78	0.03	0.0009	0.0018	0.0009	0.04
	31.75	0.00	0.0000			
	31.72	−0.03	0.0009			
0.010	33.05	0.04	0.0016	0.0241	0.0121	0.11
	32.89	−0.12	0.0144			
	33.10	0.09	0.0081			

Table 9.10 Margin of error calculations for *GeneA* C_T values at different amounts of input RNA at a 99% confidence level. The standard deviation (s) values were calculated in Table 9.8. The sample size (n) for each input RNA amount is three.

Input RNA (ng)	*GeneA* s	$\frac{s}{\sqrt{n}}$	$2.58 \times \frac{s}{\sqrt{n}}$
1.000	0.01	0.0058	0.01
0.500	0.01	0.0058	0.01
0.250	0.04	0.0231	0.06
0.100	0.03	0.0173	0.04
0.050	0.03	0.0173	0.04
0.025	0.02	0.0116	0.03
0.010	0.05	0.0289	0.07

Table 9.11 Margin of error calculations for β-actin C_T values at different amounts of input RNA at a 99% confidence level. The standard deviation (s) values were calculated in Table 9.9. The sample size (n) for each input RNA amount is three.

Input RNA (ng)	β-actin s	$\frac{s}{\sqrt{n}}$	$2.58 \times \frac{s}{\sqrt{n}}$
1.000	0.04	0.0231	0.06
0.500	0.03	0.0173	0.04
0.250	0.03	0.0173	0.04
0.100	0.04	0.0231	0.06
0.050	0.03	0.0173	0.04
0.025	0.04	0.0231	0.06
0.010	0.11	0.0636	0.16

when we report the difference between two numbers? The method we use is called **propagation of errors**.

When numbers having margins of error are either added or subtracted, the new margin of error is calculated as the square root of the sum of each margin of error squared:

$$\text{margin of error of sum or difference} = \sqrt{(\text{margin of error}_1)^2 + (\text{margin of error}_2)^2 + \ldots + (\text{margin of error}_n)^2}$$

When quantities having margins of error are multiplied or divided, the new margin of error is calculated as the square root of the sum of each margin of error divided by the quantity to which it is attached squared:

$$\text{margin of error of product or quotient} = \sqrt{\left(\frac{\text{margin of error}_1}{\text{quantity}_1}\right)^2 + \left(\frac{\text{margin of error}_2}{\text{quantity}_2}\right)^2 + \ldots + \left(\frac{\text{margin of error}_n}{\text{quantity}_n}\right)^2}$$

For the 1 ng input RNA sample, we will be subtracting the β-actin average C_T (26.59 ± 0.06) from the *GeneA* average C_T (28.71 ± 0.01). To propagate the error, we have

$$\begin{aligned}\text{propagated error} &= \sqrt{(0.01)^2 + (0.06)^2}\\ &= \sqrt{0.0001 + 0.0036}\\ &= \sqrt{0.0037}\\ &= 0.06\end{aligned}$$

The margins of error for the ΔC_T values for the remaining samples shown in Table 9.12 are calculated in the same manner.

We can now plot the ΔC_T values determined in Table 9.12 using Microsoft Excel as described in Appendix A. However, we must also include each

Table 9.12 ΔC_T values for each input RNA amount. The average C_T value of β-actin is subtracted from the average C_T value of GeneA. The attending margin of error for each ΔC_T is calculated by propagation of errors (see explanation above).

Input RNA (ng)	*GeneA* Average C_T	β-actin Average C_T	ΔCT (Average *GeneA* C_T – Average (β-actin C_T)
1.000	28.71 0.01	26.59 0.06	2.12 0.06
0.500	29.63 0.01	27.57 0.04	2.06 0.04
0.250	30.62 0.06	28.60 0.04	2.02 0.07
0.100	31.82 0.04	29.76 0.06	2.06 0.07
0.050	32.82 0.04	30.78 0.04	2.04 0.06
0.025	33.76 0.03	31.75 0.06	2.01 0.07
0.010	33.01 0.07	33.01 0.16	2.00 0.17

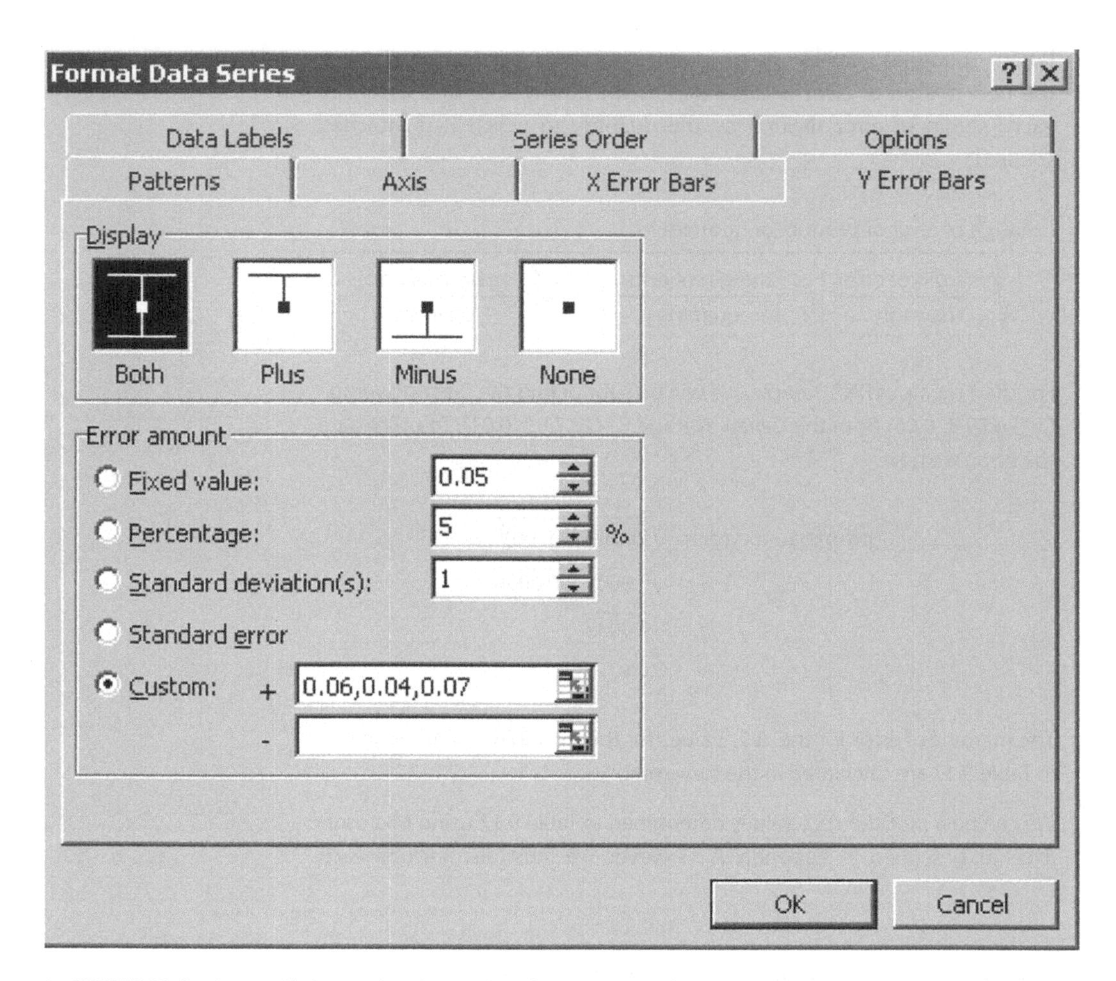

■ **FIGURE 9.12** Error bars are added to a graph produced in Microsoft Excel by using the **Format Data Series** dialogue box (obtained by clicking on one of the data points in the plot).

value's attending margin or error through the use of error bars. These can be added to a graph by completing the Excel dialog box shown in Figure 9.12.

In Figure 9.13, the ΔC_T values with their margins of error (calculated in Table 9.12) are plotted using Microsoft Excel. For this experiment, since the absolute value of the slope of the line of best fit is close to zero (0.0463), the target (*GeneA*) and reference (β-actin) genes have similar amplification efficiencies and they can be used accordingly in a gene expression assay using the $2^{-\Delta\Delta C_T}$ method.

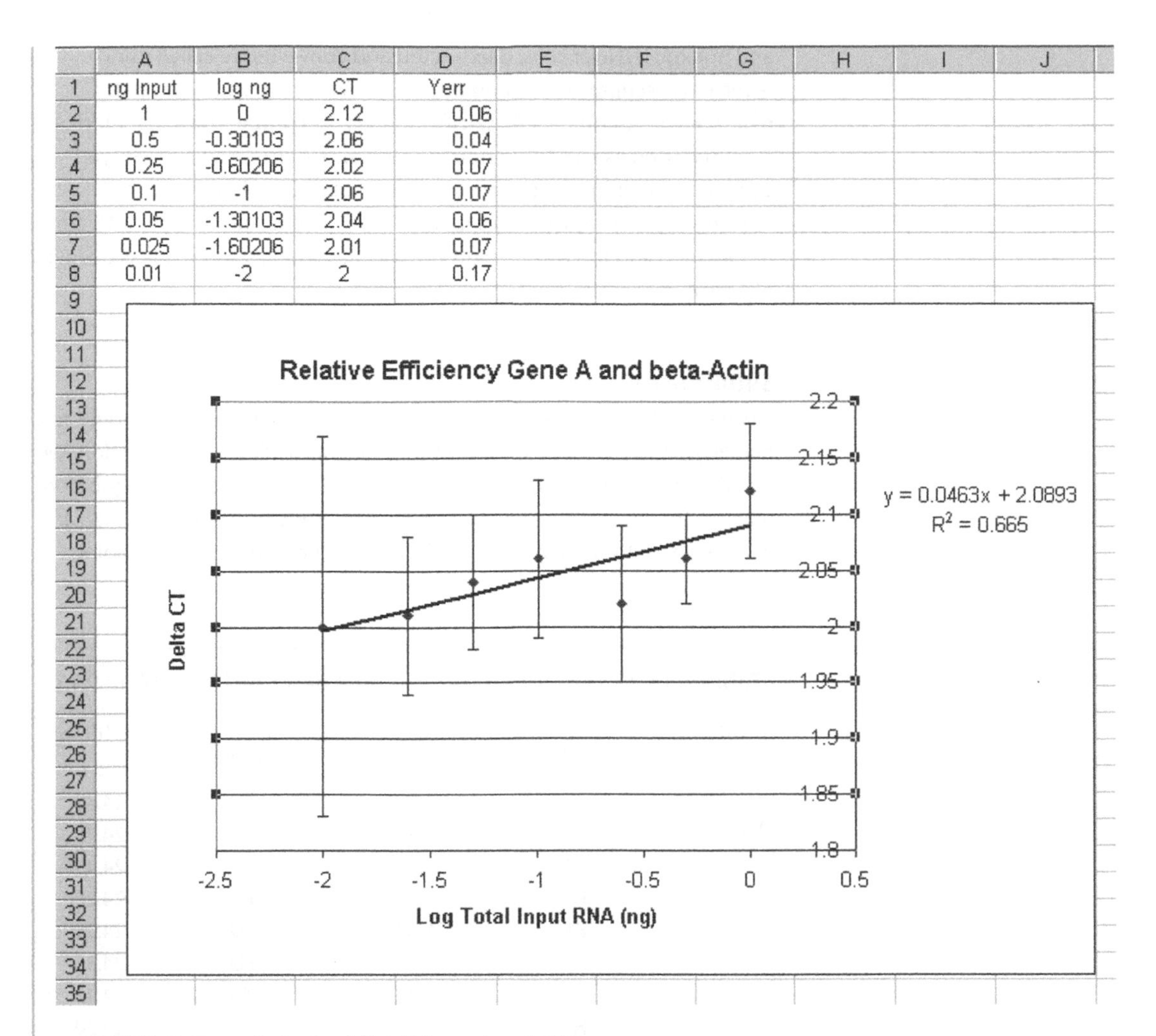

	A	B	C	D
1	ng Input	log ng	CT	Yerr
2	1	0	2.12	0.06
3	0.5	-0.30103	2.06	0.04
4	0.25	-0.60206	2.02	0.07
5	0.1	-1	2.06	0.07
6	0.05	-1.30103	2.04	0.06
7	0.025	-1.60206	2.01	0.07
8	0.01	-2	2	0.17

■ **FIGURE 9.13** Plotting the data from Table 9.12 shows a slope of 0.0463.

9.6.3 The $2^{-\Delta\Delta C_T}$ method – is the reference gene affected by the experimental treatment?

Problem 9.13 A series of 10 liver tissue culture samples are exposed to the Hepatitis C virus. As a control, a second series of 10 liver cell cultures remain unexposed. The uninfected cells serve as the calibrator. Two hours after exposure, total RNA from both sets of samples, exposed liver cells and unexposed liver cells, is extracted and converted to cDNA using reverse transcriptase. The amount of *GeneA* expression is assayed by real-time PCR using *GeneA*-specific primers and probe. The expression of β-actin is used as an endogenous control and is amplified and assayed using its own unique set of primers and probe. *GeneA* and β-actin cDNAs are amplified and assayed in separate tubes. The results in Table 9.13 are obtained. Using the $2^{-\Delta\Delta C_T}$ method, calculate the relative expression level of *GeneA* in hepatitis C-infected cells.

Solution 9.13

Since *GeneA* and β-actin are assayed in separate reactions, we should not pair any particular *GeneA* reaction with any particular β-actin reaction. We will therefore average the C_T values of the two genes separately before making the ΔC_T calculation. The average, or mean (designated as $\bar{x}$), is calculated as the sum of the data divided by the number of data points.

Table 9.13 C_T values for the experiment described in Problem 9.13.

Cells	Well number	*GeneA* C_T	Well number	β-actin C_T
Uninfected liver cells	A1	31.65	C2	24.32
	B1	31.37	D2	24.35
	C1	31.72	E2	24.45
	D1	31.51	F2	24.28
	E1	31.80	G2	24.25
	F1	31.68	H2	24.36
	G1	31.47	A3	24.42
	H1	31.59	B3	24.20
	A2	31.78	C3	24.52
	B2	31.55	D3	24.30
Hepatitis C-infected liver cells	E3	23.15	G4	25.87
	F3	22.89	H4	25.93
	G3	22.58	A5	25.75
	H3	23.05	B5	25.90
	A4	22.64	C5	25.72
	B4	22.78	D5	25.81
	C4	23.21	E5	26.03
	D4	22.92	F5	25.85
	E4	22.84	G5	25.87
	F4	22.73	H5	25.95

Average C_T of *GeneA* in uninfected liver cells:

$$\frac{\left(\begin{array}{l}31.65+31.37+31.72+31.51+31.80+31.68\\+31.47+31.59+31.78+31.55\end{array}\right)}{10}=31.61$$

Average C_T of β-actin in uninfected liver cells:

$$\frac{\left(\begin{array}{l}24.32+24.35+24.45+24.28+24.25+24.36\\+24.42+24.20+24.52+24.30\end{array}\right)}{10}=24.35$$

Average C_T of *GeneA* in hepatitis C-infected liver cells:

$$\frac{\left(\begin{array}{l}23.15+22.89+22.58+23.05+22.64+22.78\\+23.21+22.92+22.84+22.73\end{array}\right)}{10}=22.88$$

Average C_T of β-actin in hepatitis C-infected liver cells:

$$\frac{\left(\begin{array}{l}25.78+25.93+25.75+25.90+25.72+25.81\\+26.03+25.85+25.87+25.95\end{array}\right)}{10}=25.87$$

As we did in Problem 9.12, we will determine the standard error for this dataset by first calculating the standard deviation.

Again, the standard deviation (s) is given by the formula

$$s=\sqrt{\frac{\sum(x-\overline{x})^2}{n-1}}$$

where the Σ symbol represents the sum, x is a sample, $\overline{x}$ is the average (mean), and n is the number of measurements in the dataset.

The standard deviation is used to calculate the standard error by the following formula:

$$\text{standard error}=\frac{s}{\sqrt{n}}$$

where s is the standard deviation and n is the sample size.

The margin of error of the sample average with a defined degree of confidence is calculated using the following equation:

$$\text{margin of error} = Z \times \frac{s}{\sqrt{n}}$$

where Z (the Z-value) is the number of standard errors associated with a particular confidence level, s is the standard deviation, and n is the number of samples in the dataset. For a margin of error containing 99% of all possible results, the equation becomes

$$\text{margin of error} = \pm 2.58 \times \frac{s}{\sqrt{n}}$$

To arrive at our margin of error for each dataset, we will first calculate their standard deviations. Tables 9.14–9.17 show the derivation involved in these calculations.

We can now calculate the standard deviation values. Each dataset has a sample size (n) of 10.

Table 9.14 Components of the standard deviation calculation for C_T values obtained for *GeneA* in uninfected liver cells. The mean value for this dataset (calculated previously) is 31.61.

***GeneA* C_T in uninfected cells (x)**	**$(x - \bar{x})$**	**$(x - \bar{x})^2$**
31.65	0.04	0.0016
31.37	−0.24	0.0576
31.72	0.11	0.0121
31.51	−0.10	0.0100
31.80	0.19	0.0361
31.68	0.07	0.0049
31.47	−0.14	0.0196
31.59	−0.02	0.0004
31.78	0.17	0.0289
31.55	−0.06	0.0036
Total		**0.1748**

Table 9.15 Components of the standard deviation calculation for C_T values obtained for β-actin in uninfected liver cells. The mean value for this dataset (calculated previously) is 24.35.

β-actin C_T in uninfected cells (x)	**$(x - \bar{x})$**	**$(x - \bar{x})^2$**
24.32	−0.03	0.0009
24.35	0.00	0.0000
24.45	0.10	0.0100
24.28	−0.07	0.0049
24.25	−0.10	0.0100
24.36	0.01	0.0001
24.42	0.07	0.0049
24.20	−0.15	0.0225
24.52	0.17	0.0289
24.30	−0.05	0.0025
Total		**0.0847**

Table 9.16 Components of the standard deviation calculation for C_T values obtained for *GeneA* in hepatitis C virus-infected liver cells. The mean value for this dataset (calculated previously) is 22.88.

***GeneA* C_T in virus-infected liver cells (x)**	**$(x - \bar{x})$**	**$(x - \bar{x})^2$**
23.15	0.27	0.0729
22.89	0.01	0.0001
22.58	−0.30	0.0900
23.05	0.17	0.0289
22.64	−0.24	0.0576
22.78	−0.10	0.0100
23.21	0.33	0.1089
22.92	0.04	0.0016
22.84	−0.04	0.0016
22.73	−0.15	0.0225
Total		**0.3941**

Table 9.17 Components of the standard deviation calculation for C_T values obtained for β-actin in hepatitis C virus-infected liver cells. The mean value for this dataset (calculated previously) is 25.87.

β-actin C_T in virus-infected liver cells (x)	**$(x - \bar{x})$**	**$(x - \bar{x})^2$**
25.87	0.00	0.0000
25.93	0.06	0.0036
25.75	−0.12	0.0144
25.90	0.03	0.0009
25.72	−0.15	0.0225
25.81	−0.06	0.0036
26.03	0.16	0.0256
25.85	−0.02	0.0004
25.87	0.00	0.0000
25.95	0.08	0.0064
Total		**0.0774**

GeneA in uninfected liver cells:

$$s = \sqrt{\frac{\sum(x - \bar{x})^2}{n-1}} = \sqrt{\frac{0.1748}{10-1}} = \sqrt{\frac{0.1748}{9}} = \sqrt{0.0194} = 0.14$$

β-actin in uninfected liver cells:

$$s = \sqrt{\frac{\sum(x - \bar{x})^2}{n-1}} = \sqrt{\frac{0.0847}{10-1}} = \sqrt{\frac{0.0847}{9}} = \sqrt{0.0094} = 0.10$$

GeneA in virus-infected liver cells:

$$s = \sqrt{\frac{\sum(x - \bar{x})^2}{n-1}} = \sqrt{\frac{0.3941}{10-1}} = \sqrt{\frac{0.3941}{9}} = \sqrt{0.0438} = 0.21$$

β-actin in virus-infected liver cells:

$$s = \sqrt{\frac{\sum(x - \bar{x})^2}{n-1}} = \sqrt{\frac{0.0774}{10-1}} = \sqrt{\frac{0.0774}{9}} = \sqrt{0.0086} = 0.09$$

With the standard deviation values in hand, the margin of errors can be calculated. We will use a confidence level of 99% ($Z = 2.58$). The sample size (n) for each dataset is 10.

GeneA in uninfected liver cells:

$$\text{margin of error} = Z \times \frac{s}{\sqrt{n}} = 2.58 \times \frac{0.14}{\sqrt{10}} = 2.58 \times \frac{0.14}{3.16} = 0.11$$

β-actin in uninfected liver cells:

$$\text{margin of error} = Z \times \frac{s}{\sqrt{n}} = 2.58 \times \frac{0.10}{\sqrt{10}} = 2.58 \times \frac{0.10}{3.16} = 0.08$$

GeneA in virus-infected liver cells:

$$\text{margin of error} = Z \times \frac{s}{\sqrt{n}} = 2.58 \times \frac{0.21}{\sqrt{10}} = 2.58 \times \frac{0.21}{3.16} = 0.17$$

β-actin in virus-infected liver cells:

$$\text{margin of error} = Z \times \frac{s}{\sqrt{n}} = 2.58 \times \frac{0.09}{\sqrt{10}} = 2.58 \times \frac{0.09}{3.16} = 0.07$$

We can now attach the margin of error values to the average C_T values for each dataset:

Average C_T of *GeneA* in uninfected liver cells = 31.61 ± 0.11.
Average C_T of β-actin in uninfected liver cells = 24.35 ± 0.08.
Average C_T of *GeneA* in hepatitis C-infected liver cells = 22.88 ± 0.17.
Average C_T of β-actin in hepatitis C-infected liver cells = 25.87 ± 0.07.

The ΔC_T of the calibrator sample is calculated as the average C_T of *Gene A* in uninfected liver cells minus the average C_T of β-actin in uninfected liver cells. We will first calculate the margin of error for the calibrator ΔC_T. The average C_T value of *Gene A* in uninfected liver cells has a margin of error of ± 0.11. The average C_T of β-actin in the uninfected liver cells has a margin of error of ± 0.08. Using propagation of errors (see Problem 9.12), the margin of error of the difference between these two average C_T values is

$$\text{margin of error} = \sqrt{(0.11)^2 + (0.08)^2} = \sqrt{0.0121 + 0.0064}$$
$$= \sqrt{0.0185} = 0.14$$

Therefore, we will attach a margin of error of ±0.14 to the ΔC_T value for the calibrator. It is calculated as:

$$\Delta C_{T,\,\text{calibrator}} = 31.61 \pm 0.11 - 24.35 \pm 0.08 = 7.26 \pm 0.14$$

The $\Delta\Delta C_T$ for the uninfected cells (the calibrator) is calculated as the average ΔC_T for this dataset minus the average ΔC_T for the uninfected cells. We are essentially subtracting this value from itself. The result will be zero, by definition:

$$7.26 - 7.26 = 0.00$$

However, we perform this step because of the margin of error values attached to those numbers. When we calculate the $2^{-\Delta\Delta CT}$ value for the calibrator, the propagated margin of error will show us the range of normalized values. The ΔC_T for the uninfected cells has a margin of error of ± 0.14. To propagate the error, we have

$$\sqrt{(0.14)^2 + (0.14)^2} = \sqrt{0.0196 + 0.0196}\sqrt{0.0392} = 0.20$$

Therefore, our $\Delta\Delta C_T$ value for uninfected cells is expressed as 0.00 ± 0.20.

The normalized *GeneA* amount relative to itself in uninfected cells using $2^{-\Delta\Delta CT}$, by definition, is unity (equal to 1.0):

$$2^{-(0.00)} = 1.0$$

However, we can calculate a range of normalized values using the propagated error of ± 0.20. We will take 2 to the exponent $-(-0.2)$ (derived from $0.00 - 0.20 = -0.2$) and 2 to the exponent $-(0.2)$ (derived from $0.00 + 0.20 = 0.2$). We have

$$2^{-\Delta\Delta C_T} = 2^{-(-0.2)} = 2^{0.2} = 1.1$$

and

$$2^{-\Delta\Delta C_T} = 2^{-(0.2)} = 0.9$$

Therefore, the normalized *GeneA* amount in uninfected cells relative to itself has a range of from 0.9 to 1.1.

The ΔC_T of the target (*GeneA*) in infected cells is calculated as the average C_T of *GeneA* in hepatitis C-infected liver cells minus the average C_T of β-actin in hepatitis C-infected liver cells. The average C_T value of *GeneA* in virus-infected cells (22.88) has a margin of error of ± 0.17. The average C_T value of β-actin in virus-infected cells (25.87) has a margin of error of ± 0.07. The margin of error of the ΔC_T will be

$$\begin{aligned} \text{margin of error} &= \sqrt{(0.17)^2 + (0.07)^2} \\ &= \sqrt{0.0289 + 0.0049} = \sqrt{0.0338} = 0.18 \end{aligned}$$

Therefore, we will attach a margin of error of ± 0.18 to the value of ΔC_T for the target gene. It is calculated as

$$\Delta C_{T.\,target} = 22.88 \pm 0.17 - 25.87 \pm 0.07 = -2.99 \pm 0.18$$

We can now calculate the $\Delta\Delta C_T$ as the average ΔC_T of the target minus the average ΔC_T of the calibrator:

$$\Delta\Delta C_T = -2.99 - 7.26 = -10.25$$

We propagate the error as follows:

$$\text{margin of error} = \sqrt{(0.18)^2 + (0.14)^2}$$
$$= \sqrt{0.0324 + 0.0196} = \sqrt{0.0520} = 0.23$$

Therefore, we can report the $\Delta\Delta C_T$ value as -10.25 ± 0.23.

We take the change in expression of *GeneA* in uninfected cells as 2^0, which is equivalent to 1. The change in expression of our target gene in hepatitis C-infected liver cells can now be calculated as $2^{-\Delta\Delta C_T}$ where $\Delta\Delta C_T$ is -10.25.

$$2^{-\Delta\Delta C_T} = 2^{-(-10.25)} = 1218$$

Therefore, there is a 1218-fold change in the expression of *GeneA* in hepatitis C-infected liver cells compared to its expression level in uninfected liver cells. We still, however, need to account for the margin of error. We can use the margin of error to determine a range of values for $2^{-\Delta\Delta C_T}$. In this experiment, the $\Delta\Delta C_T$ values, based on the margin of error calculations, will range from -10.48 (-10.25 minus the margin of error of 0.23) to -10.02 (-10.25 plus 0.23):

$$2^{-\Delta\Delta C_T} = 2^{-(-10.48)} = 1428$$
$$2^{-\Delta\Delta C_T} = 2^{-(-10.02)} = 1038$$

Therefore, we can report that *GeneA* is expressed 1218-fold more in hepatitis C-infected liver cells than in uninfected cells with a range from 1038- to 1428-fold difference. Notice, however, that this range is not symmetrical about the average $\Delta\Delta C_T$ value of 1218 (1428 is 210-fold differences away from 1218 while 1038 is 180-fold differences away). Since C_T is related to copy number exponentially, the asymmetry of range is an artifact of calculating a linear comparison of amounts from results derived from an exponential process.

Note: If the $\Delta\Delta C_T$ value is positive (+4, for example), then

$$2^{-\Delta\Delta CT} = 2^{-(4.0)} = 0.0625$$

and we can say that the sample has 0.0625, or 1/16, the amount of target RNA as the calibrator.

Problem 9.14 We are conducting an experiment to measure the difference in expression of *GeneA* (the target gene) in two different tissue types, liver and muscle. β-actin will serve as the reference gene and muscle cells will serve as the calibrator. Total RNA is recovered from both types of tissue and 2 μg of total RNA from each tissue type is converted to cDNA by reverse transcriptase. Five replicate wells from each tissue type contain cDNA derived from 20 ng total RNA. A probe for *GeneA* is labeled with a FAM (blue) dye and a probe for β-actin is labeled with VIC (a green dye). Since each gene can be assayed for using their own distinguishable reporter dyes, *GeneA* and β-actin are amplified and assayed simultaneously in each well. The results shown in Table 9.18 are obtained. Calculate the fold difference in the expression of *GeneA* in liver cells as compared to the expression of *GeneA* in muscle cells using the $2^{-\Delta\Delta CT}$ method.

Table 9.18 C_T values for cDNA samples made from total RNA isolated from liver and muscle cells. *GeneA* and β-actin were amplified and assayed in the same well for five replicate samples from each tissue type.

Tissue	Well number	*GeneA* C_T	β-actin C_T
Liver	A1	28.24	26.44
	B1	28.32	26.37
	C1	28.28	26.56
	D1	28.36	26.52
	E1	28.34	26.48
Muscle	F1	34.81	25.48
	G1	34.65	25.45
	H1	34.72	25.62
	A2	34.58	25.54
	B2	34.84	25.56

Solution 9.14

In Problem 9.13, the target gene and the reference gene are amplified and assayed in different wells of a microtiter plate in replicate. We therefore average the C_T values of the target and reference before calculating the ΔC_T. In this problem, however, the target and reference are amplified and assayed in the same well in replicate. Because each well has the same amount of cDNA (made from 20 ng of input RNA per well), we will calculate the ΔC_T for each well separately. We will then take the average of these C_T values to calculate the $\Delta\Delta C_T$ and the value for $2^{-\Delta\Delta CT}$. We will also calculate the standard error for each average ΔC_T and for the $\Delta\Delta C_T$ value.

The ΔC_T values are calculated as the *GeneA* C_T minus the β-actin C_T for each well. These results are shown in Table 9.19. This table also shows the average ΔC_T for each tissue type.

To arrive at our margin of error for each dataset, we will first calculate their standard deviations. Tables 9.20 and 9.21 show the calculations for the '$(x - \bar{x})^2$' component of the standard deviation calculations shown in Table 9.22.

We can now calculate the standard deviations for the average ΔC_T values. Each dataset has a sample size (n) of 5. With the standard deviation values in hand, the margins of error can be calculated. We will use a confidence level of 99% ($Z = 2.58$). The sample size (n) for each dataset is 10 (Table 9.23).

Table 9.19 ΔC_T calculations for the Problem 9.14 data. The average (mean) ΔC_T is calculated as the sum of the individual ΔC_T values divided by five (the number of samples for each tissue type).

Tissue type	Well number	*GeneA* C_T	β-actin C_T	ΔC_T
Liver	A1	28.24	26.44	1.80
	B1	28.32	26.37	1.95
	C1	28.28	26.56	1.72
	D1	28.36	26.52	1.84
	E1	28.34	26.48	1.86
			Average =	**1.83**
Muscle	F1	34.81	25.48	9.33
	G1	34.65	25.45	9.20
	H1	34.72	25.62	9.10
	A2	34.58	25.54	9.04
	B2	34.84	25.56	9.28
			Average =	**9.19**

Table 9.20 Components of the standard deviation calculation for ΔC_T values obtained for liver cells. The mean value for this dataset is 1.83 (calculated in Table 9.19).

ΔC_T from Liver Cell Samples (x)	$(x - \bar{x})$	$(x - \bar{x})^2$
1.80	−0.03	0.0009
1.95	0.12	0.0144
1.72	−0.11	0.0121
1.84	0.01	0.0001
1.86	0.03	0.0009
Total		**0.0284**

Table 9.21 Components of the standard deviation calculation for C_T values obtained for muscle cells. The mean value for this dataset is 9.19 (calculated in Table 9.19).

ΔC_T from Muscle Cell Samples (x)	$(x - \bar{x})$	$(x - \bar{x})^2$
9.33	0.23	0.0529
9.20	0.01	0.0001
9.10	−0.09	0.0081
9.04	−0.15	0.0225
9.28	0.09	0.0081
Total		**0.0917**

Table 9.22 Calculations for the standard deviation values for the ΔC_T associated with the two different tissue types.

ΔC_T from Liver Cells:

$$s = \sqrt{\frac{\sum(x - \bar{x})^2}{n-1}}$$

$$= \sqrt{\frac{0.0284}{5-1}} = \sqrt{\frac{0.0284}{4}} = \sqrt{0.0071} = 0.08$$

ΔC_T from Muscle Cells:

$$s = \sqrt{\frac{\sum(x - \bar{x})^2}{n-1}}$$

$$= \sqrt{\frac{0.0917}{5-1}} = \sqrt{\frac{0.0917}{4}} = \sqrt{0.0229} = 0.15$$

Table 9.23 Calculations for the margin of error for the ΔC_T values associated with the two different tissue types.

ΔC_T from Liver Cells:

$$\text{margin of error} = Z \times \frac{s}{\sqrt{n}} = 2.58 \times \frac{0.08}{\sqrt{5}} = 2.58 \times \frac{0.08}{2.24} = 0.04$$

ΔC_T from Muscle Cells:

$$\text{margin of error} = Z \times \frac{s}{\sqrt{n}} = 2.58 \times \frac{0.15}{\sqrt{5}} = 2.58 \times \frac{0.15}{2.24} = 0.07$$

We can now attach the margin of error values to the average ΔC_T values for each tissue type dataset:

Average ΔC_T for liver cells $= 1.83 \pm 0.04$.
Average ΔC_T for muscle cells $= 9.19 \pm 0.07$.

The $\Delta\Delta C_T$ value can now be calculated by subtracting the average ΔC_T of the reactions from muscle (the calibrator) from the average ΔC_T value associated with each tissue type. For liver, we have

$$1.83 - 9.19 = -7.36$$

and for muscle, we have

$$9.19 - 9.19 = 0.00$$

Since our average ΔC_T values, however, have associated margin of error values, we must propagate those values and include them in the answer. As described in Problem 9.12, we propagate the error for the ΔC_T liver minus ΔC_T muscle as

$$\text{margin of error} = \sqrt{(0.04)^2 + (0.07)^2} = \sqrt{0.0016 + 0.0049} = \sqrt{0.0065} = 0.08$$

The $\Delta\Delta C_T$ value for liver cells is therefore expressed as

$$-7.36 \pm 0.08$$

The margin of error associated with the $\Delta\Delta C_T$ for ΔC_T muscle minus ΔC_T muscle is

$$\begin{aligned}\text{margin of error} &= \sqrt{(0.07)^2 + (v)^2} \\ &= \sqrt{0.0049 + 0.0049} = \sqrt{0.0098} = 0.10\end{aligned}$$

The $\Delta\Delta C_T$ value for muscle cells is therefore expressed as

$$0.00 \pm 0.10$$

We can now calculate the normalized *GeneA* amounts relative to their expression level in muscle using the $2^{-\Delta\Delta CT}$ equation. For *GeneA* expression in liver cells, we have

$$2^{-\Delta\Delta C_T} = 2^{-(-7.36)} = 2^{7.36} = 164.3$$

Since we have included a margin of error measurement (± 0.08), we will find the range of fold change that is bounded by this margin of error by taking 2 to the exponent $-(-7.44)$ (since $-(-7.36 - 0.08) = -(-7.44)$) and 2 to the exponent $-(-7.28)$ (since $-(-7.36 + 0.08) = -(-7.28)$). We have

$$2^{-\Delta\Delta C_T} = 2^{-(-7.44)} = 2^{7.44} = 173.6$$

and

$$2^{-\Delta\Delta C_T} = 2^{-(-7.28)} = 2^{7.28} = 155.4$$

Therefore, *GeneA* is expressed some 164.3-fold more in liver cells than in muscle cells and has a range of from 155.4- to 173.6-fold more.

GeneA expression relative to itself in muscle cells (the calibrator) will have a fold change of unity ($2^0 = 1.0$), by definition. However, since there is a margin of error associated with this value, we will also calculate its range by taking 2 to the exponent $-(-0.1)$ (since $-(0.00 - 0.10) = -(-0.1)$) and 2 to the exponent $-(0.1)$ (since $-(0.00 + 0.10) = -(0.1)$):

$$2^{-(-0.1)} = 2^{0.1} = 1.1$$

and

$$2^{-0.1} = 0.9$$

Therefore, normalized *GeneA* amount in muscle relative to muscle has a range of 0.9- to 1.1-fold.

Problem 9.15 In an experiment similar to that described in Problem 9.14, the expression of *GeneA* in liver cells in response to hepatitis C infection is measured. β-actin serves as an endogenous control to which *GeneA* activity is normalized. Triplicate samples of liver cells are taken at times 0, 1, and 2 hr post infection and cDNA prepared. *GeneA* is assayed in real-time PCRs using a FAM (blue)-labeled probe. By a multiplex reaction for each sample (the reactions for each gene are performed in the same well), β-actin is assayed by real-time PCR using a VIC (green)-labeled probe. The zero time point serves as the calibrator sample. The data in Table 9.24 is collected. What is the fold change in the expression of *Gene A* normalized to the expression of β-actin and relative to time zero for 1 and 2 hr post exposure to hepatitis C virus?

Solution 9.15

The calibrator for this experiment is the expression of *GeneA* normalized to β-actin at time zero. We therefore need to calculate the mean C_T value for *GeneA* at time zero and the mean C_T value for β-actin at time zero. These values will be subtracted from each other to give the ΔC_T calibrator, which will be further subtracted from the ΔC_T of each sample to yield the $\Delta\Delta C_T$.

The Mean $2^{-\Delta\Delta C_T}$ value calculated in the last column of Table 9.25 represents the mean fold change in gene expression. At 1 hr, therefore, *GeneA* is expressed at a 28.8-fold increase over time zero. By 2 hr, *GeneA* expression has increased by 2719.7-fold.

Table 9.24 Data collected for a real-time PCR experiment to measure the effect of hepatitis C infection on the expression of *GeneA* in liver cells.

Sample	Well number	Time	*GeneA* C_T	β-actin C_T
1	A1	0	27.8	27.4
2	B1	0	28.2	27.2
3	C1	0	28.3	27.0
4	D1	1 hr	24.1	27.9
5	E1	1 hr	23.8	28.0
6	F1	1 hr	24.0	27.8
7	G1	2 hr	17.5	28.2
8	H1	2 hr	17.6	27.9
9	A2	2 hr	17.9	28.4

Table 9.25 The ΔC_T of the target at t = n is calculated as (*GeneA* C_T) − (β-actin C_T) for each well at each time point. The $\Delta\Delta C_T$ is calculated as ((*GeneA* C_T) − (β-actin C_T)$_{Time\ n}$) − ((mean C_T of *GeneA* at time 0) − (mean C_T of β-actin at time 0)).

Sample	Time	*GeneA* C_T	Mean *GeneA* C_T at t = 0	β-actin C_T	Mean β-actin C_T at t = 0	ΔC_T of the target at t = n	$\Delta\Delta C_T$	$2^{-\Delta\Delta CT}$	Mean $2^{-\Delta\Delta CT}$
1	0	27.8	28.1	27.4	27.2	0.4	−0.5	1.41	1.03
2	0	28.2	28.1	27.2	27.2	1.0	0.1	0.93	
3	0	28.3	28.1	27.0	27.2	1.3	0.4	0.76	
4	1 hr	24.1	28.1	27.9	27.2	−3.8	−4.7	26.0	28.8
5	1 hr	23.8	28.1	28.0	27.2	−4.2	−5.1	34.3	
6	1 hr	24.0	28.1	27.8	27.2	−3.8	−4.7	26.0	
7	2 hr	17.5	28.1	28.2	27.2	−10.7	−11.6	3104.2	2719.7
8	2 hr	17.6	28.1	27.9	27.2	−10.3	−11.2	2352.5	
9	2 hr	17.9	28.1	28.4	27.2	−10.5	−11.4	2702.4	

For completeness, we will calculate the standard deviations and the coefficient of variation.

Again, the standard deviation (s) is given by the formula

$$s = \sqrt{\frac{\sum (x - \bar{x})^2}{n - 1}}$$

where the Σ symbol represents the sum, x is a sample, $\bar{x}$ is the average (mean), and n is the number of measurements in the dataset. These standard deviations for the dataset are calculated in Table 9.26.

Therefore, the calculated zero time mean fold increase in gene expression has a standard deviation (s) of 0.31. The values determined for 1 and 2 hr have standard deviations of 4.79 and 376.15, respectively.

The coefficient of variation (cv) is a measure of the dispersion of the data points around a mean and is given by the equation

$$\mathrm{CV} = \frac{\text{standard deviation}}{\text{mean}} = \frac{s}{\bar{x}}$$

The CV is a dimensionless number in that it has no regard for the units of the dataset. It can be useful, therefore, when comparing the degree of variation from one dataset to another when those datasets have different units.

Table 9.26 Calculations of standard deviations for the Problem 9.15 dataset. The mean $2^{-\Delta\Delta CT}$ for zero time is 1.03, for 1 hr is 28.8, and for 2 hours is 2719.7 (determined in Table 9.25).

Time	$2^{-\Delta\Delta CT}$	$(x-\bar{x})$	$(x-\bar{x})^2$	$\sum(x-\bar{x})^2$	$\frac{\sum(x-\bar{x})^2}{n-1}$	$\sqrt{\frac{\sum(x-\bar{x})^2}{n-1}}$
0	1.41	0.38	0.1444	0.2273	0.1137	0.34
	0.93	−0.01	0.0100			
	0.76	−0.27	0.0729			
1 hour	26.0	−2.8	7.84	45.93	22.97	4.79
	34.3	5.5	30.25			
	26.0	−2.8	7.84			
2 hours	3104.2	384.5	147 840	282 975	141 488	376.15
	2352.5	−367.2	134 836			
	2702.4	−17.3	299			

It is also useful when the different datasets have vastly different means, as is the case here. The CV is often expressed as a percent. The higher the CV value, the greater the dispersion of data about the mean. However, if a mean is close to zero, the CV value can be disproportionately impacted by small changes in the mean, and, under those circumstances, is a less valuable measure of dataset variation.

The CV values for this dataset are as follows. For the zero time fold increase, we have

$$CV = \frac{0.34}{1.03} \times 100 = 33\%$$

For the 1 hr value, we have

$$CV = \frac{4.79}{28.8} \times 100 = 17\%$$

For the 2 hr value, we have

$$CV = \frac{376.15}{2719.7} \times 100 = 14\%$$

Therefore, these datasets have similar dispersions about the mean values calculated for each time measurement: 0, 1, and 2 hr.

9.7 THE RELATIVE STANDARD CURVE METHOD

9.7.1 Standard curve method for relative quantitation

In the Standard Curve method, the simplest method of quantitation, the amount of transcript for each gene in each tissue type is first calculated using a standard curve of a known nucleic acid at known concentrations. The concentrations of the experimental samples are then expressed as fractions relative to a single calibrator (usually the transcript of a different gene in some other tissue type). The concentration of the calibrator is arbitrarily set at one – at unity. The concentrations of the target transcripts are expressed as a difference to the calibrator – the quantity of the target divided by the quantity of the calibrator. That is, for example, the transcript of the HGPRT gene is made at a two-fold greater amount than the calibrator.

In the previous section, we saw that absolute quantification of an unknown sample is possible through the use of a standard curve in which the concentration of a quantified, serially diluted DNA sample is plotted against each diluted sample's C_T. The C_T value of an unknown sample can then be used to determine that sample's concentration by extrapolation from the standard curve.

A standard curve can also be used to measure gene expression when the exact concentration of the standard is not known. In this case, quantification of the target is measured relative to a reference. The reference can be a calibrator such as a housekeeping gene (β-actin, GAPDH, or a ribosomal RNA gene) or it can be the study's actual target gene but from untreated or time zero cells. The reference's concentration, by definition, is at 1X and all target quantities from the treated samples are expressed as an *n*-fold difference relative to the reference. A standard curve can be prepared from a dilution series of any independent sample of RNA or DNA carrying the desired target gene(s). The actual concentrations (in ng/μL) of the standards need not be known, just their relative dilutions.

Standard curves should be prepared for both the target and the reference. Each standard curve is then used to determine the quantities, respectively, of the target and the reference gene. For all experimental samples, target quantity (as determined from the standard curve) is divided by the target quantity of the reference to yield a relative *n*-fold difference. The actual nanogram amount of product generated for each gene is not important. It is a relative comparison.

The Relative Standard Curve method for the analysis of relative gene expression is simpler to perform than the $2^{-\Delta\Delta CT}$ method because it requires the smallest amount of optimization and validation. The two methods, however, can give very similar results.

Problem 9.16 We wish to determine the relative expression of *GeneA* in several different tissue types. We will use the housekeeping gene *HSKG*α as a reference. *GeneA* and *HSKG*α will be amplified in separate tubes. Dilutions of cDNA prepared from a stock of total RNA (purchased from a commercial source) are used to construct standard curves separately for *GeneA* and *HSKG*α. Fifty ng of cDNA prepared from the RNA stock is diluted serially in two-fold increments. Aliquots of those dilutions are distributed in triplicate into each of two 96-well optical plates (plates A and B) such that, in each plate, three wells carry 2 ng of cDNA, three wells carry 1 ng of cDNA, three wells carry 0.5 ng of cDNA, etc., down to 0.015 ng of cDNA per well in three wells. Samples of total RNA are isolated from brain, muscle, and liver cells. That RNA is converted to cDNA by reverse transcriptase, and 1 ng aliquots of the cDNA are dispensed into three wells each of plate A and plate B. *GeneA* is amplified in plate A using specific primers and a FAM-labeled probe in a real-time PCR instrument. The *HSKG*α gene is amplified and assayed in plate B. The results shown in Tables 9.27 and 9.28 are obtained. Using the Relative Standard Curve method, what are the relative levels of expression of *GeneA* in the three tissue types: brain, muscle, and liver?

Solution 9.16

We will first construct standard curves for the dataset from plates A and B. Using the equation for the curve's regression line, we will calculate the relative amounts of *GeneA* and *HSKG*α and then compare them.

If the instrument software cannot construct a standard curve, one can be made using Microsoft Excel using the following steps:

1. In an Excel workbook, add the information into columns of 'ng,' 'log ng,' and 'C_T,' as shown in Figure 9.14.
2. Highlight a box in the 'log ng' column next to a value in the 'ng' column. Calculate the log (base 10) for that ng value by
   a. Clicking open the **Paste Function** icon.
   b. Highlighting 'Math & Trig' in the dialogue box that appears, highlighting 'LOG10' in the **Function name**, and clicking the **OK** button.
   c. Entering the ng value in the **Number** field of the LOG10 dialog box and clicking **OK**. This will enter the log value in the highlighted box in the log ng column.

Table 9.27 Cycle threshold data for plate A as described in Problem 9.16.
Plate A: *GeneA* standard curve and *GeneA* expression in brain, muscle, and liver

Sample	Well	C_T
2 ng cDNA (standard)	AI	25.43
2 ng cDNA (standard)	BI	25.56
2 ng cDNA (standard)	CI	25.58
1 ng cDNA (standard)	DI	26.52
1 ng cDNA (standard)	EI	26.59
1 ng cDNA (standard)	FI	26.44
0.5 ng cDNA (standard)	GI	27.38
0.5 ng cDNA (standard)	HI	27.48
0.5 ng cDNA (standard)	A2	27.50
0.25 ng cDNA (standard)	B2	28.43
0.25 ng cDNA (standard)	C2	28.34
0.25 ng cDNA (standard)	D2	28.50
0.125 ng cDNA (standard)	E2	29.36
0.125 ng cDNA (standard)	F2	29.38
0.125 ng cDNA (standard)	G2	29.43
0.06 ng cDNA (standard)	H2	30.28
0.06 ng cDNA (standard)	A3	30.29
0.06 ng cDNA (standard)	B3	30.39
0.03 ng cDNA (standard)	C3	31.30
0.03 ng cDNA (standard)	D3	31.32
0.03 ng cDNA (standard)	E3	31.27
0.016 ng cDNA (standard)	F3	32.21
0.016 ng cDNA (standard)	G3	32.31
0.016 ng cDNA (standard)	H3	32.39
0.5 ng cDNA from Brain	A4	25.78
0.5 ng cDNA from Brain	B4	25.69
0.5 ng cDNA from Brain	C4	25.84
0.5 ng cDNA from Muscle	D4	30.74
0.5 ng cDNA from Muscle	E4	30.68
0.5 ng cDNA from Muscle	F4	31.02
0.5 ng cDNA from Liver	G4	24.98
0.5 ng cDNA from Liver	H4	25.12
0.5 ng cDNA from Liver	A5	25.26

Table 9.28 Cycle threshold data for plate B as described in Problem 9.16.
Plate B: *HSKGa* standard curve and *HSKGa* expression in brain, muscle, and liver

Sample	Well	C_T
2 ng cDNA (standard)	AI	23.98
2 ng cDNA (standard)	BI	24.04
2 ng cDNA (standard)	CI	24.02
1 ng cDNA (standard)	DI	25.01
1 ng cDNA (standard)	EI	25.02
1 ng cDNA (standard)	FI	24.96
0.5 ng cDNA (standard)	GI	26.01
0.5 ng cDNA (standard)	HI	26.04
0.5 ng cDNA (standard)	A2	25.92
0.25 ng cDNA (standard)	B2	27.01
0.25 ng cDNA (standard)	C2	27.05
0.25 ng cDNA (standard)	D2	26.89
0.125 ng cDNA (standard)	E2	28.02
0.125 ng cDNA (standard)	F2	28.04
0.125 ng cDNA (standard)	G2	27.92
0.06 ng cDNA (standard)	H2	29.03
0.06 ng cDNA (standard)	A3	29.05
0.06 ng cDNA (standard)	B3	29.01
0.03 ng cDNA (standard)	C3	30.03
0.03 ng cDNA (standard)	D3	29.90
0.03 ng cDNA (standard)	E3	30.03
0.016 ng cDNA (standard)	F3	31.04
0.016 ng cDNA (standard)	G3	31.07
0.016 ng cDNA (standard)	H3	31.09
0.5 ng cDNA from Brain	A4	27.48
0.5 ng cDNA from Brain	B4	27.02
0.5 ng cDNA from Brain	C4	27.28
0.5 ng cDNA from Muscle	D4	25.38
0.5 ng cDNA from Muscle	E4	25.27
0.5 ng cDNA from Muscle	F4	25.41
0.5 ng cDNA from Liver	G4	31.48
0.5 ng cDNA from Liver	H4	31.55
0.5 ng cDNA from Liver	A5	31.61

	A	B	C	D
1	**Well**	**ng**	**log ng**	CT
2	A1	2	0.30103	25.43
3	B1	2	0.30103	25.56
4	C1	2	0.30103	25.58
5	D1	1	0	26.52
6	E1	1	0	26.59
7	F1	1	0	26.44
8	G1	0.5	-0.30103	27.38
9	H1	0.5	-0.30103	27.48
10	A2	0.5	-0.30103	27.5
11	B2	0.25	-0.60206	28.43
12	C2	0.25	-0.60206	28.34
13	D2	0.25	-0.60206	28.5
14	E2	0.125	-0.90309	29.36
15	F2	0.125	-0.90309	29.38
16	G2	0.125	-0.90309	29.43
17	H2	0.06	-1.22185	30.28
18	A3	0.06	-1.22185	30.29
19	B3	0.06	-1.22185	30.39
20	C3	0.03	-1.52288	31.3
21	D3	0.03	-1.52288	31.32
22	E3	0.03	-1.52288	31.27
23	F3	0.016	-1.79588	32.21
24	G3	0.016	-1.79588	32.31
25	H3	0.016	-1.79588	32.39

■ **FIGURE 9.14** Data for the *GeneA* standard curve obtained from plate A as it appears in a Microsoft Excel spreadsheet.

3. Perform the following steps to construct a standard curve:
   a. Highlight all the values listed in the 'ng' and 'log ng' columns (do not include the headers).
   b. Click the **Chart Wizard** open and select 'XY (Scatter)' in the **Chart type** list.
   c. Click the **Next** button through the fields to format the chart to your satisfaction. When done, click the **Finish** button. The chart of the standard curve will appear on the spreadsheet.
   d. Click on the chart to make sure it is selected (it will highlight along its edges).
   e. From the **Chart** pull-down menu in the top tool bar, select 'Add Trendline.' In the dialogue box that appears, select 'Linear' for the **Trend/Regression type**. In the **Options** window of the **Add Trendline** box, make sure the checkboxes for 'Display equation on chart' and 'Display R-squared value on chart' are checked. Click the **OK** button. This will add the regression line's equation on the chart (Figure 9.15).

This process is repeated for plate B as shown in Figure 9.16.

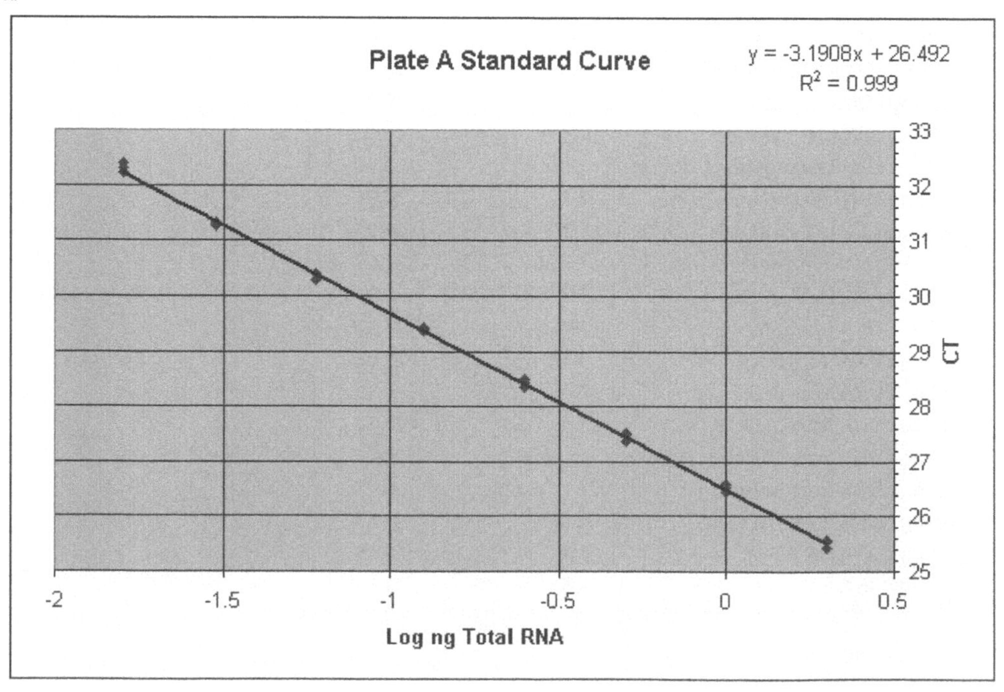

■ **FIGURE 9.15** The regression line and its equation for the standard curve plotted using the dataset of Problem 9.16 and Microsoft Excel.

	A	B	C	D
1	**Well**	**ng**	**log ng**	**CT**
2	A1	2	0.30103	23.98
3	B1	2	0.30103	24.04
4	C1	2	0.30103	24.02
5	D1	1	0	25.01
6	E1	1	0	25.02
7	F1	1	0	24.96
8	G1	0.5	-0.30103	26.01
9	H1	0.5	-0.30103	26.04
10	A2	0.5	-0.30103	25.92
11	B2	0.25	-0.60206	27.01
12	C2	0.25	-0.60206	27.05
13	D2	0.25	-0.60206	26.89
14	E2	0.125	-0.90309	28.02
15	F2	0.125	-0.90309	28.04
16	G2	0.125	-0.90309	27.92
17	H2	0.06	-1.22185	29.03
18	A3	0.06	-1.22185	29.05
19	B3	0.06	-1.22185	29.01
20	C3	0.03	-1.52288	30.03
21	D3	0.03	-1.52288	29.9
22	E3	0.03	-1.52288	30.02
23	F3	0.016	-1.79588	31.04
24	G3	0.016	-1.79588	31.07
25	H3	0.016	-1.79588	31.09

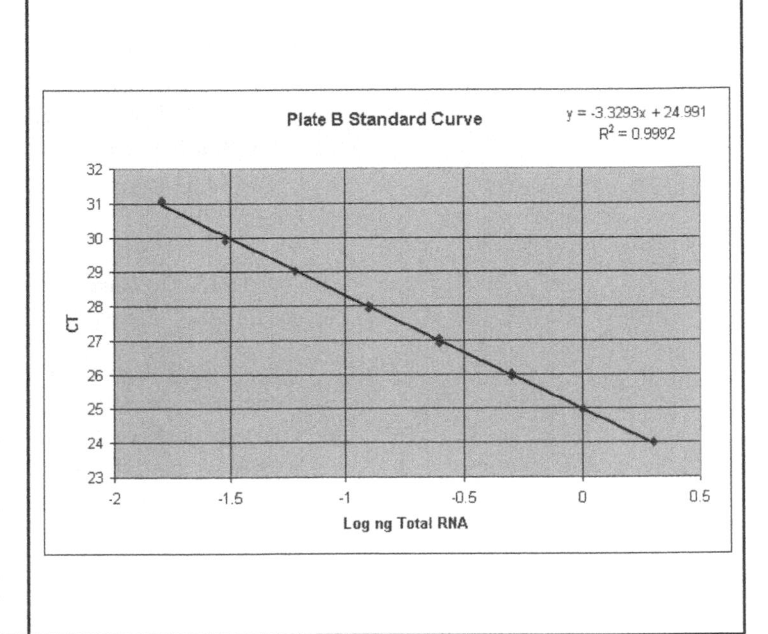

■ **FIGURE 9.16** The C_T data and *HSKGα* standard curve for the plate B dataset.

We now use the equations for the regression lines determined for the standard curves to calculate the input amounts for the unknown samples. The equations are in the general form for a line $y = mx + b$, where y is the C_T value, m is equal to the slope of the line, x is the log of the input amount of RNA in nanograms, and b is the y intercept at x equal to 0. This equation can be rearranged to solve for x as follows.

$$y = mx + b$$
$$x = \frac{y - b}{m}$$

By knowing the C_T value (y), therefore, we can use the linear regression equation to calculate the input amount of RNA. For example, the amount of *GeneA* input RNA corresponding to the sample in well A4 of plate A (see Table 9.26) is

$$x = \frac{25.78 - 26.492}{-3.1908} = \frac{-0.712}{-3.1908} = 0.223$$

This value, however, is the log of the ng amount. To convert it to ng, we must take its antilog. On a calculator, that value is obtained by using the **10^x** function. In Microsoft Excel, the antilog is obtained by using the **Paste Function** function and selecting 'Math & Trig' from the **Function category** and 'POWER' from the **Function name**. After clicking **OK**, entering '10' in the **Number** field and the *x* value in the **Power** field will give you the *x* value's antilog.

$$\text{Antilog } 0.223 = 1.67$$

Therefore, a C_T value of 25.78 corresponds to an input amount of RNA of 1.67 ng.

This protocol is used to determine the input amount of RNA for all of the unknown samples. For the *HSKG*α samples, however, the regression equation from the standard curve generated from the plate B dataset is used for the calculation. These calculations yield the results shown in Table 9.29.

Table 9.29 Calculation results for the equivalent amounts of total RNA as determined by the linear regression equations from the standard curves for genes *GeneA* and *HSKG*α.

Gene/tissue	C_T	Log ng total RNA	Ng total RNA
GeneA/brain	25.78	0.223	1.67
	25.69	0.251	1.78
	25.84	0.204	1.60
GeneA/muscle	30.74	−1.331	0.047
	30.68	−1.313	0.049
	31.02	−1.419	0.038
GeneA/liver	24.98	0.474	2.98
	25.12	0.430	2.69
	25.26	0.386	2.43
HSKGα*/brain*	27.48	−0.748	0.179
	27.02	−0.609	0.246
	27.28	−0.712	0.194
HSKGα*/muscle*	25.38	−0.117	0.764
	25.27	−0.108	0.780
	25.41	−0.126	0.748
HSKGα*/liver*	31.48	−1.949	0.011
	31.55	−1.970	0.011
	31.61	−1.988	0.010

We will now calculate the standard deviation for each ng amount for each gene and tissue type. Standard deviation (s) is given by the equation

$$s = \sqrt{\frac{\sum(x - \bar{x})^2}{n - 1}}$$

In calculating the standard deviation (see Table 9.30), we should average the *GeneA* and *HSKGα* amounts separately because, even though we are measuring the two genes in the same tissues, we have amplified and assayed them in separate tubes.

Table 9.30 Calculations for the standard deviation values of the total RNA amounts as determined from the standard curve described in Problem 9.16.

Gene/tissue	ng Total RNA	$(x - \bar{x})^2$	$\sum(x - \bar{x})^2$	$\frac{\sum(x - \bar{x})^2}{n - 1}$	$\sqrt{\frac{\sum(x - \bar{x})^2}{n - 1}}$
GeneA/brain	1.67	0.0001	0.0165	0.0083	0.091
	1.78	0.0100			
	1.60	0.0064			
Average	1.68				
GeneA /muscle	0.047	0.000004	0.00007	0.00004	0.006
	0.049	0.000016			
	0.038	0.000049			
Average	0.045				
GeneA/liver	2.98	0.0784	0.1514	0.0757	0.275
	2.69	0.0001			
	2.43	0.0729			
Average	2.70				
HSKGα/brain	0.179	0.00073	0.00247	0.00124	0.035
	0.246	0.00160			
	0.194	0.00014			
Average	0.206				
HSKGα/muscle	0.764	0.00000	0.00052	0.0003	0.016
	0.780	0.00026			
	0.748	0.00026			
Average	0.764				
HSKGα/liver	0.011	0.00000	0.00000	0.00000	0.000
	0.011	0.00000			
	0010	0.00000			
Average	0.011				

Table 9.31 The input RNA amounts with their standard deviations for *GeneA* and *HSKGα* in the different tissue types described in Problem 9.16.

Gene/tissue	Ng input RNA
GeneA/brain	1.68 ± 0.091
GeneA/muscle	0.045 ± 0.006
GeneA/liver	2.70 ± 0.275
HSKGα/brain	0.206 ± 0.035
HSKGα/muscle	0.764 ± 0.016
HSKGα/muscle	0.011 ± 0.00

These standard deviation values (Table 9.30) will be attached as a '±' number to each average input RNA amount calculated from the C_T values (Table 9.31).

The average amount of *GeneA* is now divided by the average amount of *HSKGα* to give the normalized amount of *GeneA* for each tissue type. Since there are standard deviation values associated with each ng amount of input RNA, we will need to propagate that standard deviation in our quotient. We can do that through the **coefficient of variation** (***cv***) – a unitless value that is a measure of variation (dispersion) of data points relative to the mean. The *cv* is a useful number when you are interested in the magnitude of the dispersion relative to the size of the observation. The higher the *cv*, the greater the dispersion. The *cv* is less useful to a statistical analysis, however, when the mean is close to zero and the *cv* becomes more sensitive to small changes in the mean. It is given by the expression

$$cv = \frac{\text{standard deviation}}{\text{mean}} = \frac{s}{\bar{x}}$$

The *cv* of a quotient (as will be determined in this problem) is calculated as

$$cv = \sqrt{cv_1^2 + cv_2^2}$$

For example, *GeneA* in brain tissue is equivalent, in ng of total input RNA, to 1.68 ± 0.091 and *HSKGα* is equivalent to 0.206 ± 0.035 (see Table 9.31). In dividing 1.68 ± 0.091 by 0.206 ± 0.035 to give the normalized amount of *GeneA* in brain tissue, we would calculate the quotient's *cv* as

$$cv = \sqrt{\left(\frac{0.091}{1.68}\right)^2 + \left(\frac{0.035}{0.206}\right)^2} = \sqrt{0.003 + 0.029} = \sqrt{0.032} = 0.179$$

Table 9.32 The *cv* values for *GeneA* normalized to *HSKG*α in the three different tissue types described in Problem 9.16. Calculations were performed as described in the text.

GeneA normalized to *HSKG*α	*cv* for each cell type
Brain	0.179
Muscle	0.135
Liver	0.102

Table 9.33 *GeneA* values normalized to *HSKG*α amounts in the three different cell types. The *GeneA* amount is divided by the *HSKG*α amount. Errors are propagated as described in the text.

Tissue	*GeneA*	*HSKG*α	*GeneA* normalized to *HSKG*α
Brain	1.68 ± 0.091	0.206 ± 0.035	8.16 ± 1.46
Muscle	0.045 ± 0.006	0.764 ± 0.016	0.06 ± 0.008
Liver	2.70 ± 0.275	0.01 ± 10.00	245.45 ± 25

A *cv* value calculated in this manner for each tissue type is shown in Table 9.32.

Since the *cv* equals the standard deviation divided by the mean, we can rearrange and solve for *s*, the standard deviation, as

$$cv = \frac{s}{\overline{x}}$$

$$s = (cv)(\overline{x})$$

The standard deviation for the quotient for the example above, then, is

$$s = 0.179 \times 8.16 = 1.46$$

where the 8.16 value in this calculation is from 1.68 (the average ng total RNA for *GeneA* in brain tissue; see Table 9.30) divided by 0.206 (the average ng total RNA for *HSKG* α in brain tissue; see Table 9.30). Therefore the quotient (8.16) has a standard deviation of ±1.46. In Table 9.33, showing normalized *GeneA* values, the quotients' standard deviations are calculated in this manner.

The normalized *GeneA* amount has no units (the ng quantities cancel out during the division step to obtain the normalized value). When we compare the values between cell types, therefore, we are examining them as *n*-fold differences. To give the values a little more meaning, we can designate one of the samples as a calibrator – the sample to which the others are directly compared. The decision as to which cell type should be the calibrator is, for the most part, arbitrary since we are examining relative differences between the different cell types. However, here, since muscle cells show the lowest expression of *GeneA*, we will choose those cells as the calibrator. To get an *n*-fold difference, then, we divide each *GeneA* value by the muscle *GeneA* value.

Since the designation of the calibrator is arbitrary, the standard errors should not change from those of the normalized values for each cell type. We simply multiply each one by the new normalized value. For example, the standard deviation for the sample from brain tissue has a standard deviation of

$$136 \times 0.179 = 24.34$$

The normalized values for the quotients shown in Table 9.34 are calculated in this manner.

Therefore, these results indicate that brain cells have roughly 136 times as much *GeneA* mRNA as muscle cells. Likewise, liver cells carry some 4091 times as much *GeneA* mRNA as muscle cells.

Table 9.34 Calculations for the amounts of *GeneA* in each tissue normalized to the amount of *GeneA* in muscle cells as described in Problem 9.16. This includes a calculation in which *GeneA* in muscle is normalized to itself.

Tissue	Normalized *GeneA* relative to muscle
Brain	$\frac{8.16}{0.06} = 136 \pm 24.34$
Muscle	$\frac{0.06}{0.06} = 1.00 \pm 0.135$
Liver	$\frac{245.45}{0.06} = 4091 \pm 417$

Table 9.35 Results obtained for the experiment described in Problem 9.17.

Sample	Well	C_T *GeneA* (FAM)	C_T *HSKG*α (VIC)
2 ng cDNA (standard)	Al	24.38	23.57
2 ng cDNA (standard)	Bl	24.65	23.44
2 ng cDNA (standard)	Cl	24.67	23.68
1 ng cDNA (standard)	Dl	25.60	24.40
1 ng cDNA (standard)	El	25.55	24.60
1 ng cDNA (standard)	Fl	25.48	24.78
0.5 ng cDNA (standard)	Gl	26.43	25.74
0.5 ng cDNA (standard)	Hl	26.52	25.64
0.5 ng cDNA (standard)	A2	26.61	25.56
0.25 ng cDNA (standard)	B2	27.54	26.49
0.25 ng cDNA (standard)	C2	27.39	26.81
0.25 ng cDNA (standard)	D2	27.62	26.68
0.125 ng cDNA (standard)	E2	28.69	27.72
0.125 ng cDNA (standard)	F2	28.43	27.68
0.125 ng cDNA (standard)	G2	28.53	27.76
0.06 ng cDNA (standard)	H2	29.62	28.71
0.06 ng cDNA (standard)	A3	29.42	28.62
0.06 ng cDNA (standard)	B3	29.42	28.83
0.03 ng cDNA (standard)	C3	30.55	29.72
0.03 ng cDNA (standard)	D3	30.42	29.78
0.03 ng cDNA (standard)	E3	30.51	29.85
0.016 ng cDNA (standard)	F3	31.52	30.84
0.016 ng cDNA (standard)	G3	31.37	30.71
0.016 ng cDNA (standard)	H3	31.65	30.98
0.5 ng cDNA from Lung	A4	25.12	28.57
0.5 ng cDNA from Lung	B4	24.89	28.69
0.5 ng cDNA from Lung	C4	24.96	28.52
0.5 ng cDNA from Heart	D4	28.33	30.67
0.5 ng cDNA from Heart	E4	28.42	30.63
0.5 ng cDNA from Heart	F4	28.36	30.59
0.5 ng cDNA from Adipose	G4	30.67	24.03
0.5 ng cDNA from Adipose	H4	30.56	24.11
0.5 ng cDNA from Adipose	A5	30.61	24.07

Problem 9.17 We wish to measure the relative expression of *GeneA* in lung, heart, and adipose tissue. We use a FAM (blue)-labeled probe to assay for *GeneA* during real-time PCR. In the same wells of a 96-well optical plate, we assay for the *HSKG*α housekeeping gene using a VIC (green)-labeled probe. We prepare a standard curve as described in

Problem 9.16; however, we assay for *GeneA* and *HSKGα* simultaneously in each well of the diluted standard. RNA prepared from lung, heart, and adipose tissue is converted to cDNA by reverse transcription. Ten nanograms of the RNA sample (converted to cDNA) is placed in each of three wells per tissue type and assayed for both *GeneA* and *HSKGα*. The results shown in Table 9.35 are obtained. What is the relative expression of *GeneA* normalized to adipose tissue?

Solution 9.17

We will first construct standard curves for *GeneA* based on the FAM (blue) signal and for *HSKGα* based on the VIC (green) signal. These standard curves and the equations for their regression lines (Figures 9.17 and 9.18) are constructed using Microsoft Excel as described in Appendix A and in Problem 9.16.

We now use the linear regression equations from the standard curves to solve for the ng of total RNA equivalent to the amounts needed to generate the C_T values obtained for *GeneA* and *HSKGα* in the RNA of the different tissue types. From the standard curve for *GeneA* (Figure 9.17), we have the equation

$$y = -3.2889x + 25.542$$

where y is the C_T value and x is the log of the total nanograms of RNA. We can substitute C_T for y and rearrange the equation to solve for x:

$$x = \frac{C_T - 25.542}{-3.2889}$$

Placing the C_T values obtained for *GeneA* in each tissue type and solving for x gives us the log of the total nanograms of RNA. To get nanogram values, we take the antilog of x as described in Problem 9.16. These calculations yield the results shown in Table 9.35.

From the standard curve for *HSKGα* (Figure 9.18), we have the equation

$$y = -3.4378x + 24.593$$

Substituting C_T for y and rearranging the equation to solve for x gives

$$x = \frac{C_T - 24.593}{-3.4378}$$

This equation is then used to calculate the log ng amounts of total RNA equivalent to the amount giving the C_T value for *HSKGα* seen in the

	A	B	C	D
1	**Well**	**ng**	**log ng**	CT
2	A1	2	0.30103	24.38
3	B1	2	0.30103	24.65
4	C1	2	0.30103	24.67
5	D1	1	0	25.6
6	E1	1	0	25.55
7	F1	1	0	25.48
8	G1	0.5	-0.30103	26.43
9	H1	0.5	-0.30103	26.52
10	A2	0.5	-0.30103	26.61
11	B2	0.25	-0.60206	27.54
12	C2	0.25	-0.60206	27.39
13	D2	0.25	-0.60206	27.62
14	E2	0.125	-0.90309	28.69
15	F2	0.125	-0.90309	28.43
16	G2	0.125	-0.90309	28.53
17	H2	0.06	-1.22185	29.62
18	A3	0.06	-1.22185	29.42
19	B3	0.06	-1.22185	29.52
20	C3	0.03	-1.52288	30.55
21	D3	0.03	-1.52288	30.42
22	E3	0.03	-1.52288	30.51
23	F3	0.016	-1.79588	31.52
24	G3	0.016	-1.79588	31.37
25	H3	0.016	-1.79588	31.65

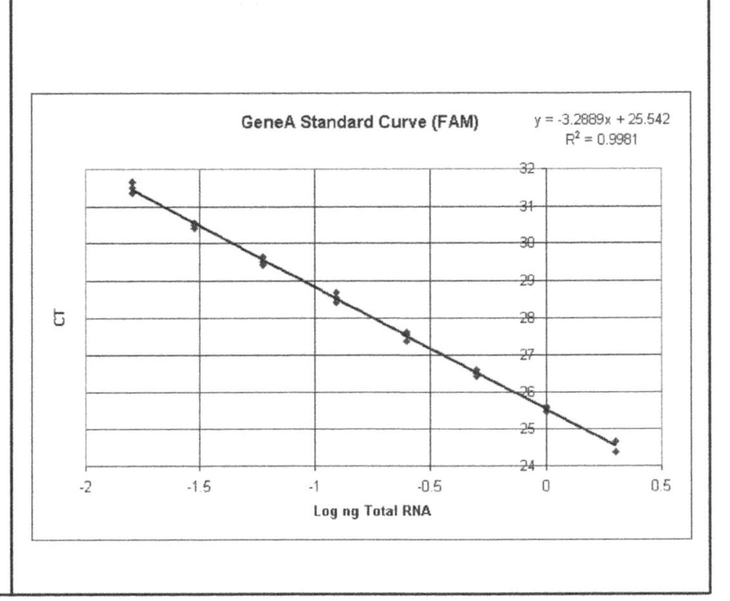

■ **FIGURE 9.17** The standard curve for *GeneA* constructed using Microsoft Excel and the dataset for Problem 9.17.

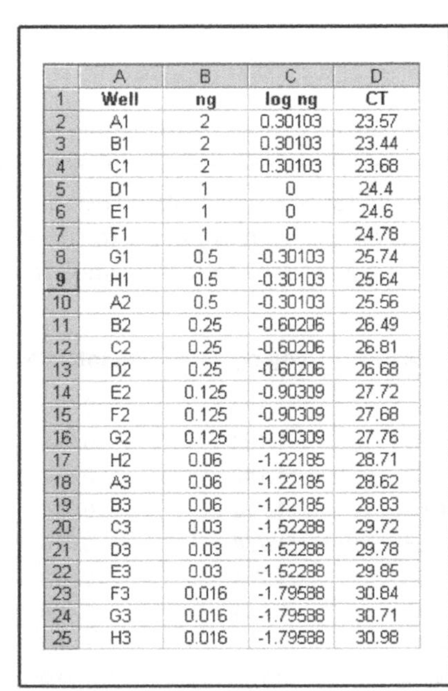

	A	B	C	D
1	**Well**	**ng**	**log ng**	CT
2	A1	2	0.30103	23.57
3	B1	2	0.30103	23.44
4	C1	2	0.30103	23.68
5	D1	1	0	24.4
6	E1	1	0	24.6
7	F1	1	0	24.78
8	G1	0.5	-0.30103	25.74
9	H1	0.5	-0.30103	25.64
10	A2	0.5	-0.30103	25.56
11	B2	0.25	-0.60206	26.49
12	C2	0.25	-0.60206	26.81
13	D2	0.25	-0.60206	26.68
14	E2	0.125	-0.90309	27.72
15	F2	0.125	-0.90309	27.68
16	G2	0.125	-0.90309	27.76
17	H2	0.06	-1.22185	28.71
18	A3	0.06	-1.22185	28.62
19	B3	0.06	-1.22185	28.83
20	C3	0.03	-1.52288	29.72
21	D3	0.03	-1.52288	29.78
22	E3	0.03	-1.52288	29.85
23	F3	0.016	-1.79588	30.84
24	G3	0.016	-1.79588	30.71
25	H3	0.016	-1.79588	30.98

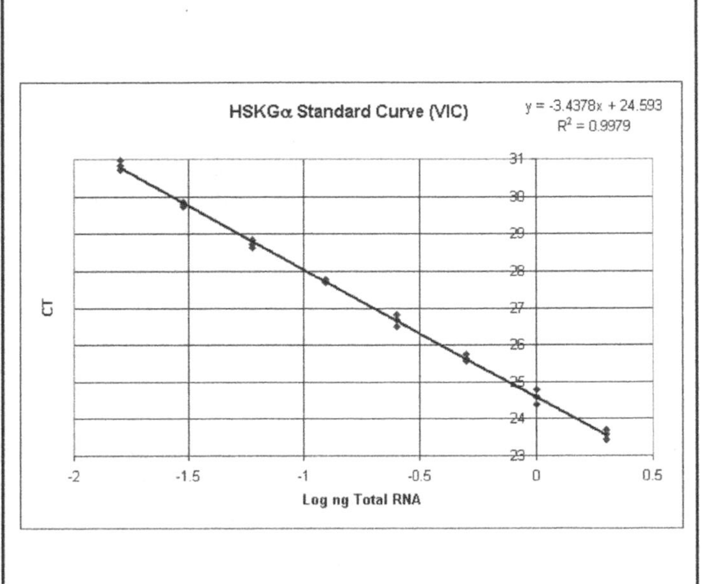

■ **FIGURE 9.18** The standard curve for *HSKGα* constructed using Microsoft Excel and the dataset for Problem 9.17

tissue samples. The antilog of these values gives the ng amounts shown in Table 9.36).

We then normalize the amount of *GeneA* by dividing its amount by the amount of *HSKGα* (calculated in Tables 9.36 and 9.37). Since both *GeneA* and *HSKGα* are multiplexed (amplified in the same well), we normalize within, rather than across, wells.

$$\text{normalized } GeneA = \frac{\text{ng } GeneA}{\text{ng } HSKG\alpha}$$

Since the ng units cancel out, normalized *GeneA* has no units. The normalized values for *GeneA* are shown in Table 9.38.

Table 9.36 Nanograms of total RNA for *GeneA* calculated as described in the text for Problem 9.17.

Tissue	Well	*GeneA* C_T	Log ng	ng total RNA
Lung	A4	25.12	0.128	1.344
	B4	24.89	0.198	1.579
	C4	24.96	0.177	1.503
Heart	D4	28.33	−0.848	0.142
	E4	28.42	−0.875	0.133
	F4	28.36	−0.857	0.139
Adipose	G4	30.67	−1.559	0.028
	H4	30.56	−1.526	0.030
	A5	30.61	−1.541	0.029

Table 9.37 Nanograms of total RNA for *HSKGα* calculated using the equation for the linear regression line from the standard curve. The antilog of x in that equation is equal to the total RNA amount. Data is for Problem 9.17.

Tissue	Well	*GeneA* C_T	Log ng	ng total RNA
Lung	A4	28.57	−1.157	0.070
	B4	28.69	−1.192	0.064
	C4	28.52	−1.142	0.072
Heart	D4	30.67	−1.768	0.017
	E4	30.63	−1.756	0.018
	F4	30.59	−1.744	0.018
Adipose	G4	24.03	0.164	1.458
	H4	24.11	0.141	1.382
	A5	24.07	0.152	1.419

Table 9.38 *GeneA* normalized to *HSKG*α. The normalized values are obtained by dividing the amount of *GeneA* by the amount of *HSKG*α for each tissue type. Since the ng units cancel out during this division step, normalized *GeneA* is a value without units. This data is for Problem 9.17.

Tissue	Well	ng *GeneA*	ng *HSKG*α	*GeneA* normalized to *HSKG*α
Lung	A4	1.344	0.070	19.20
	B4	1.579	0.064	24.67
	C4	1.503	0.072	20.88
Heart	D4	0.142	0.017	8.35
	E4	0.133	0.018	7.39
	F4	0.139	0.018	7.72
Adipose	G4	0.028	1.458	0.019
	H4	0.030	1.382	0.022
	A5	0.029	1.419	0.020

We will now determine the average normalized *GeneA* value for each tissue type and assign a standard deviation (as a '±' number) to each of those values. The standard deviation is given by the equation

$$s = \sqrt{\frac{\sum(x - \bar{x})^2}{n - 1}}$$

where n is the number of samples used to obtain data for each tissue type. The averages and standard deviations for *GeneA* in the different tissue types are shown in Table 9.39.

We now have the normalized average *GeneA* values with their standard deviations for each tissue type:

Lung: 21.58 ± 2.80.
Heart: 7.82 ± 0.48.
Adipose: 0.02 ± 0.00.

We will now calculate the normalized *GeneA* values relative to a calibrator. In this case, we will designate *GeneA* activity in adipose tissue as the calibrator to which *GeneA* activity in the other tissues (lung and heart) will be compared. To obtain a relative value, we will divide each number by the value associated with adipose tissue. We will also attach a standard deviation to each relative gene expression value we calculate.

Table 9.39 Mean normalized *GeneA* values and calculation of their standard deviations. Data is for Problem 9.17.

Tissue	Normalized *GeneA*	$(x - \bar{x})$	$(x - \bar{x})^2$	$\sum(x - \bar{x})^2$	$\frac{\sum(x - \bar{x})^2}{n-1}$	$\sqrt{\frac{\sum(x - \bar{x})^2}{n-1}}$
Lung	19.20	−2.38	5.66	15.70	7.85	2.80
	24.67	3.09	9.55			
	20.ES	−0.70	0.49			
Average	21.58					
Heart	8.35	0.53	0.28	0.47	0.24	0.48
	7.39	−0.43	0.18			
	7.72	−0.10	0.01			
Average	7.82					
Adipose	0.02	0.00	0.00	0.00	0.00	0.00
	0.02	0.00	0.00			
	0.02	0.00	0.00			
Average	0.02					

As in Problem 9.16, we will use the following relation of *cv* to standard deviation:

$$cv = \frac{s}{\bar{x}}$$

Since the designation of which sample serves as the calibrator is an arbitrary decision and its value, therefore, is an arbitrary constant, we will use the standard deviation of each normalized value from each respective tissue type for the calculation. The *cv* value we calculate is then multiplied by each relative value (normalized *GeneA* in lung divided by normalized *GeneA* in adipose, for example) to give us its standard deviation. In other words, we already calculated standard deviations (above) but those numbers are for their specific means. We need to know the standard deviation for the new mean value – the relative expression level value obtained by comparing each *GeneA* level to the calibrator.

Since *cv* is equal to the standard deviation divided by the mean, it is also true, by rearranging the equation, that

$$s = cv \times \bar{x}$$

The relative expression of *GeneA* in lung cells relative to its expression in adipose is

$$\text{relative } GeneA_{\text{lung}} \text{ to } GeneA_{\text{adipose}} = \frac{21.58}{0.02} = 1079$$

We calculate normalized *GeneA*'s *cv* as

$$cv = \frac{2.80}{21.58} = 0.13$$

The relative amount of *GeneA* has a standard deviation, therefore, of

$$\bar{x} \times cv = s$$
$$1079 \times 0.13 = 140$$

GeneA RNA, therefore, is 1079 ± 140 times more abundant in lung cells than it is in adipose cells.

Likewise, *GeneA* RNA amount in heart cells relative to its abundance in adipose is calculated as

$$\frac{7.82}{0.02} = 391$$

This value's standard deviation is calculated from the *cv* as

$$cv = \frac{0.48}{7.82} = 0.06$$

and the standard deviation, therefore, is

$$391 \times 0.06 = 24$$

By this experiment, *GeneA* RNA is 391 ± 24 times more abundant in heart cells than in adipose cells.

Since *GeneA* in adipose is the calibrator, its relative amount is one (unity).

$$\frac{0.02}{0.02} = 1.0$$

Its standard deviation is

$$cv = \frac{0.00}{0.02} = 0.00 \text{ and } 1.0 \times 0.00 = 0.00$$

The expression of GeneA in adipose relative to its expression in adipose is 1.0 ± 0.00.

9.8 RELATIVE QUANTIFICATION BY REACTION KINETICS

Relative quantification by the standard curve method requires that a number of wells carry a dilution series of a quantified nucleic acid independently prepared from that isolated during the gene expression study. Although not necessarily difficult to prepare, the standard curve does monopolize much of the plate real estate and, because of the additional pipetting steps, can potentially introduce additional error into the experiment as well as require longer time spent in the lab.

The $\Delta\Delta C_T$ method requires a generous helping of up-front validation work designed to convince the experimenter that the approach can yield unambiguous results. It demands, for example, that the reference gene has an amplification efficiency equal to that of the target and that both efficiencies are close to 100%. If this is not the case, the experimenter needs to spend some effort on optimizing one or both of the reactions. In addition, a validation experiment must be performed to see what effect the experimental treatment might have on the reference gene used for normalization.

In a method described by Liu and Saint (2002), the kinetic curve of the amplification is used to calculate an amplification efficiency and that number is used to derive the initial amount of transcript present in the reaction. That value can be normalized to a reference gene and relative levels of RNA (gene expression) can be extrapolated. In this approach, there is no need for a standard curve and the validation experiments that might discourage the use of the $\Delta\Delta C_T$ method.

We have seen that exponential amplification by PCR is described by the equation

$$X_n = X_0 \times (1 + E)^n$$

where X_n is the number of target molecules at cycle n, X_0 is the initial number of template molecules, E is the amplification efficiency of the target, and n is the number of cycles.

The foundation of the real-time PCR assay is the assumption that X_n is proportional to the amount of reporter fluorescence, R. The equation for exponential amplification can be written, therefore, as

$$R_n = R_0 \times (1 + E)^n$$

where R_n is the amount of reporter fluorescence at cycle n, R_0 is the amount of initial reporter fluorescence, E is the amplification of the target, and n is the number of cycles. The amplification efficiency, E, is equal to one when the product is doubling with each cycle. If no product is generated at any cycle, the amplification efficiency would be equal to zero.

Although, as we have seen earlier, we can calculate an overall amplification efficiency for the reactions that generate a standard curve based on the slope of the regression line for that curve (using the equation $E = 10^{(-1/\text{slope})} - 1$), the efficiency of amplification is not constant over all the cycles of the PCR. In the latter stages of the reaction, components become limiting and exponential growth can no longer be sustained. The synthesis rate of new, full-length products slows down and eventually ceases. The equations for X_n and R_n shown above, therefore, can only actually apply to those cycles tucked within the exponential phase of amplification.

We can derive E, the amplification efficiency, by rearranging the above equation but with a focus on the logarithmic phase of the reaction. We will arbitrarily assign two points, A and B, to the exponential section of the amplification curve. E, then, is calculated as

$$E = \left(\frac{R_{n,A}}{R_{n,B}}\right)^{\frac{1}{C_{T,A} - C_{T,B}}} - 1 \qquad \text{Eq. 9.1}$$

where $R_{n,A}$ and $R_{n,B}$ represent R_n (reporter fluorescence at cycle n) at arbitrary thresholds A and B in the exponential part of the amplification curve, respectively, and $C_{T,A}$ and $C_{T,B}$ are the threshold cycles at these arbitrary thresholds.

For the amplification of any gene, we can write the amount of reporter signal at any cycle n as

$$R_n = R_0 \times (1 + E)^n$$

We can use this equation to normalize the amplification of a target gene to a reference gene in a relative quantification experiment by using the equation

$$\frac{R_{0,\text{Target}}}{R_{0,\text{Reference}}} = \frac{R_{n,\text{Target}} \times (1 + E_{\text{Reference}})^n}{R_{n,\text{Reference}} \times (1 + E_{\text{Target}})^n} \qquad \text{Eq. 9.2}$$

Likewise, since the amount of reporter signal at any cycle can be written as

$$R_n = R_0 \times (1 + E)^{C_T}$$

normalizing one gene to another, a target to a reference, can be expressed as

$$\frac{R_{0,\text{Target}}}{R_{0,\text{Reference}}} = \frac{(1 + E_{\text{Reference}})^{C_{T,\text{Reference}}}}{(1 + E_{\text{Target}})^{C_{T,\text{Target}}}} \qquad \text{Eq. 9.3}$$

In the following problem, we will use Equation 1 and Equation 3 to calculate the relative expression level of two genes in liver cells.

Problem 9.18 We wish to know to what level *GeneA* (the target) is expressed in relation to the housekeeping gene *HSKG*α (designated as the reference) in liver cells. Total RNA is isolated from a cell sample, it is reverse transcribed into cDNA, and aliquots dispensed into six separate wells of a 96-well optical microtiter plate. Primers and probes specific for *GeneA* and *HSKG*α in a PCR master mix are added to the six wells (and, of course, to the usual controls (wells devoted to NTC and positive control samples)). *GeneA* is assayed for using a blue dye-labeled probe and *HSKG*α is assayed for using a green dye-labeled probe. The plate is cycled in a real-time PCR instrument and the results shown in Table 9.40 are obtained. Using the Reaction Kinetics method for quantifying gene expression, calculate the relative expression level of *GeneA* normalized to *HSKG*α.

Solution 9.18

We first need to calculate the amplification efficiency (*E*) for the two reactions. In this experiment, all six wells give superimposable curves for the R_n vs. cycle number plot. We will look at the curve generated for just one of the wells (Figure 9.19) and, from that plot, derive the values needed to calculate the amplification efficiency for the *GeneA* and *HSKG*α reactions using Equation 1.

Table 9.40 Real-time polymerase chain reaction (PCR) data for the experiment described in Problem 9.18.

Well position	*GeneA* C_T (blue)	*HSKG*αC_T (green)
A1	23.46	24.81
B1	23.39	24.79
C1	23.42	24.72
D1	23.35	24.86
E1	23.38	24.81
F1	23.36	24.71

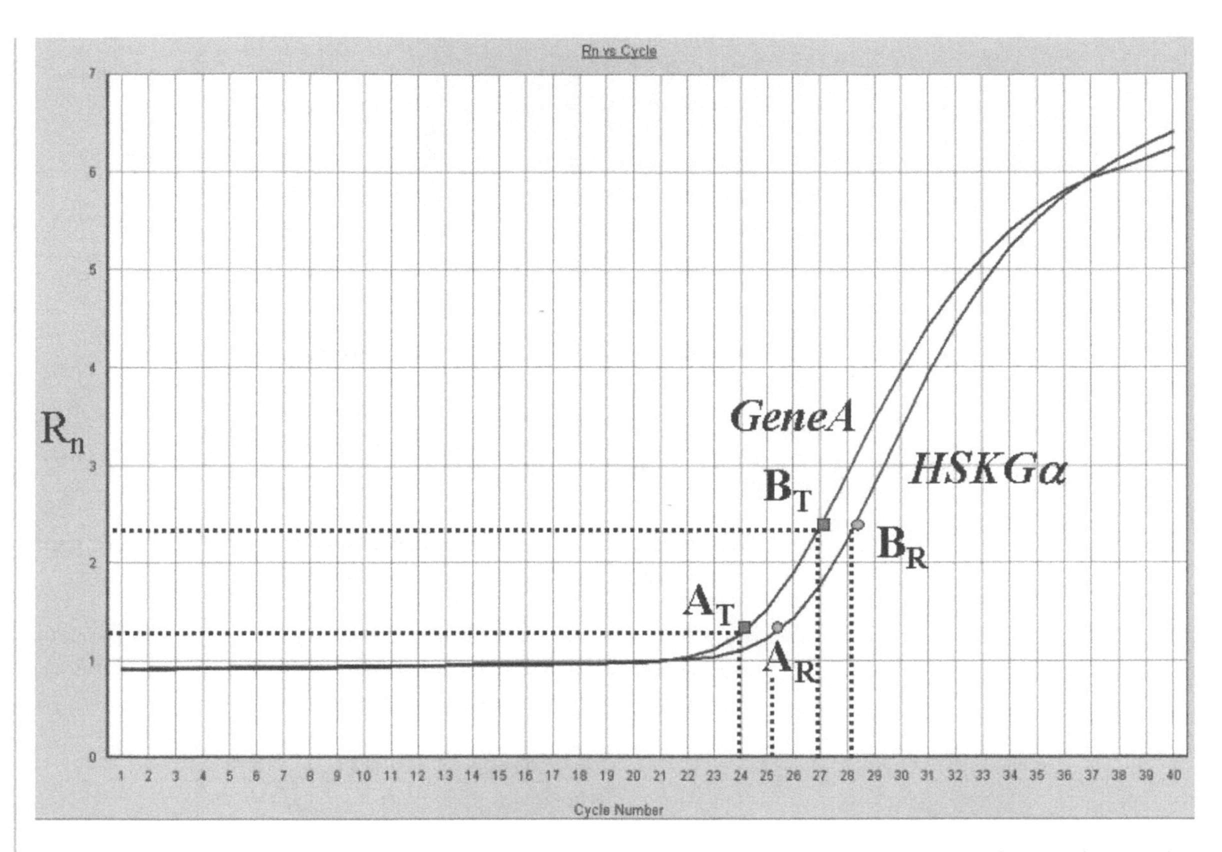

■ **FIGURE 9.19** Amplification plot for well C1 of the experiment described in Problem 9.18. The plot shows R_n (normalized reporter) vs. cycle number with curves both for *GeneA* and *HSKGα* shown on the same plot. Two points, A and B, are arbitrarily chosen within the exponential amplification part of the curves. $_T$ denotes 'target' and $_R$ subscript denotes 'reference.' Values for R_n and C_T are derived from this plot for each gene (dashed lines).

Note: The values needed to calculate the amplification efficiency using Equation 1 are not readily available through the software of most real-time PCR instruments. Therefore, these values are read from the plot of R_n vs. cycle number, as shown in Figure 9.19, which is an easily retrievable form of the analyzed data.

From the plot shown in Figure 9.19, the values shown in Table 9.41 are derived.

With the values in Table 9.41, we can calculate the efficiency of the *GeneA* and the *HSKGα* reactions using the following equation (Equation 1):

$$E = \left(\frac{R_{n,A}}{R_{n,B}}\right)^{\frac{1}{C_{T,A} - C_{T,B}}} - 1$$

Amplification efficiency of the reaction for the target, *GeneA*, is

$$E = \left(\frac{1.3}{2.4}\right)^{\frac{1}{24-27}} - 1 = 0.54^{\frac{1}{-3}} - 1 = 0.54^{-0.33} - 1 = 1.23 - 1 = 0.23$$

Table 9.41 Values for normalized reporter and C_T derived for *GeneA* and *HSKG*α, extrapolated from Figure 9.19.

Gene	$R_{n,A}$	$R_{n,B}$	$C_{T,A}$	$C_{T,B}$
GeneA	1.3	2.4	24.0	27.0
HSKGα	1.3	2.4	25.3	28.3

Amplification efficiency of the reaction for the reference, *HSKG*α, is

$$E = \left(\frac{1.3}{2.4}\right)^{\frac{1}{25.3-28.3}} - 1 = 0.54^{\frac{1}{-3}} - 1 = 0.54^{-0.33} - 1 = 1.23 - 1 = 0.23$$

From these calculations, the amplification efficiencies of the two genes, the target and the reference, appear to be equal.

We will now use this amplification efficiency in the following equation (Equation 3) to determine a relative difference in the amount of mRNA of the two genes within the cells being tested.

$$\frac{R_{0,\text{Target}}}{R_{0,\text{Reference}}} = \frac{(1+E_{\text{Reference}})^{C_{T,\text{Reference}}}}{(1+E_{\text{Target}})^{C_{T,\text{Target}}}}$$

We will do this calculation for each well and then average the results. For well A1, as an example, we have

$$\frac{R_{0,\text{GeneA}}}{R_{0,\text{HSKG}\alpha}} = \frac{(1+0.23)^{24.81}}{(1+0.23)^{23.46}} = \frac{170.04}{128.58} = 1.32$$

Therefore, in well A1, *GeneA* RNA is present in liver cells at a 1.32-fold higher level than *HSKG*α RNA. The fold difference values for the other wells are calculated in this manner. They appear in Table 9.42.

Therefore, from Table 9.42, we see that, in the liver cells tested, *GeneA* is expressed at a 1.33-fold higher level than *HSKG*α. However, we should also attach a standard deviation to this value. The standard deviation calculations for this problem are shown in Table 9.43.

Therefore, from Table 9.43, we can attach a standard deviation of 0.02 to the fold difference value of 1.33. We can say that *GeneA* RNA is 1.33 $\pm$ 0.02-fold higher than *HSKG*α RNA in the liver cells tested.

Table 9.42 Calculation results for the fold difference of *GeneA* to *HSKGα*, as determined using Equation 3 using the Problem 9.18 dataset.

Well position	*GeneA* C_T (target)	*HSKGα* C_T (Reference)	Fold difference *GeneA* to *HSKGα*
A1	23.46	24.81	1.32
B1	23.39	24.79	1.34
C1	23.42	24.72	1.31
D1	23.35	24.86	1.37
E1	23.38	24.81	1.34
F1	23.36	24.71	1.32
		Average	1.33

Table 9.43 Standard deviation for the average fold difference in *GeneA* expression above *HSKGα*. The mean of 1.33 was calculated in Table 9.41. Calculations are for the Problem 9.18 dataset.

Well	Fold difference	$(x-\bar{x})$	$(x-\bar{x})^2$	$\sum(x-\bar{x})^2$	$\frac{\sum(x-\bar{x})^2}{n-1}$	$\sqrt{\frac{\sum(x-\bar{x})^2}{n-1}}$
A1	1.32	−0.01	0.0001	0.0024	0.00048	0.02
B1	1.34	0.01	0.0001			
C1	1.31	−0.02	0.0004			
D1	1.37	0.04	0.0016			
E1	1.34	0.01	0.0001			
F1	1.32	−0.01	0.0001			

Note: The designation of the A and B points on the amplification plot as needed for the determination of an amplification efficiency, though arbitrary, must be done within the exponential section of the plot. An underestimation of the efficiency will be carried forward into an underestimation of the relative difference between the target and reference genes.

9.9 THE R_0 METHOD OF RELATIVE QUANTIFICATION

Polymerase chain reaction amplifies DNA at an exponential rate. With a 100% efficient reaction, product doubles every cycle. In real-time PCR, the increase in fluorescent signal generated during the amplification reaction is proportional to the DNA concentration. The term R_0 is used to designate the theoretical amount of initial fluorescence at time zero. It can be expressed mathematically as

$$R_0 = R_n \times 2^{-C_T}$$

where C_T is the cycle threshold and R_n is the reporter signal at that threshold cycle.

Pierson (2006) describes the use of this equation for measuring relative gene expression. Knowing the initial amount of fluorescence, it is a simple matter to normalize a target to a reference by dividing the individual target gene's R_0 value by its respective internal control reference gene's R_0 value. Manually moving the threshold R_n value within the exponential region of the amplification curve should not change the R_0 value since, in so doing, there will be a corresponding change in the C_T.

An amplification efficiency (if it is known from a standard curve) can be incorporated into the equation as follows

$$R_0 = R_n \times (1 + E)^{-C_T}$$

The R_0 method can be used in a manner similar to that described for the $\Delta\Delta C_T$ method whereby a target gene, normalized to a reference, can be compared to a calibrator sample.

Problem 9.19 Total RNA is isolated from a sample of lung cells and is converted to cDNA. An aliquot of the cDNA reaction is placed into a well of an optical microtiter plate. Within that well, a target gene, *GeneA*, and a reference gene, *HSKG*α, are assayed for using their own unique primers and probes. *GeneA* is detected using a blue dye-labeled probe and *HSKG*α is detected using a green dye-labeled probe. When the run is completed, the threshold is set at 1.5 R_n. At this level of fluorescent signal, *GeneA* has a C_T value of 25.0 and *HSKG*α has a C_T of 26.35 (Figure 9.20). What is the fold difference in their amounts in this one well?

Solution 9.19

We will use the following equation to determine the R_0 values for the two genes:

$$R_0 = R_n \times 2^{-C_T}$$

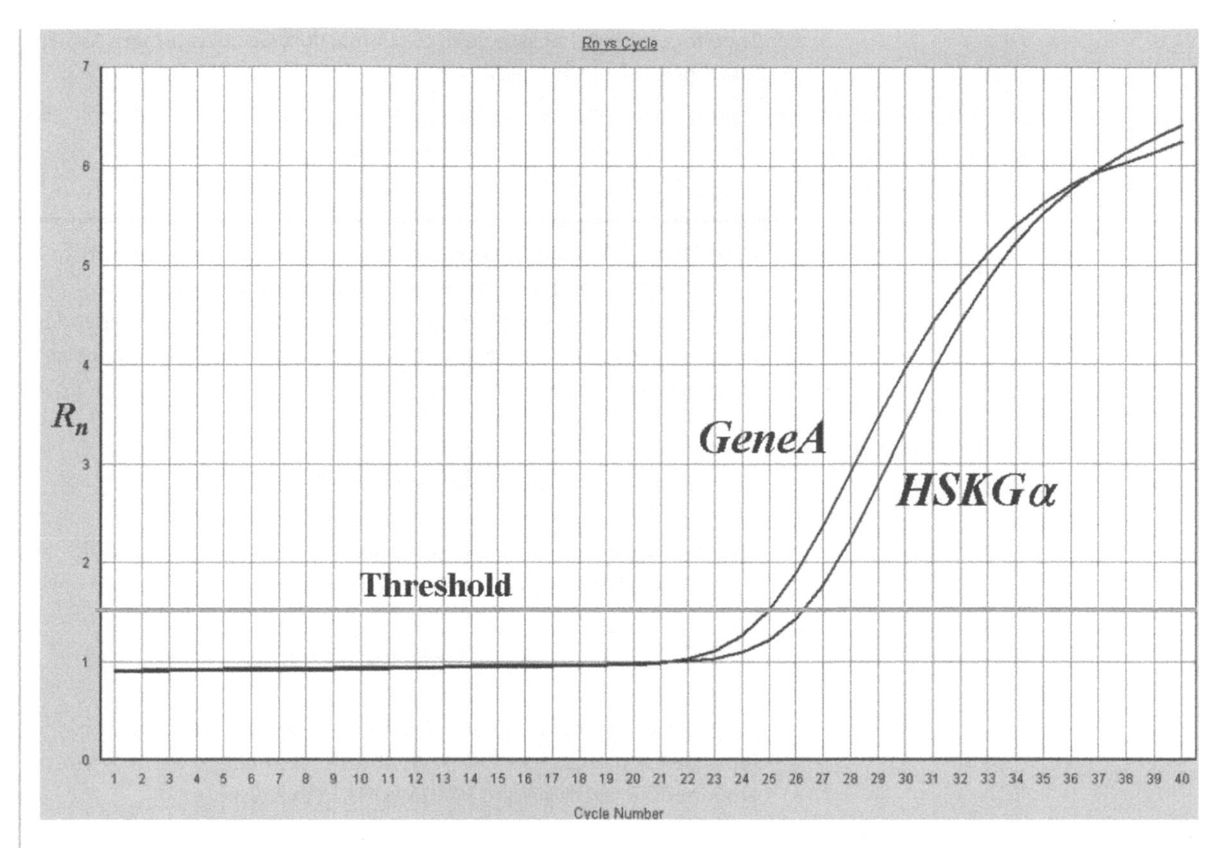

■ **FIGURE 9.20** Amplification curve for *GeneA* and *HSKGα* expression by the real-time PCR experiment described in Problem 9.19.

For *GeneA*, we have

$$\begin{aligned} R_0 &= 1.5 \times 2^{-25} \\ &= 1.5 \times (2.98 \times 10^{-8}) \\ &= 4.5 \times 10^{-8} \end{aligned}$$

For *HSKGα*, we have

$$R_0 = 1.5 \times 2^{-26.3} = 1.5 \times (1.21 \times 10^{-8}) = 1.8 \times 10^{-8}$$

Dividing the *GeneA* R_0 value by the *HSKGα* R_0 value yields

$$\frac{R_{0,\text{GeneA}}}{R_{0,\text{HSKG}\alpha}} = \frac{4.5 \times 10^{-8}}{1.8 \times 10^{-8}} = 2.5$$

Therefore, within that sample of cDNA reaction, *GeneA* is present at a 2.5-fold higher level than *HSKG*α.

Problem 9.20 A one-well experiment similar to that described in Problem 9.19 is conducted. However, this time, a standard curve is generated for *GeneA* and *HSKG*α using a dilution series of an independent and quantified DNA sample. The regression line for the *GeneA* standard curve yields the following equation:

$$y = -3.026349x + 28.535219$$

The regression line from the standard curve for *HSKG*α amplification derived from the standard curve has the equation

$$y = -3.056389x + 27.387451$$

Setting the threshold line at $1.5\,R_n$ for data analysis yields a C_T of 25.0 for *GeneA* and 26.35 for *HSKG*α. What is the fold difference between the amount of *GeneA* cDNA and *HSKG*α cDNA in that well?

Solution 9.20

We will first calculate the amplification efficiency of each gene. The regression lines equation has the general formula of $y = mx + b$ where m is the line's slope. Amplification efficiency is calculated using the following equation:

$$E = 10^{(-1/\text{slope})} - 1$$

GeneA is therefore amplified with an efficiency of

$$E = 10^{\left(\frac{-1}{-3.026349}\right)} - 1 = 10^{0.3304} - 1 = 2.14 - 1 = 1.14$$

HSKGα is amplified with an efficiency of

$$E = 10^{\left(\frac{-1}{-3.056389}\right)} - 1 = 10^{0.3272} - 1 = 2.12 - 1 = 1.12$$

We now use the following equation to determine R_0 for each gene:

$$R_0 = R_n \times (1 + E)^{-C_T}$$

For *GeneA*, we have

$$\begin{aligned} R_0 &= 1.5 \times (1 + 1.14)^{-25} \\ &= 1.5 \times (2.14)^{-25} = 1.5 \times (5.49 \times 10^{-9}) = 8.24 \times 10^{-9} \end{aligned}$$

and for *HSKG*α, we have

$$\begin{aligned} R_0 &= 1.5 \times (1 + 1.12)^{-26.3} \\ &= 1.5 \times (2.12)^{-26.3} = 1.5 \times (2.61 \times 10^{-9}) = 3.92 \times 10^{-9} \end{aligned}$$

The ratio of *GeneA* to *HSKG*α is

$$\frac{R_{0,GeneA}}{R_{0,HSKG\alpha}} = \frac{8.24 \times 10^{-9}}{3.92 \times 10^{-9}} = 2.1$$

Therefore, cDNA for *GeneA* is 2.1-fold more abundant in that sample than cDNA for *HSKG*α.

9.10 THE PFAFFL MODEL

Pfaffl (2001) describes a method for measuring relative gene expression based on the C_T values and amplification efficiencies of the target and reference genes. Normalization of the target to the reference and consideration for differences in their amplification efficiencies are incorporated into the calculation presented as a ratio of the change in expression of the target to the change in the expression of the reference. This ratio is calculated as

$$\text{ratio} = \frac{(E_{\text{target}})^{\Delta C_{T,\,\text{target}}}}{(E_{\text{reference}})^{\Delta C_{T,\,\text{reference}}}}$$

where E is the amplification efficiency derived from the slope of a standard curve and calculated as $E = 10^{(-1/\text{slope})}$ and ΔC_T is calculated as the C_T of the untreated sample minus the C_T of the treated sample for either the target or the reference gene.

The E values for the target and reference genes are an integral part of the calculation and so their real-time PCR amplification efficiencies should be well characterized. Although this method does not require that a standard dilution curve be run at the same time and under the same reaction conditions as those used for the amplification of the experimental samples, since the equation has amplification efficiency as its central feature, running such a set of reactions in parallel would help the experimenter sleep at night. It should also be noted that efficiency is not a percentage. Rather, a 99% efficient reaction, for this method, should be expressed as 1.99 (see note in Section 9.4).

Problem 9.21 An experiment is performed to access the relative expression of *GeneA* (the target) to the housekeeping gene *HSKG*α (the reference) following 2 hr of heat-shock treatment (a shift from 37 to 44°C). Total RNA isolated from liver cells incubated at 37°C (t_0 for the untreated samples) is reverse transcribed into cDNA. Ten nanograms of the cDNA reaction are added to replicate wells of a 96-well optical plate. Following the recovery of the total RNA sample, the liver cells are shifted to 44°C and incubated for 2 hr at this elevated temperature. Total RNA isolated from the heat-shocked cells is converted to cDNA and 10 ng of the reaction is added to replicate wells of the microtiter plate. *GeneA* and *HSKG*α are assayed simultaneously in each well using their own unique primers and dye-labeled probes. *GeneA* is determined from a standard dilution curve to amplify with an efficiency of 1.97. *HSKG*α is determined to amplify with an efficiency of 1.99. Figure 9.21 shows the amplification curve for the *GeneA* and *HSKG*α genes from both untreated (37°C at t_0) and treated (44°C incubation for 2 hr) samples (one well each condition). Two wells give the results shown in Table 9.43. Using the Pfaffl method, calculate the relative gene expression of *GeneA* compared to *HSKG*α.

Note: In any experiment of this kind, it is important to perform replicates of each sample. In this experiment, like samples should be averaged (all GeneA/untreated C_T values averaged, all *HSKG*α/untreated C_T values averaged, etc.) and those averages used to calculate a relative expression ratio. However, since this approach has been demonstrated in previous problems in this section, for this example and to demonstrate the math, we will only examine a single set of C_T values.

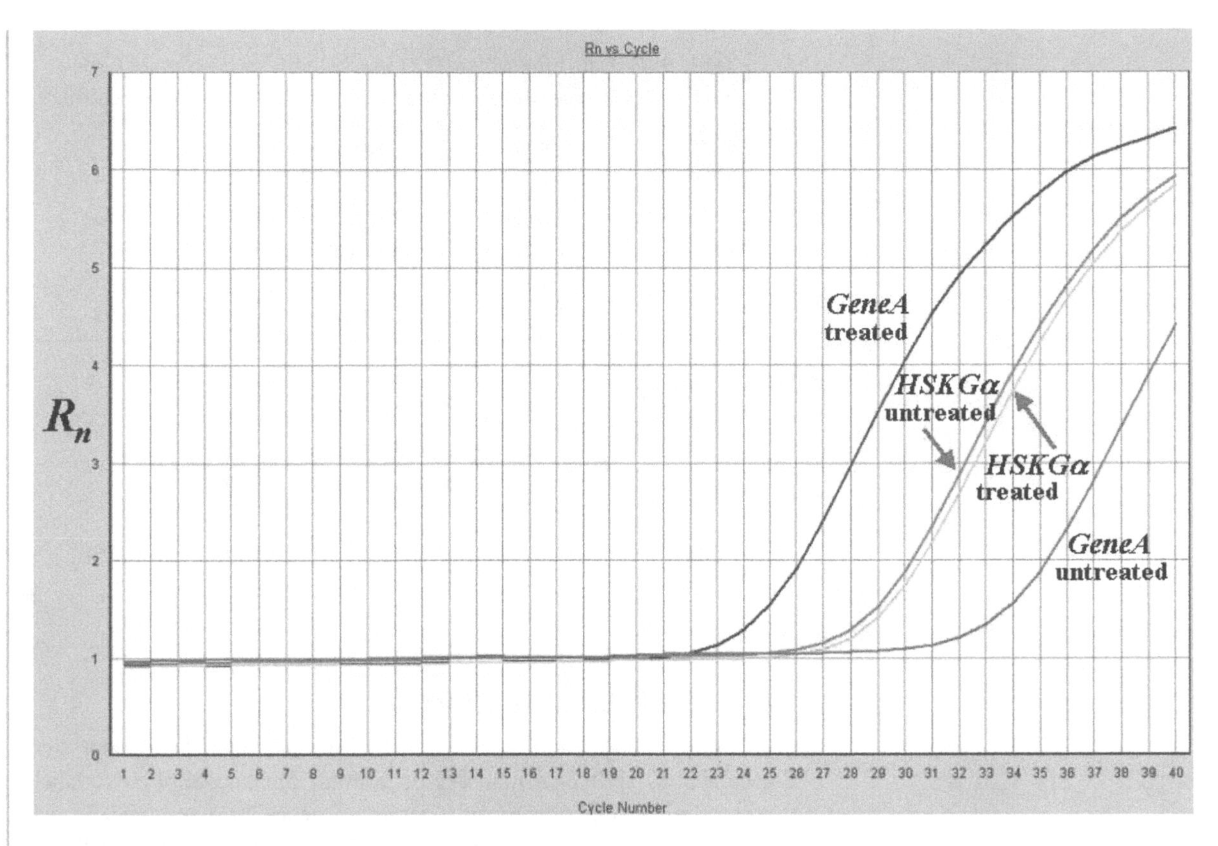

■ **FIGURE 9.21** Amplification curve for the samples described in Problem 9.21. R_n represents the normalized reporter signal. Note that the expression of the housekeeping gene *HSKGα* is changed very little by the heat-shock treatment, as is desirable for a reference gene in a relative expression study.

Solution 9.21

We will first calculate the fold increase in *GeneA*. Untreated (before heat shock) *GeneA* gives a C_T of 32.54. After two hours at 44°C, *GeneA* gives a C_T of 23.41. With an amplification efficiency of 1.97, its fold increase is

$$\begin{aligned}\text{increase} &= (E)^{C_{T,\,\text{untreated}} - C_{T,\,\text{treated}}}\\ &= (1.97)^{32.54-23.41}\\ &= (1.97)^{9.13} = 488\text{-fold increase}\end{aligned}$$

HSKGα amplifies with an efficiency of 1.99. Untreated, it generated a C_T value of 27.72. Treated (heat-shocked), it generates a C_T value of 27.93. Its fold increase is calculated as

$$\text{increase} = (1.99)^{27.72-27.93} = (1.99)^{-0.21} = 0.87$$

Table 9.44 Data for Problem 9.21 in which the expression of *GeneA* and *HSKG*α are monitored for their reaction to a heat-shock treatment.

Well position	Gene	Experimental condition	C_T
A1	*GeneA*	Untreated (37°C)	32.54
A1	*HSKG*α	Untreated (37°C)	27.72
A6	*GeneA*	Treated (44°C)	23.41
A6	*HSKG*α	Treated (44°C)	27.93

Our ratio is

$$\text{ratio} = \frac{\text{change in the target}}{\text{change in the reference}} = \frac{488}{0.87} = 561$$

Therefore, *GeneA* expression is increased 561-fold relative to *HSKG*α.

CHAPTER SUMMARY

Real-time PCR uses an increase in the intensity of a fluorescent signal generated by an intercalating dye or from the breakdown of a dye-labeled probe during amplification of a target sequence to detect nucleic acids either for their presence or absence or for their amount. The PCR cycle number at which fluorescent signal becomes discernible above background noise is called the C_T value. Between two samples, a decrease in a CT value (ΔC_T) of one cycle represents a doubling of the amount of target. This relationship can be expressed as

$$\text{The fold increase in the amount of amplified product} = 2^{-\Delta C_T}$$

Standard deviation, a measure of the degree of variability in a dataset, is calculated to determine to what amount the C_T values of replicate real-time PCRs vary from the mean. It is given by the equation

$$s = \sqrt{\frac{\sum(x - \bar{x})^2}{n - 1}}$$

DNA quantification using real-time PCR is performed by generating a standard curve from a dilution series of an independently quantified supply of reference DNA, plotting their C_T values vs. the log of the DNA concentration, and determining the amount of DNA contained in the unknown sample from the plot's regression line.

Real-time PCR and the generation of fluorescent signal proceeds at a rate defined by the equation

$$R_f = R_i(1 + E)^n$$

where R_f is equivalent to the final fluorescence signal of the reporter dye, R_i is equivalent to the reporter dye's initial fluorescent signal, E is the amplification efficiency, and n is the number of cycles. Amplification efficiency (E) is determined from a standard curve using the following formula:

$$EXPONENTIAL\ AMPLIFICATION = 10^{(-1/slope)}$$

$$EFFICIENCY = \left(10^{(-1/slope)} - 1\right) \times 100$$

where the slope in this equation is that of the standard curve's regression line. A 100% efficient reaction will have a slope of -3.32. During exponential amplification, a 100% efficient reaction will double every cycle and will produce a 10-fold increase in PCR product every 3.32 cycles. In mathematical terms, that is

$$\log_2 10 = 3.3219$$

or

$$2^{3.3219} = 10$$

A slope with a more negative value will represent a PCR having efficiency less than 100% and in need of optimization. A slope having a more positive value can indicate a problem with the reaction.

Gene expression can be assessed using real-time PCR by converting a cell's mRNA into cDNA and using that as the reaction's substrate. Most real-time PCR gene expression studies involve comparing the amount of one gene's (the target's) amount of transcript in a cell to another gene's (the reference's, or calibrator's) amount of transcript. A gene's expression can be compared across tissue types and in response to an experimental treatment. Several methods have been reported for evaluating real-time PCR data. One of the first is known as the $\Delta\Delta CT$ or Comparative C_T method. The change in the target gene's C_T value relative to the change in a calibrator's C_T is given as

$$\text{Relative Fold Change} = 2^{-\Delta\Delta C_T}$$

In the $\Delta\Delta C_T$ method, ΔC_T is equivalent to the change in the target relative to the endogenous control (reference):

$$C_{T,\ target} - C_{T,\ reference} = \Delta C_T$$

The $\Delta\Delta C_T$ value is calculated as the ΔC_T of each sample minus the ΔC_T of the calibrator (the sample representing zero time or the untreated control against which all samples are compared).

Use of the comparative C_T method requires that the reference gene is amplified at a similar efficiency to that of the target gene and that it is not dramatically affected by the experimental treatment. An initial set of experiments should be conducted to test whether or not this is the case. ANOVA (analysis of variance) can be used to compare differences in amplification efficiencies between genes. It can assess the variability between a set of samples within an individual group and between entire groups of samples. Variance between samples within a group is calculated using the general equation for variance:

$$\text{variance} = s^2 = \frac{\sum(x - \bar{x})^2}{n - 1}$$

where x is a sample's C_T value within a group, $\bar{x}$ is the average C_T value for all the samples within that individual group, and n is the number of samples. The between-groups variance s^{*2} is calculated as

$$s^{*2} = \frac{(\bar{a} - \bar{\bar{x}})^2 + (\bar{b} - \bar{\bar{x}})^2 + (\bar{c} - \bar{\bar{x}})^2 + \ldots}{m - 1}$$

where $\bar{a}$, $\bar{b}$, and $\bar{c}$ are the averages of each different group, $\bar{\bar{x}}$ is the average of all the averages from each group, and m is the number of groups. The average of all sample variances is calculated as

$$s^2 = \frac{s_a^{\ 2} + s_b^{\ 2} + s_c^{\ 2} + \ldots}{m}$$

These values are used to calculate an F statistic, the ratio of the between-groups variance to the within-groups variance:

$$F = \frac{n \times s^{*2}}{s^2}$$

where n is the number of measurements in each individual group. If the F value is large, the variability between groups (the numerator) must be large in comparison to the variability within each group (the denominator). The F statistic is used to determine whether the sample means of several groups are equal. If they are, the null hypothesis, that the experimental treatment has no effect on the expression of the reference gene, can be accepted. The cutoff value where that determination can be made is found in an F distribution table. A p-value associated with the F statistic will indicate how probable it is that you could get an F statistic more extreme than the one you determined if the null hypothesis were true. The smaller the p-value, the more likely it is that the null hypothesis can be rejected.

To test whether or not the target and reference genes have similar amplification efficiencies and can, therefore, be used with the $\Delta\Delta C_T$ method involves preparing standard curves for each gene from a dilution series of cDNA. The average C_T values at each diluted concentration are found by subtracting reference gene from target gene and these C_T values are plotted against the log of the input RNA used to make the diluted cDNA for the standard curve. If the regression line to that plot has a slope less than 0.1, then it can be assumed that the two genes have amplification efficiencies similar enough that they can be paired target and reference.

The Standard Curve method for measuring relative gene expression uses the equation for the regression lines derived from standard curves for the target and reference genes to calculate the amount of the target gene's transcript in an unknown sample. The amount of target is normalized to the reference gene and compared to the amount of reference RNA in a calibrator sample.

Real-time PCR kinetics can be used to calculate an amplification efficiency and, from that, an initial amount of target. Efficiency is derived from the equation

$$E = \left(\frac{R_{n,A}}{R_{n,B}}\right)^{\frac{1}{C_{T,A} - C_{T,B}}} - 1$$

where $R_{n,A}$ and $R_{n,B}$ represent R_n (reporter fluorescence at cycle n) at arbitrary thresholds A and B in the exponential part of the amplification curve, respectively, and $C_{T,A}$ and $C_{T,B}$ are the threshold cycles at these arbitrary thresholds. In a relative quantification experiment, one gene can be normalized to a reference by using the following equation:

$$\frac{R_{0,\text{Target}}}{R_{0,\text{Reference}}} = \frac{(1 + E_{\text{Reference}})^{C_{T,\text{Reference}}}}{(1 + E_{\text{Target}})^{C_{T,\text{Target}}}}$$

where R_0 is the initial amount of fluorescence.

In the R_0 method for measuring relative gene expression, the following equation can be used to determine an initial amount of fluorescence at zero time (when the temperature is shifted, when cells are infected, when cells are exposed to a drug, etc.):

$$R_0 = R_n \times 2^{-C_T}$$

where C_T is the cycle threshold and R_n is the reporter signal at that threshold cycle. When R_0 is determined for a target gene and a reference gene, the two can be compared relative to each other. If an amplification efficiency (E) is known, that value can be incorporated as follows:

$$R_0 = R_n \times (1+E)^{-C_T}$$

The Pfaffl Model is used to measure relative gene expression based on C_T values and amplification efficiencies according to the equation

$$\text{ratio} = \frac{(E_{\text{target}})^{\Delta C_{T,\text{target}}}}{(E_{\text{reference}})^{\Delta C_{T,\text{reference}}}}$$

where E is the amplification efficiency derived from the slope of a standard curve and calculated as $E = 10^{(-1/\text{slope})}$ and ΔC_T is calculated as the C_T of the untreated sample minus the C_T of the treated sample for either the target or the reference gene.

REFERENCES

Liu, W., and D.A. Saint (2002). A new quantitative method of real time reverse transcription polymerase chain reaction assay based on simulation of polymerase chain reaction kinetics. *Anal. Biochem.* 302:52–59.

Livak, K.J., and T.D. Schmittgen (2001). Analysis of relative gene expression data using real-time quantitative PCR and the $2^{-\Delta\Delta C_T}$ method. *Methods* 25:402–408.

Pfaffl, M.W. (2001). A new mathematical model for relative quantification in real-time RT-PCR. *Nucleic Acids Res.* 29:2002–2007.

Pierson, S.N. (2006). Quantitative analysis of ocular gene expression. In *Real-time PCR* (M. Tevfik Dorak, ed.). Taylor & Francis Group, New York, NY, pp. 107–126.

FURTHER READING

Freeman, W.M., S.J. Walker, and K.E. Vrana (1999). Quantitative RT-PCR: pitfalls and potential. *BioTechniques* 26:112–125.

O'Connor, W., and E.A. Runquist (2008). Error minimization algorithm for comparative quantitative PCR analysis: Q-Anal. *Anal. Biochem.* 378:96–98.

Peirson, S.N., J.N. Butler, and R.G. Foster (2003). Experimental validation of novel and conventional approaches to quantitative real-time PCR data analysis. *Nucleic Acids Res.* 31(14):e73.

Qiu, H., K. Durand, H. Rabinovitch-Chable, M. Rigaud, V. Gazaille, P. Clavere, and F.G. Sturtz (2007). Gene expression of *HIF-1α* and *XRCC4* measured in human samples by real-time RT-PCR using the sigmoidal curve-fitting method. *BioTechniques* 42:355–362.

Rutledge, R.G. (2004). Sigmoidal curve-fitting redefines quantitative real-time PCR with the prospective of developing automated high-throughput applications. *Nucleic Acids Res.* 32(22):e178 (doi:10.1093/nar/gnh177).

Schmittgen, T.D., and B.A. Zahrajsek (2000). Effect of experimental treatment on housekeeping gene expression: validation by real-time, quantitative RT-PCR. *J. Biochem. Biophys. Methods* 46:69–81.

User Bulletin #2: ABI PRISM 7700 Sequence Detection System. Applied Biosystems publication, 2001.

Wong, M.L., and J.F. Medrano (2005). Real-time PCR for mRNA quantitation. *BioTechniques* 39:75–85.

Chapter 10

Recombinant DNA

INTRODUCTION

The development of **recombinant DNA** technology (also called **gene cloning** or **gene splicing**), the methodology used to join DNA from different biological sources for the purpose of determining the sequence or for manipulating its expression, spawned an era of gene discovery and launched an industry. In its simplest terms, gene cloning is achieved in four steps:

1. A particular DNA is first fragmented into smaller pieces.
2. A specific segment of the fragmented DNA is then joined to another DNA, termed a **vector**, containing the genetic elements required for intracellular replication.
3. The recombinant molecule is introduced into a host cell, where it can replicate, producing multiple copies of the cloned DNA fragment.
4. Cells containing recombinant clones are identified and propagated.

10.1 RESTRICTION ENDONUCLEASES

Restriction endonucleases are enzymes that recognize a specific DNA sequence, called a **restriction site**, and cleave the DNA within or adjacent to that site. For example, the restriction endonuclease *Eco*R I, isolated from the bacterium *E. coli*, recognizes the following sequence:

$$^{5'}\mathrm{GAATTC}^{3'}$$

$$_{3'}\mathrm{CTTAAG}_{5'}$$

The broken strands of DNA depicted here should line up such that the C and G are right across from each other as shown here:

$$^{5'}\text{---------G}^{3'} \quad \text{and} \quad ^{5'}\text{AATTC--------}^{3'}$$

$$_{3'}\text{---------CTTAA}_{5'} \qquad\qquad _{3'}\text{G--------}_{5'}$$

Calculations for Molecular Biology and Biotechnology. DOI: 10.1016/B978-0-12-375690-9.00010-3
© 2010 Elsevier Inc. All rights reserved.

The termini produced by *EcoR* I, since they are complementary at their single-stranded overhangs, are said to be **cohesive** or **sticky**.

A number of restriction enzymes have been isolated from a variety of microbial sources. Recognition sites for specific enzymes range in size from 4 to 13 bp, and, for most restriction enzymes used in gene cloning, are palindromes: sequences that read in the 5′-to-3′ direction on one strand are the same as those in the 5′-to-3′ direction on the opposite strand. Restriction enzyme activity is usually expressed in terms of units, in which one unit is that amount of enzyme that will cleave all specific sites in 1 μg of a particular DNA sample (usually λ DNA) in one hour at 37°C.

Problem 10.1 A stock of human genomic DNA has a concentration of 275 μg/mL. You wish to cut 5 μg with 10 units of the restriction endonuclease *Hind* III. The *Hind* III enzyme has a concentration of 2500 units/mL. What volumes of DNA and enzyme should be used?

Solution 10.1

Restriction digests are typically performed in small volumes (20 to 100 μL). Such reactions are prepared with a micropipettor delivering μL amounts. The concentrations of both the human DNA and the enzyme stocks are given as a quantity per mL. A conversion factor, therefore, must be used to convert mL to μL. The volumes needed for each reaction component can be written as follows.

The amount of input DNA is calculated as follows:

$$\frac{275\,\mu\text{g DNA}}{\text{mL}} \times \frac{1\text{ mL}}{1000\,\mu\text{L}} \times x\,\mu\text{L} = 5\,\mu\text{g DNA}$$

Solving the equation for x yields the following result:

$$\frac{(275\,\mu\text{g DNA})x}{1000} = 5\,\mu\text{g DNA}$$

$$(275\,\mu\text{g DNA})x = 5000\,\mu\text{g DNA}$$

$$x = \frac{5000\,\mu\text{g DNA}}{275\,\mu\text{g DNA}} = 18.2$$

Therefore, 18.2 μL of the human DNA stock contains 5 μg of DNA.

The amount of *Hin*d III enzyme is calculated as follows:

$$\frac{2500\ \text{units}}{\text{mL}} \times \frac{1\ \text{mL}}{1000\ \mu\text{L}} \times x\ \mu\text{L} = 10\ \text{units}$$

Solving this equation for x yields the following result:

$$\frac{(2500\ \text{units})x}{1000} = 10\ \text{units}$$

$$(2500\ \text{units})x = 10\,000\ \text{units}$$

$$x = \frac{10\,000\ \text{units}}{2500\ \text{units}} = 4$$

Therefore, 4 μL of the enzyme stock is needed to deliver 10 units.

10.1.1 The frequency of restriction endonuclease cut sites

The frequency with which a particular restriction site occurs in any DNA depends on its base composition and the length of the recognition site. For example, in mammalian genomes, a G base follows a C in a much lower frequency than would be expected by chance. Therefore, a restriction endonuclease such as *Nru* I, which recognizes the sequence TCGCGA, cuts mammalian DNA less frequently than it does DNA from bacterial sources having a more random distribution of bases. Also, as might be expected, restriction sites that are 4 bp in length will occur more frequently than restriction sites that are 6 bp in length. The average size of a restriction fragment produced from a random sequence by a particular endonuclease can be estimated by the method illustrated in the following problem.

Problem 10.2 The restriction endonuclease *Hin*d III recognizes the sequence AAGCTT. If genomic DNA of random sequence is cleaved with *Hin*d III, what will be the average size of the fragments produced?

Solution 10.2

The chance that any one base, A, C, G, or T, will occur at any particular position in DNA of random sequence is one in four. The number of cutting sites in random DNA can be estimated by raising ¼ to the n ($(1/4)^n$), where the exponent n is the number of bps in the recognition sequence.

A 'six-cutter' such as the restriction enzyme *Hin*d III, which recognizes a 6 bp sequence, will cut at a frequency of $(1/4)^6$:

$$\left(\frac{1}{4}\right)^6 = \frac{1}{4096}$$

Therefore, in DNA of random sequence, a *Hin*d III site might be expected, on average, every 4096 bp along the molecule. (This value reflects only an estimate, and actual sizes of restriction fragments generated from any DNA may vary considerably. For example, bacteriophage λ DNA, 48 502 bp in length, generates *Hin*d III fragments ranging in size from 125 to over 23 000 bp.)

10.2 CALCULATING THE AMOUNT OF FRAGMENT ENDS

DNA is digested with restriction enzymes to create fragments for cloning, for gene mapping, or for the production of genetic probes. For some applications, such as fragment end labeling, it is important for the experimenter to estimate the amount of DNA fragment ends. For a given amount of a linear fragment having a certain length, the moles of fragment ends are given by the following equation:

$$\text{moles of DNA ends} = \frac{2 \times (\text{grams of DNA})}{(\text{number of bp}) \times \left(\frac{660\ \text{g/mol}}{\text{bp}}\right)}$$

(There is a 2 in the numerator of this equation because a linear DNA fragment has two ends.)

Problem 10.3 You have 4 μg of a 3400 bp fragment. How many picomoles of ends does this represent?

Solution 10.3

First, calculate the moles of DNA ends and then convert this amount to picomoles. The amount of fragment is expressed in μg. Use of the preceding equation requires that the fragment amount be converted to g. This is done as follows:

$$4\,\mu g \times \frac{1 g}{1 \times 10^{6}\,\mu g} = 4 \times 10^{-6}\,g$$

Substituting the given values into the equation for calculating moles of DNA ends yields the following result:

$$\text{moles of DNA ends} = \frac{2 \times (\text{grams of DNA})}{(\text{number of bp}) \times \left(\frac{660\ \text{g/mol}}{\text{bp}}\right)}$$

$$\text{moles of DNA ends} = \frac{2 \times (4 \times 10^{-6}\ \text{g})}{(3400\ \text{bp}) \times \left(\frac{660\ \text{g/mol}}{\text{bp}}\right)} = \frac{8 \times 10^{-6}\ \text{g}}{2.2 \times 10^{6}\ \text{g/mol}}$$

$$= 3.6 \times 10^{-12}\ \text{mol}$$

This value is converted to picomoles by multiplying by the conversion factor 1×10^{12} pmol/mol.

$$3.6 \times 10^{-12}\ \text{mol} \times \frac{1 \times 10^{12}\ \text{pmol}}{\text{mol}} = 3.6\ \text{pmol}$$

Therefore, 4 μg of a 3400 bp fragment is equivalent to 3.6 pmol of DNA ends.

■

10.2.1 The amount of ends generated by multiple cuts

The number of moles of ends created by a restriction digestion of a DNA containing multiple recognition sites can also be determined. For a circular DNA (such as a plasmid or cosmid), the moles of ends created by a restriction digest can be calculated using the following equation:

$$\text{moles ends} = 2 \times (\text{moles DNA}) \times (\text{number of restriction sites})$$

The 2 multiplier at the beginning of this expression is necessary since each restriction enzyme cut generates two ends.

The moles of ends created by multiple-site digestion of a linear DNA molecule are given by the following expression:

$$\text{moles ends} = [(2 \times (\text{moles DNA}) \times (\text{number of restriction sites})] + [2 \times (\text{moles DNA})]$$

In this equation, the function at the end of the second expression in brackets is necessary to account for the two ends already on the linear molecule.

Problem 10.4 A 4200 bp plasmid contains five sites for the restriction endonuclease *Eco*R II. If 3 μg are cut with this restriction enzyme, how many picomoles of ends will be produced?

Solution 10.4

Since a circular plasmid is being digested, we will use the following equation:

$$\text{moles ends} = 2 \times (\text{moles DNA}) \times (\text{number of restriction sites})$$

This equation requires that we determine the number of moles of plasmid DNA. To determine that value, the plasmid's molecular weight must first be determined. Moles of plasmid can then be calculated from the amount of plasmid DNA to be digested (3 μg). The molecular weight of the plasmid is calculated as follows:

$$4200 \text{ bp} \times \frac{660 \text{ g/mol}}{\text{bp}} = 2.8 \times 10^6 \text{ g/mol}$$

The number of moles of plasmid represented by 3 μg is determined by multiplying the amount of plasmid (converted μg to g) by the plasmid's molecular weight:

$$3\,\mu\text{g} \times \frac{1\text{ g}}{1 \times 10^6\,\mu\text{g}} \times \frac{1\text{ mol}}{2.8 \times 10^6\text{ g}} = \frac{3\text{ mol}}{2.8 \times 10^{12}} = 1.1 \times 10^{-12}\text{ mol}$$

Therefore, 3 μg of a 4200 bp plasmid are equivalent to 1.1×10^{-12} moles. Placing this value into the equation for moles of ends created from digestion of a circular molecule gives the following result:

$$\text{moles of ends} = 2 \times (1.1 \times 10^{-12}\text{ mol}) \times 5 = 1.1 \times 10^{-11}\text{ mol}$$

Therefore, 1.1×10^{-11} moles of ends are generated by an *Eco*R II digest of the 4200 bp plasmid. This value is equivalent to 11 picomoles of ends, as shown here:

$$1.1 \times 10^{-11}\text{ mol} \times \frac{1 \times 10^{12}\text{ pmol}}{\text{mol}} = 11\text{ pmol}$$

Problem 10.5 Four micrograms of a purified 6000 bp linear DNA fragment are to be cut with *Alu* I. There are 12 *Alu* I sites in this fragment. How many picomoles of ends will be created by digesting the fragment with this enzyme?

Solution 10.5

This problem will be approached in the same manner as Problem 9.4. The molecular weight of the 6000 bp fragment is calculated first:

$$6000\,\text{bp} \times \frac{660\,\text{g/mol}}{\text{bp}} = 4 \times 10^{6}\,\text{g/mol}$$

The number of moles of 6000 bp fragment in 4 μg is then determined:

$$4\,\mu\text{g} \times \frac{1\,\text{g}}{1 \times 10^{6}\,\mu\text{g}} \times \frac{1\,\text{mol}}{4 \times 10^{6}\,\text{g}} = \frac{4\,\text{mol}}{4 \times 10^{12}} = 1 \times 10^{-12}\,\text{mol DNA}$$

Placing this value into the equation for calculating moles of ends created from digestion of a linear molecule gives the following relationship:

$$\text{moles of ends} = [2 \times (\text{moles DNA}) \times (\text{number of restriction sites})] + [2 \times (\text{moles DNA})]$$

$$\begin{aligned}\text{moles of ends} &= [2 \times (1 \times 10^{-12}\,\text{mol}) \times (12)] + [2 \times (1 \times 10^{-12}\,\text{mol})] \\ &= [2.4 \times 10^{-11}\,\text{mol}] + [2 \times 10^{-12}\,\text{mol}] = 2.6 \times 10^{-11}\,\text{mol}\end{aligned}$$

This value is converted to picomoles of ends in the following manner:

$$2.6 \times 10^{-11}\,\text{mol} \times \frac{1 \times 10^{12}\,\text{pmol}}{\text{mol}} = 26\,\text{pmol}$$

Therefore, digesting 4 μg of the 6000 bp fragment with *Alu* I will generate 26 pmol of ends.

10.3 LIGATION

Once the DNA to be cloned exists as a defined fragment (called the **target** or **insert**), it can be joined to the vector by the process called **ligation**. The tool used for this process is DNA ligase, an enzyme that catalyzes the formation of a phosphodiester bond between juxtaposed 5′-phosphate and 3′-hydroxyl termini in duplex DNA. The desired product in a ligation reaction is a functional hybrid molecule that consists exclusively of the vector plus the insert. However, a number of other events can occur in a ligation reaction, including circularization of the target fragment, ligation of vector without the addition of an insert, ligation of a target molecule to another target molecule, and any number of other possible combinations.

Two vector types are used in recombinant DNA research: circular molecules such as plasmids and cosmids, and linear cloning vectors such as those derived from the bacteriophage λ. In either case, for joining to a target DNA fragment, a vector is first cut with a restriction enzyme that produces compatible ends with those of the target. Circular vectors, therefore, are converted to a linear form prior to ligation to target. The insert fragment is then ligated to the prepared vector to create a recombinant molecule capable of replication once introduced into a host cell.

Dugaiczyk et al. (1975) described the events occurring during ligation of *Eco* R I fragments, observations that can be applied to any ligation reaction. They identified the two factors that play the biggest role in determining the outcome of any particular ligation. First, the readiness with which two DNA molecules join is dependent on the concentration of their ends: the higher the concentration of compatible ends (those capable of being joined), the greater the likelihood that two termini will meet and be ligated. This parameter is designated by the term i and is defined as the *total* concentration of complementary ends in the ligation reaction. Second, the amount of circularization occurring within a ligation reaction is dependent on the parameter j, the concentration of same-molecule ends in close enough proximity to each other that they could potentially and effectively interact. For any DNA fragment, j is a constant value dependent on the fragment's length, not on its concentration in the ligation reaction.

For a linear fragment of duplex DNA with self-complementary, cohesive ends (such as those produced by *Eco* R I), i is given by the formula

$$i = 2N_0M \times 10^{-3} \text{ ends/mL}$$

where N_0 is Avogadro's number (6.023×10^{23}) and M is the molar concentration of the DNA.

Some DNA fragments to be used for cloning are generated by a double digest, say by using both *Eco* R I and *Hin*d III in the same restriction reaction. In such a case, one end of the fragment will be cohesive with other *Eco* R I-generated termini while the other end of the DNA fragment will be cohesive with other *Hin*d III-generated termini. For such DNA fragments with ends that are not self-complementary, the concentration of each end is given by

$$i = N_0 M \times 10^{-3} \text{ ends/mL}$$

Dugaiczyk et al. (1975) determined that a value for j, the concentration of one end of a molecule in the immediate vicinity of its other end, can be calculated using the equation

$$j = \left(\frac{3}{2\pi lb}\right)^{3/2} \text{ ends/mL}$$

where l is the length of the DNA fragment, b is the minimal length of DNA that can bend around to form a circle, and π is the number pi, the ratio of the circumference of a circle to its diameter (approximately 3.14, a number often seen when performing calculations relating to measurements of the area of a circle). The authors specifically determined the value of j for bacteriophage λ DNA. λ DNA is 48 502 bp in length, and they assigned a value of 13.2×10^{-4} cm to l. For λ DNA in ligation buffer, they assigned a value of 7.7×10^{-6} cm to b. With these values, the j value for λ (j_λ) is calculated to be 3.22×10^{11} ends/mL.

The j value for any DNA molecule can be calculated in relation to j_λ using the equation

$$j = j_\lambda \left(\frac{MW_\lambda}{MW}\right)^{1.5} \text{ ends/mL}$$

where j_λ is equal to 3.22×10^{11} ends/mL and MW represents molecular weight. Since the λ genome is 48 502 bp in length, its molecular weight can be calculated using the conversion factor 660 g/mol/bp, as shown in the following equation:

$$48\,502 \text{ bp} \times \frac{660 \text{ g/mol}}{\text{bp}} = 3.2 \times 10^7 \text{ g/mol}$$

Therefore, the molecular weight of lambda (MW_λ) is 3.2×10^7 g/mol.

Under circumstances in which j is equal to i ($j = i$, or $j/i = 1$), the end of any particular DNA molecule is just as likely to join with another molecule as it is to interact with its own opposite end. If j is greater than i ($j > i$), intramolecular ligation events predominate and circles are the primary product.

If i is greater than j ($i > j$), intermolecular ligation events are favored and hybrid linear structures predominate.

10.3.1 Ligation using λ-derived vectors

The vectors derived from phage λ carry a cloning site within the linear genome. Cleavage at the cloning site by a restriction enzyme produces two fragments, each having a λ cohesive end sequence (*cos*) at one end and a restriction enzyme-generated site at the other. Successful ligation of target with λ vector actually requires three primary events.

1. Joining of the target segment to one of the two λ fragments.
2. Ligation of the other λ fragment to the other end of the target fragment to create a full-length λ genome equivalent.
3. Joining of the *cos* ends between other full-length λ molecules to create a **concatemer**, a long DNA molecule in which a number of λ genomes are joined in a series, end-to-end. Only from a concatemeric structure can individual λ genomes be efficiently packaged into a protein coat allowing for subsequent infection and propagation.

In ligation of a fragment to the arms of a λ vector, both ends of the fragment will be compatible with the cloning site of either λ arm, but each arm has only one compatible end with the insert. For the most efficient ligation, the reaction should contain equimolar amounts of the restriction enzyme-generated ends of each of the three fragments, the λ left arm, the insert, and the λ right arm. However, since there are two fragments of λ and one insert fragment per recombinant genome, when preparing the ligation reaction, the molar ratio of annealed arms to insert should be 2:1 so that the three molecules have a ratio of 2:1:2 (left arm:insert:right arm). Furthermore, since linear ligation products need to predominate, the concentration of ends (i) and the molecule's length (j) must be manipulated such that j is less than i, thereby favoring the production of linear hybrid molecules over circular ones.

Problem 10.6 You have performed a partial *Sau*3A I digest of human genomic DNA, producing fragments having an average size of 20 000 bp. You will construct a **genomic library** (a collection of recombinant clones representing the entire genetic complement of a human) by ligating these fragments into a λEMBL3 vector digested with *Bam*H I. (*Bam*H I creates ends compatible with those generated by *Sau*3A I; both enzymes produce a GATC 5′ overhang.) The λEMBL3 vector is approximately 42 360 bp in length. Removal of the 13 130 bp 'stuffer' fragment (the segment between the *Bam*H I sites, which is replaced by the target insert), leaves two arms, which, if annealed, would give a linear genome of 29 230 bp.

a) If you wish to prepare a 150 µL ligation reaction, how many µg of insert and vector arms should be combined to give optimal results?

b) How many micrograms of λEMBL3 vector should be digested with *Bam*H I to yield the desired amount of arms?

Solution 10.6(a)

Since we wish to ensure the formation of linear concatemers to optimize the packaging/infection step, j should be less than i. The first step then is to calculate the j values for the λ arms (j_{arms}) and the inserts ($j_{inserts}$). The following equation can be used:

$$j = j_{\lambda}\left(\frac{MW_{\lambda}}{MW}\right)^{1.5} \text{ ends/mL}$$

where j_{λ} is 3.22×10^{11} ends/mL and, as calculated previously, $MW_{\lambda} = 3.2 \times 10^{7}$ g/mol.

The molecular weight of the λEMBL3-annealed arms is calculated as follows:

$$29\,230 \text{ bp} \times \frac{660 \text{ g/mol}}{\text{bp}} = 1.93 \times 10^{7} \text{ g/mol}$$

The j_{arms} value can now be calculated by substituting the proper values into the equation for determining j:

$$j_{arms} = (3.22 \times 10^{11} \text{ ends/mL})\left(\frac{3.2 \times 10^{7} \text{ g/mol}}{1.93 \times 10^{7} \text{ g/mol}}\right)^{1.5}$$

$$j_{arms} = (3.22 \times 10^{11} \text{ ends/mL})\left(\frac{1.8 \times 10^{11}}{8.48 \times 10^{10}}\right)$$

$$j_{arms} = (3.22 \times 10^{11} \text{ ends/mL})(2.12)$$

$$j_{arms} = 6.83 \times 10^{11} \text{ ends/mL}$$

Therefore, j for the annealed arms is equal to 6.83×10^{11} ends/mL.

The molecular weight of a 20 000 bp insert is calculated as follows:

$$20\,000 \text{ bp} \times \frac{660 \text{ g/mol}}{\text{bp}} = 1.3 \times 10^{7} \text{ g/mol}$$

The $j_{inserts}$ value can now be calculated.

$$j_{inserts} = (3.22 \times 10^{11} \text{ ends/mL})\left(\frac{3.2 \times 10^{7} \text{ g/mol}}{1.3 \times 10^{7} \text{ g/mol}}\right)^{1.5}$$

$$j_{inserts} = (3.22 \times 10^{11} \text{ ends/mL})\left(\frac{1.8 \times 10^{11}}{4.7 \times 10^{10}}\right)$$

$$j_{inserts} = (3.22 \times 10^{11} \text{ ends/mL})(3.8)$$

$$j_{inserts} = 1.22 \times 10^{12} \text{ ends/mL}$$

So that the production of linear ligation products can be ensured, j should be decidedly less than i. Maniatis et al. (1982) suggested that j should be 10-fold less than i, such that

$$(i = 10j)$$

The j values for both the arms and the insert have now been calculated. To further ensure that j will be less than i, we will use the larger of the j values, which in this case is $j_{inserts}$. If we use $i = 10j$, we then have

$$i = (10)(1.22 \times 10^{12} \text{ ends/mL}) = 1.22 \times 10^{13} \text{ ends/mL}$$

Since i represents the sum of all cohesive termini in the ligation reaction ($i = i_{inserts} + i_{arms}$), the following relationship is true:

$$i = (2N_0M_{insert} \times 10^{-3} \text{ ends/mL}) + (2N_0M_{arms} \times 10^{-3} \text{ ends/mL})$$

As discussed earlier, when using a λ-derived vector, the molar concentration of annealed arms should be twice the molar concentration of the insert ($M_{arms} = 2M_{insert}$). The equation above then becomes

$$i = (2N_0M_{insert} \times 10^{-3} \text{ ends/mL}) + (2N_0 2M_{insert} \times 10^{-3} \text{ ends/mL})$$

This equation can be rearranged and simplified to give the following relationship:

$$i = (2N_0M_{insert} + 2N_0 2M_{insert}) \times 10^{-3} \text{ ends/mL}$$

$$i = (2N_0M_{insert} + 4N_0M_{insert}) \times 10^{-3} \text{ ends/mL}$$

$$i = 6N_0M_{insert} \times 10^{-3} \text{ ends/mL}$$

The equation can be solved for M_{insert} by dividing each side of the equation by $6N_0 \times 10^{-3}$ ends/mL:

$$M_{insert} = \frac{i}{6N_0 \times 10^{-3} \text{ ends/mL}}$$

Substituting the value we have previously calculated for i (1.22×10^{13} ends/mL) and using Avogadro's number for N_0 gives the following relationship:

$$M_{\text{insert}} = \frac{1.22 \times 10^{13} \text{ ends/mL}}{6(6.023 \times 10^{23}) \times (1 \times 10^{-3} \text{ ends/mL})}$$

$$M_{\text{insert}} = \frac{1.22 \times 10^{13} \text{ ends/mL}}{3.6 \times 10^{21} \text{ ends/mL}} = 3.4 \times 10^{-9}\ M$$

By multiplying the molarity value by the average size of the insert (20000 bp) and the average molecular weight of a base pair (660 g/mol/bp), it is converted into a concentration expressed as μg/mL. Conversion factors are also used to convert liters to milliliters and to convert g to μg. (Remember also that molarity (M) is a term for a concentration given in moles/liter; $3.4 \times 10^{-9} M = 3.4 \times 10^{-9}$ mol/L.)

$$x \ \mu\text{g insert/mL} = \frac{3.4 \times 10^{-9} \text{moles}}{\text{liter}} \times \frac{660 \text{ g/mol}}{\text{bp}} \times 20\,000 \text{ bp}$$
$$\times \frac{\text{liter}}{1000 \text{ mL}} \times \frac{1 \times 10^{6} \ \mu\text{g insert}}{\text{g}}$$

$$x = \frac{44\,880 \ \mu\text{g insert}}{1000 \text{ mL}} = 44.9 \ \mu\text{g insert/mL}$$

The ligation reaction, therefore, should contain insert DNA at a concentration of 44.9 mg/mL. To calculate the amount of insert to place in a 150 mL reaction, the following relationship can be used. (Remember that one mL is equivalent to 1000 μL).

$$\frac{x \ \mu\text{g insert}}{150 \ \mu\text{L}} = \frac{44.9 \ \mu\text{g insert}}{1000 \ \mu\text{L}}$$

$$x \ \mu\text{g insert} = \frac{(44.9 \ \mu\text{g insert})(150 \ \mu\text{L})}{1000 \ \mu\text{L}} = \frac{6735 \ \mu\text{g insert}}{1000}$$
$$= 6.74 \ \mu\text{g insert}$$

Therefore, adding 6.74 μg of insert DNA to a 150 μL reaction will yield a concentration of 44.9 μg insert/mL.

It was determined previously that the required molarity of λ vector arms should be equivalent to twice the molarity of the inserts ($M_{\text{arms}} = 2M_{\text{insert}}$). The molarity of the insert fragment was calculated to be $3.4 \times 10^{-9} M$ (as determined earlier). The molarity of the arms, therefore, is

$$M_{\text{arms}} = (2)(3.4 \times 10^{-9}\ M) = 6.8 \times 10^{-9}\ M$$

Therefore, the λEMBL3 arms should be at a concentration of $6.8 \times 10^{-9} M$. This value is converted to a μg arms/mL concentration, as shown in the following calculation. The 29 230 bp term represents the combined length of the two λ arms (see Problem 9.6):

$$x\ \mu g\ \text{arms/mL} = \frac{6.8 \times 10^{-9}\ \text{moles}}{\text{liter}} \times \frac{660\ \text{g/mol}}{\text{bp}} \times 29\,230\ \text{bp} \times \frac{\text{liter}}{1000\ \text{mL}} \times \frac{1 \times 10^{6}\ \mu g\ \text{arms}}{\text{g}}$$

$$x\ \mu g\ \text{arms/mL} = \frac{1.3 \times 10^{5}\ \mu g\ \text{arms}}{1000\ \text{mL}} = 130\ \mu g\ \text{arms/mL}$$

Therefore, the λ arms should be at a concentration of 130 μg/mL. The amount of λEMBL3 arms to add to a 150 μL ligation reaction can be calculated by using a relationship of ratios, as follows:

$$\frac{x\ \mu g\ \lambda\text{EMBL3 arms}}{150\ \mu L} = \frac{130\ \mu g\ \lambda\text{EMBL3 arms}}{1000\ \mu L}$$

$$x\ \mu g\ \lambda\text{EMBL3 arms} = \frac{(130\ \mu g\ \lambda\text{EMBL3 arms})(150\ \mu L)}{1000\ \mu L} = 19.5\ \mu g\ \lambda\text{EMBL3 arms}$$

Therefore, 19.5 μg of λEMBL3 arms should be added to 6.74 μg of insert DNA in the 150 μL ligation reaction to give optimal results.

Solution 10.6(b)

The amount of λEMBL3 vector DNA to yield 19.5 μg of λEMBL3 arms can now be calculated. We will first determine the weight of a single λ genome made up of only the two annealed arms. We will then calculate how many genomes are represented by 19.5 μg of λEMBL3 arms.

As stated in the original problem, λEMBL3 minus the stuffer fragment is 29 230 bp, and this represents one genome. The weight of one 29 230 bp genome is calculated as follows:

$$x\ \mu g/\text{genome} = 29\,230\ \text{bp} \times \frac{660\ \text{g/mol}}{\text{bp}} \times \frac{1 \times 10^{6}\ \mu g}{\text{g}} \times \frac{1\ \text{mol}}{6.023 \times 10^{23}\ \text{genomes}}$$

$$x\ \mu g/\text{genome} = \frac{1.93 \times 10^{13}\ \mu g}{6.023 \times 10^{23}\ \text{genomes}} = 3.2 \times 10^{-11}\ \mu g/\text{genome}$$

Therefore, each 29 230 bp genome weighs 3.2×10^{-11} μg.

The number of genomes represented by 19.5 mg can now be calculated:

$$19.5\,\mu g \times \frac{1\text{ genome}}{3.2 \times 10^{-11}\,\mu g} = 6.1 \times 10^{11}\text{ genomes}$$

Since one genome of λEMBL3 vector will give rise to one 29 230 bp genome of λ arms, we now need to know how many micrograms of uncut λEMBL3 DNA is equivalent to 6.1×10^{11} genomes. λEMBL3 is 42 360 bp in length. We can determine how much each λEMBL3 genome weighs by the following calculation:

$$x\,\mu g/\text{genome} = 42\,360\text{ bp} \times \frac{660\text{ g/mol}}{\text{bp}} \times \frac{1 \times 10^{6}\,\mu g}{\text{g}} \times \frac{1\text{ mol}}{6.023 \times 10^{23}\text{ genomes}}$$

$$x\,\mu g/\text{genome} = \frac{2.8 \times 10^{13}\,\mu g}{6.023 \times 10^{23}\text{ genomes}} = 4.6 \times 10^{-11}\,\mu g/\text{genome}$$

Therefore, each λEMBL3 genome weighs 4.6×10^{-11} μg. The amount of λEMBL3 DNA equivalent to 6.1×10^{11} genomes can now be calculated:

$$\frac{4.6 \times 10^{-11}\,\mu g}{\text{genome}} \times 6.1 \times 10^{11}\text{ genomes} = 28.1\,\mu g\ \lambda\text{EMBL3}$$

Therefore, digestion of 28.1 μg of λEMBL3 vector DNA will yield 19.5 μg of vector arms.

■

10.3.2 **Packaging of recombinant λ genomes**

Following ligation of an insert fragment to the arms of a λ vector and concatemerization of recombinant λ genomes, the DNA must be packaged into phage protein head and tail structures, rendering them fully capable of infecting sensitive *E. coli* cells. This is accomplished by adding ligated recombinant λ DNA to a prepared extract containing the enzymatic and structural proteins necessary for complete assembly of mature virus particles. The packaging mixture is then plated with host cells that allow the formation of plaques on an agar plate (see Chapter 4). Plaques can then be analyzed for the presence of the desired clones. The measurement of the effectiveness of these reactions is called the **packaging efficiency** and can be expressed as the number of plaques (PFUs) generated from a given number of phage genomes or as the number of plaques per microgram of phage DNA.

Packaging extracts purchased commercially can have packaging efficiencies for wild-type concatemeric λ DNA of between 1×10^7 and 2×10^9 PFU/μg DNA.

A method for calculating this value is demonstrated in the following problem.

Problem 10.7 The 150 μL ligation reaction described in Problem 9.6 is added to 450 μL of packaging extract. Following a period to allow for recombinant genome packaging, 0.01 mL is withdrawn and diluted to a total volume of 1 mL of λ dilution buffer. From that tube, 0.1 mL is withdrawn and diluted to a total volume of 10 mL. One mL is withdrawn from the last dilution and is diluted further to a total volume of 10 mL dilution buffer. One-tenth mL of this last tube is plated with 0.4 mL of *E. coli* K-12 sensitive to λ infection. Following incubation of the plate, 813 plaques are formed. What is the packaging efficiency?

Solution 10.7

The packaging reaction is diluted in the following manner:

$$\frac{0.01\,\text{mL}}{1\,\text{mL}} \times \frac{0.1\,\text{mL}}{10\,\text{mL}} \times \frac{1\,\text{mL}}{10\,\text{mL}} \times 0.1\,\text{mL plated}$$

Multiplying all numerators together and all denominators together yields the total dilution:

$$\frac{0.01 \times 0.1 \times 1 \times 0.1}{1 \times 10 \times 10} = \frac{1 \times 10^{-4}\,\text{mL}}{100} = 1 \times 10^{-6}\,\text{mL}$$

The concentration of viable phage in the packaging mix (given as PFU/mL) is calculated as the number of plaques counted on the assay plate divided by the total dilution:

$$\frac{813\,\text{PFU}}{1 \times 10^{-6}\,\text{mL}} = 8.13 \times 10^{8}\,\text{PFU/mL}$$

Therefore, the packaging mix contains 8.13×10^8 PFU/mL.

The total number of PFU in the packaging mix is calculated by multiplying the foregoing concentration by the volume of the packaging reaction (packaging mix volume = 150 μL ligation reaction + 450 μL packaging extract = 600 μL (equivalent to 0.6 mL)):

$$\frac{8.13 \times 10^8 \text{ PFU}}{\text{mL}} \times 0.6\,\text{mL} = 4.88 \times 10^8 \text{ PFU total}$$

In Problem 10.6, it was determined that 6.1×10^{11} genomes could potentially be constructed from the λEMBL3 arms placed in the reaction. The percentage of these potential genomes actually packaged into viable virions can now be calculated by dividing the titer of the packaging mix by the number of potential genomes and multiplying by 100:

$$\frac{4.88 \times 10^8}{6.1 \times 10^{11}} \times 100 = 0.08\%$$

Therefore, 0.08% of the potential λ genomes were packaged.

The packaging efficiency can be calculated as PFU per microgram of recombinant λ DNA. The average recombinant phage genome should be the length of the λEMBL3 arms (29 230 bp) + the average insert (20 000 bp) = 49 230 bp. The weight of each 49 230 bp recombinant genome can be calculated by multiplying the length of the recombinant by the molecular weight of each bp and a conversion factor that includes Avogadro's number, as follows:

$$49\,230\,\text{bp} \times \frac{660\,\text{g/mol}}{\text{bp}} \times \frac{1 \times 10^6\,\mu\text{g}}{\text{g}} \times \frac{1\,\text{mol}}{6.023 \times 10^{23}\text{ genomes}}$$

$$= \frac{3.25 \times 10^{13}\,\mu\text{g}}{6.023 \times 10^{23}\text{ genomes}} = 5.4 \times 10^{-11}\,\mu\text{g/genome}$$

Therefore, each 49 230 bp recombinant genome weighs 5.4×10^{-11} μg. If all 6.1×10^{11} possible genomes capable of being constructed from the λEMBL3 arms were made into 49 230 bp genomes, the total amount of recombinant genomes (in μg) would be

$$6.1 \times 10^{11}\text{ genomes} \times \frac{5.4 \times 10^{-11}\,\mu\text{g}}{\text{genome}} = 32.9\,\mu\text{g}$$

Therefore, if all possible genomes are made and are recombinant with 20 000 bp inserts, they would weigh a total of 32.9 μg. In the packaging mix, a total of 4.88×10^8 infective viruses were made. A PFU/μg concentration is then obtained by dividing the total number of phage in the packaging mix by the total theoretical amount of recombinant DNA:

$$\frac{4.88 \times 10^8 \text{ PFU}}{32.9\,\mu\text{g}} = 1.5 \times 10^7 \text{ PFU}/\mu\text{g}$$

Therefore, the packaging efficiency for the recombinant genomes is 1.5×10^7 PFU/μg.

Packaging extracts from commercial sources can give packaging efficiencies as high as 2×10^9 PFU/μg. These efficiencies are usually determined by using wild-type λ DNA made into concatemers, the proper substrate for packaging. There are any number of reasons why a packaging experiment using recombinant DNA from a ligation reaction will not be as active. First, the ligation reaction may not be 100% efficient because of damaged cohesive ends or reduced ligase activity. Second, not all ligated products can be packaged; only genomes larger than 78% or smaller than 105% of the size of wild-type λ can be packaged. This means that λEMBL3 can accommodate only fragments from about 8600 to 21 700 bp. Though ligation with the λEMBL3 arms may occur with fragments of other sizes, they will not be packaged into complete phage particles capable of infecting a cell.

10.3.3 Ligation using plasmid vectors

Ligation of insert DNA into a bacteriophage λ-derived cloning vector requires the formation of linear concatemers of the form λ left arm : insert : λ right arm : λ left arm : insert : λ right arm : λ left arm : insert : λ right arm, etc. For this reaction to be favored, i needs to be much higher than j ($i \gg j$). In contrast, ligation using a plasmid vector requires that two types of event occur. First, a linear hybrid molecule must be formed by ligation of one end of the target fragment to one complementary end of the linearized plasmid. Second, the other end of the target fragment must be ligated to the other end of the vector to create a circular recombinant plasmid. The circularization event is essential since only circular molecules transform *E. coli* efficiently. Ligation conditions must be chosen, therefore, that favor intermolecular ligation events followed by intramolecular ones.

It was suggested by Maniatis et al. (1982) that optimal results will be achieved with plasmid vectors when i is greater than j by two- to threefold. Such a ratio will favor intermolecular ligation but will still allow for circularization of the recombinant molecule. Furthermore, Maniatis et al. recommend that the concentration of the termini of the insert (i_{insert})

be approximately twice the concentration of the termini of the linearized plasmid vector ($i_{insert} = 2i_{vector}$). The following problem will illustrate the mathematics involved in calculating optimal concentrations of insert and plasmid vector to use in the construction of recombinant clones.

Problem 10.8 The plasmid cloning vector pUC19 (2686 bp) is to be cut at its single *Eco*R I site within the polylinker cloning region. This cut will linearize the vector. The insert to be cloned is a 4250 bp *Eco*R I-generated fragment. What concentration of vector and insert should be used to give an optimal yield of recombinants?

Solution 10.8

We will first calculate the values of j for both the vector and the vector + insert. These are determined by using the following equation:

$$j = j_\lambda \left(\frac{MW_\lambda}{MW} \right)^{1.5} \text{ends/mL}$$

To use this equation for the vector, we must first calculate its molecular weight. This is accomplished by multiplying its length (in bp) by the conversion factor 660 g/mol/bp, as shown in the following calculation:

$$2686 \text{ bp} \times \frac{660 \text{ g/mol}}{\text{bp}} = 1.77 \times 10^6 \text{ g/mol}$$

Therefore, the pUC19 vector has a molecular weight of 1.77×10^6 g/mol.

It was determined previously that j_λ has a value of 3.22×10^{11} ends/mL and that the molecular weight of lambda (MW_λ) is equal to 3.2×10^7 g/mol. Placing these values into the earlier equation for calculating j gives the following result:

$$j = j_\lambda \left(\frac{MW_\lambda}{MW} \right)^{1.5} \text{ends/mL}$$

$$j_{vector} = 3.22 \times 10^{11} \text{ ends/mL} \left(\frac{3.2 \times 10^7 \text{ g/mol}}{1.77 \times 10^6 \text{ g/mol}} \right)^{1.5}$$

$$j_{vector} = 3.22 \times 10^{11} \text{ ends/mL } (18.1)^{1.5}$$

$$= 3.22 \times 10^{11} \text{ ends/mL } (77) = 2.5 \times 10^{13} \text{ ends/mL}$$

Therefore, the j value for this vector is 2.5×10^{13} ends/mL. The j value for the linear ligation intermediate formed by the joining of one end of the insert fragment to one end of the linearized plasmid vector can also be calculated. The molecular weight of the ligation intermediate (vector + insert) is calculated as follows:

$$(2686\ \text{bp} + 4250\ \text{bp}) \times \frac{660\ \text{g/mol}}{\text{bp}} = 4.58 \times 10^{6}\ \text{g/mol}$$

Using the molecular weight of the intermediate ligation product, the j value for the intermediate can be calculated:

$$j_{\text{insert+vector}} = 3.22 \times 10^{11}\ \text{ends/mL}\left(\frac{3.2 \times 10^{7}\ \text{g/mol}}{4.58 \times 10^{6}\ \text{g/mol}}\right)^{1.5}$$

$$j_{\text{vector+insert}} = 3.22 \times 10^{11}\ \text{ends/mL}\ (6.99)^{1.5}$$

$$= 3.22 \times 10^{11}\ \text{ends/mL}\ \ (18.48) = 5.95 \times 10^{12}\ \text{ends/mL}$$

Therefore, the ligation intermediate has a j value of 5.95×10^{12} ends/mL. Note that the j value for the ligation intermediate ($j = 5.95 \times 10^{12}$ ends/mL) is less than that for the linearized vector alone ($j = 2.5 \times 10^{13}$ ends/mL); shorter molecules have larger j values and can circularize more efficiently than longer DNA molecules.

For optimal results, we will prepare a reaction such that i is three-fold greater than j_{vector} ($i = 3j_{\text{vector}}$). j_{vector} has been calculated to be 2.5×10^{13} ends/mL. Therefore, we want i to equal three times 2.5×10^{13} ends/mL:

$$i = 3 \times (2.5 \times 10^{13}\ \text{ends/mL}) = 7.5 \times 10^{13}\ \text{ends/mL}$$

Furthermore, to favor the creation of the desired recombinant, i_{insert} should be equal to twice i_{vector} ($i_{\text{insert}}:i_{\text{vector}} = 2$, or $i_{\text{insert}} = 2i_{\text{vector}}$). Since i is the *total* concentration of complementary ends in the ligation reaction, $i_{\text{insert}} + i_{\text{vector}}$ must equal 7.5×10^{13} ends/mL. Since $i_{\text{insert}} = 2i_{\text{vector}}$, we have

$$2i_{\text{vector}} + i_{\text{vector}} = 7.5 \times 10^{13}\ \text{ends/mL}$$

which is equivalent to

$$3i_{\text{vector}} = 7.5 \times 10^{13}\ \text{ends/mL}$$

$$i_{\text{vector}} = \frac{7.5 \times 10^{13}\ \text{ends/mL}}{3} = 2.5 \times 10^{13}\ \text{ends/mL}$$

Therefore, i_{vector} is equivalent to 2.5×10^{13} ends/mL. Since we want i_{insert} to be two times greater than i_{vector}, we have

$$i_{insert} = 2 \times (2.5 \times 10^{13} \text{ ends/mL}) = 5.0 \times 10^{13} \text{ ends/mL}$$

Therefore i_{insert} is equivalent to 5.0×10^{13} ends/mL. To calculate the molarity of the DNA molecules we require in the ligation reaction, we will use the equation

$$i = 2N_0M \times 10^{-3} \text{ ends/mL}$$

This expression can be rearranged to give us molarity (M), as follows:

$$M = \frac{i}{2N_0 \times 10^{-3} \text{ ends/mL}}$$

For the vector, M is calculated as follows:

$$M_{vector} = \frac{2.5 \times 10^{13} \text{ ends/mL}}{2(6.023 \times 10^{23})(1 \times 10^{-3} \text{ ends/mL})}$$

$$M_{vector} = \frac{2.5 \times 10^{13}}{1.2 \times 10^{21}} = 2.1 \times 10^{-8}\ M$$

Therefore, the molarity of the vector in the ligation reaction should be $2.1 \times 10^{-8}\,M$. Converting molarity to moles/liter gives the following result:

$$2.1 \times 10^{-8}\ M = 2.1 \times 10^{-8} \text{ mol/L}$$

An amount of vector (in μg/mL) must now be calculated from the moles/liter concentration.

$$x\ \mu\text{g vector/mL} = \frac{2.1 \times 10^{-8} \text{ mol}}{\text{L}} \times \frac{660 \text{ g/mol}}{\text{bp}} \times 2686 \text{ bp} \times \frac{\text{L}}{1000 \text{ mL}} \times \frac{1 \times 10^{6}\ \mu\text{g vector}}{\text{g}}$$

$$x\ \mu\text{g vector/mL} = \frac{3.7 \times 10^{4}\ \mu\text{g vector}}{1000 \text{ mL}} = 37\ \mu\text{g vector/mL}$$

Therefore, in the ligation reaction, the vector should be at a concentration of 37 μg/mL. The ligation reaction is going to have a final volume of 50 μL. A relationship of ratios can be used to determine how many micrograms of vector to place into the 50 μL reaction.

$$\frac{x\ \mu\text{g vector}}{50\ \mu\text{L}} = \frac{37\ \mu\text{g vector}}{1000\ \mu\text{L}}$$

$$x = \frac{(50\,\mu\text{L})(37\,\mu\text{g vector})}{1000\,\mu\text{L}} = 1.9\,\mu\text{g vector}$$

Therefore, in the 50 μL ligation reaction, 1.9 μg of cut vector should be added.

It has been calculated that i_{insert} should be at a concentration of 5.0×10^{13} ends/mL. This can be converted to a molarity value by the same method used for i_{vector} using the following equation:

$$M = \frac{i}{2N_0 \times 10^{-3}\ \text{ends/mL}}$$

Substituting the calculated values into this equation yields the following result:

$$M = \frac{5.0 \times 10^{13}\ \text{ends/mL}}{2 \times (6.023 \times 10^{23}) \times (1 \times 10^{-3}\ \text{ends/mL})}$$

$$= \frac{5.0 \times 10^{13}}{1.2 \times 10^{21}} = 4.2 \times 10^{-8}\ M$$

Therefore, the insert should have a concentration of $4.2 \times 10^{-8}\ M$. This is equivalent to 4.2×10^{-8} moles of insert/L. Converting moles of insert/L to μg insert/mL is performed as follows:

$$x\ \mu\text{g insert/mL} = \frac{4.2 \times 10^{-8}\ \text{mol}}{\text{liter}} \times \frac{660\ \text{g/mol}}{\text{bp}} \times 4250\ \text{bp} \times \frac{1\text{L}}{1000\ \text{mL}} \times \frac{1 \times 10^{6}\ \mu\text{g insert}}{\text{g}}$$

$$x\ \mu\text{g insert/mL} = \frac{1.2 \times 10^{5}\ \mu\text{g insert}}{1000\ \text{mL}} = 120\ \mu\text{g insert/mL}$$

This is equivalent to 120 μg insert/1000 μL. We can determine the amount of insert to add to a 50 μL reaction by setting up a relationship of ratios, as shown here:

$$\frac{x\ \mu\text{g insert}}{50\ \mu\text{L}} = \frac{120\ \mu\text{g insert}}{1000\ \mu\text{L}}$$

$$x = \frac{(50\ \mu\text{L}) \times (120\ \mu\text{g insert})}{1000\ \mu\text{L}} = 6\ \mu\text{g insert}$$

Therefore, 6 μg of the 4250 bp fragment should be added to a 50 μL ligation reaction.

10.3.4 **Transformation efficiency**

Following ligation of a DNA fragment to a plasmid vector, the recombinant molecule, in a process called **transformation**, must be introduced into a host bacterium in which it can replicate. Plasmids used for cloning carry a gene for antibiotic resistance that allows for the selection of transformed cells. **Transformation efficiency** is a quantitative measure of how many cells take up plasmid. It is expressed as transformants per μg of plasmid DNA. Its calculation is illustrated in the following problem.

Problem 10.9 A 25 μL ligation reaction contains 0.5 μg of DNA of an Ap^R (ampicillin resistance) plasmid cloning vector. 2.5 μL of the ligation reaction is diluted into sterile water to a total volume of 100 μL. Ten μL of the dilution are added to 200 μL of competent cells (cells prepared for transformation). The transformation mixture is heat shocked briefly to increase plasmid uptake, and then 1300 μL of growth medium are added. The mixture is incubated for 60 min to allow for phenotypic expression of the ampicillin resistance gene. Twenty μL are then spread onto a plate containing 50 μg/mL ampicillin and the plate is incubated overnight at 37°C. The following morning, 220 colonies appear on the plate. What is the transformation efficiency?

Solution 10.9

The first step in calculating the transformation efficiency is to determine how many micrograms of plasmid DNA were in the 20 μL sample spread on the ampicillin plate. The original 25 μL ligation reaction contained 0.5 μg of plasmid DNA. We will multiply this concentration by the dilutions and the amount plated to arrive at the amount of DNA in the 20 μL used for spreading.

$$x\ \mu\text{g plasmid DNA} = \frac{0.5\ \mu\text{g plasmid DNA}}{25\ \mu\text{L}} \times \frac{2.5\ \mu\text{L}}{100\ \mu\text{L}} \times \frac{10\ \mu\text{L}}{1500\ \mu\text{L}} \times 20\ \mu\text{L}$$

$$x\ \mu\text{g plasmid DNA} = \frac{250\ \mu\text{g plasmid DNA}}{3750000} = 6.7 \times 10^{-5}\ \mu\text{g plasmid DNA}$$

Therefore, the 20 μL volume spread on the ampicillin plate contained 6.7×10^{-5} μg of plasmid DNA. Transformation efficiency is then calculated as the number of colonies counted divided by the amount of plasmid DNA contained within the spreading volume:

$$\text{Transformation efficiency} = \frac{220 \text{ transformants}}{6.7 \times 10^{-5}\ \mu\text{g DNA}} = \frac{3.3 \times 10^{6} \text{ transformants}}{\mu\text{g DNA}}$$

Therefore, the transformation efficiency is 3.3×10^6 transformants/μg DNA. Competent cells prepared for transformation can be purchased from commercial sources and may exhibit transformation efficiencies greater than 1×10^9 transformants/μg DNA, as measured by using supercoiled plasmid. It should be noted that adding more than 10 ng of plasmid to an aliquot of competent cells can result in saturation and can actually decrease the transformation efficiency.

10.4 GENOMIC LIBRARIES – HOW MANY CLONES DO YOU NEED?

The haploid human genome is approximately 3×10^9 bp, contained on 23 chromosomes. The challenge in constructing a recombinant library from a genome of this size is in creating clones in large enough number that the experimenter will be provided with some reasonable expectation that that library will contain the DNA segment carrying the particular gene of interest. How many clones need to be created to satisfy this expectation? The answer depends on two properties: (1) the size of the genome being fragmented for cloning and (2) the average size of the cloned fragments. Clark and Carbon (1976) derived an expression based on the Poisson distribution that can be used to estimate the likelihood of finding a particular clone within a randomly generated recombinant library. The relationship specifies the number of independent clones, N, needed to isolate a specific DNA segment with probability P. It is given by the equation

$$N = \frac{\ln(1 - P)}{\ln\left[1 - \left(\frac{I}{G}\right)\right]}$$

where I is the size of the average cloned insert, in bp, and G is the size of the target genome, in bp.

Problem 10.10 In Problem 9.6, λEMBL3 is used as a vector to clone 20 000 bp fragments generated from a partial *Sau*3A1 digest of the human genome (3×10^9 bp). We wish to isolate a gene contained completely on a 20 000 bp fragment. To have a 99% chance of isolating this gene in the λEMBL3 recombinant genomic library, how many independent clones must be examined?

Solution 10.10

For this problem, $P = 0.99$ (99% probability), $I = 20000$ bp, and $G = 3 \times 10^9$ bp. Placing these values into the preceding equation yields the following result:

$$N = \frac{\ln(1 - 0.99)}{\ln\left[1 - \left(\frac{2 \times 10^4 \text{ bp}}{3 \times 10^9 \text{ bp}}\right)\right]}$$

$$N = \frac{\ln(0.01)}{\ln(1 - 6.7 \times 10^{-6})} = \frac{\ln 0.01}{\ln 0.9999933} = \frac{-4.61}{-6.7 \times 10^{-6}} = 688\,000$$

Therefore, there is a 99% chance that the gene of interest will be found in 688 000 independent λEMBL3 clones of the human genome.

This estimation, of course, is a simplification. It assumes that all clones are equally represented. For this to occur, the genome must be randomly fragmented into segments of uniform sizes, all of which can be ligated to vectors with equal probability, and all of which are equally viable when introduced into host cells. When using restriction endonuclease to generate the insert fragments, this is not always possible; some sections of genomic DNA may not contain the recognition site for the particular enzyme being used for many thousands of bps.

10.5 cDNA LIBRARIES – HOW MANY CLONES ARE ENOUGH?

Different tissues express different genes. There are over 500 000 mRNA molecules in most mammalian cells, representing the expression of as many as 34 000 different genes. The mRNA present in a particular tissue type can be converted to cDNA copies by the actions of reverse transcriptase (to convert mRNA to ssDNA) then by DNA polymerase (to convert the ssDNA to double-stranded molecules). Preparing a recombinant library from a cDNA preparation provides the experimenter with a cross section of the genes expressed in a specific tissue type.

Some genes in a cell may be expressed at a high level and be responsible for as much as 90% of the total mRNA in a cell. It can be fairly straightforward to isolate cDNA clones of the more abundant mRNA species. Up to 30% of some cells' mRNA, however, may be composed of transcripts made from genes expressed at only a very low level, such that each type may be present at quantities less than 14 copies per cell. A cDNA library is not complete unless it contains enough clones that each mRNA type in the cell is represented.

The number of clones that need to be screened to obtain the recombinant of a rare mRNA at a certain probability is given by the equation

$$N = \frac{\ln(1-P)}{\ln\left[1-\left(\frac{n}{T}\right)\right]}$$

where N is the number of cDNA clones in the library; P is the probability that each mRNA type will be represented in the library at least once (P is usually set to 0.99; a 99% chance of finding the rare mRNA represented in the cDNA library); n is the number of molecules of the rarest mRNA in a cell; and T is the total number of mRNA molecules in a cell.

Problem 10.11 The rarest mRNA in a cell of a particular tissue type has a concentration of five molecules per cell. Each cell contains 450 000 mRNA molecules. A cDNA library is made from mRNA isolated from this tissue. How many clones will need to be screened to have a 99% probability of finding at least one recombinant containing a cDNA copy of the rarest mRNA?

Solution 10.11

For this problem, $P = 0.99$, $n = 5$, and $T = 450\,000$. Placing these values into the preceding equation gives

$$N = \frac{\ln(1-0.99)}{\ln\left[1-\left(\frac{5}{450\,000}\right)\right]}$$

$$N = \frac{\ln(0.01)}{\ln(1-1.1\times10^{-5})} = \frac{-4.61}{-1.1\times10^{-5}} = 420\,000$$

Therefore, in a cDNA library of 420 000 independent clones, one clone should represent the rarest mRNA in that particular cell type.

10.6 EXPRESSION LIBRARIES

A number of cloning vectors have been designed with genetic elements that allow for the expression of foreign genes. These elements usually include the signals for transcription and translation initiation. For many expression vectors, the cloning site is downstream from an ATG start signal. For example, a truncated version of the beta-galactosidase gene from *E. coli* is often used as a source for translation signals. Insertion of a target fragment in the proper orientation and translation frame results in the production of a fusion protein between the truncated gene on the vector and the foreign inserted gene.

Expression libraries can be made directly from bacterial genomic DNA since microbial genes characteristically lack introns. Mammalian gene expression libraries, however, are best constructed from cDNA, which, since it is made from processed mRNA, is free of intron sequences.

No matter what the source of target fragment, expression relies on the positioning of the insert in the correct orientation and the correct reading frame with the vector's expression elements. The probability of obtaining a recombinant clone with the insert fragment positioned correctly in the expression vector is given by the equation

$$P = \frac{\text{coding sequence size (in kb)}}{\text{genome size (in kb)} \times 6}$$

The denominator in this equation contains the factor 6 since there are six possible positions the target fragment can insert in relation to the vector's open reading frame. The target fragment can be inserted in either the sense or antisense orientation and in any one of three possible reading frames. One of the six possible positions should be productive.

The proportion of positive recombinants expressing the gene on the target fragment is given by the expression $1/P$.

This equation provides only an estimate. For randomly sheared genomic fragments, optimal results will be achieved if the cloning fragments are roughly ½ the size of the actual gene's coding sequence, because smaller fragments are less likely to carry upstream termination signals. For the cloning of fragments from large genomes containing many introns, such as those from humans, a large number of clones must be examined.

Problem 10.12 An *E. coli* gene has a coding sequence spanning 2000 bp. It is to be recovered from a shotgun cloning experiment in which sheared fragments of genomic DNA are inserted into an expression vector. What proportion of recombinants would be expected to yield a fusion protein with the vector?

Solution 10.12

Two thousand bp is equivalent to 2 kb:

$$2000\text{ bp} \times \frac{1\text{ kb}}{1000\text{ bp}} = 2\text{ kb}$$

The *E. coli* genome is 4 640 000 bp in length. This is equivalent to 4640 kb. Placing these values into the equation gives

$$P = \frac{2\text{ kb}}{(4640\text{ kb}) \times 6} = \frac{2}{2.8 \times 10^{4}} = 7.1 \times 10^{-5}$$

Taking the reciprocal of this value gives the following result:

$$\frac{1}{P} = \frac{1}{7.1 \times 10^{-5}} = 1.4 \times 10^{4}$$

Therefore, one out of approximately 14 000 clones will have the desired insert in the proper orientation to produce a fusion protein.

10.7 SCREENING RECOMBINANT LIBRARIES BY HYBRIDIZATION TO DNA PROBES

Either plaques on a bacterial lawn or colonies spread on an agar plate can be screened for the presence of clones carrying the desired target fragment by hybridization to a labeled DNA probe. In this technique, plaques or colonies are transferred to a nitrocellulose or nylon membrane. Once the DNA from the recombinant clones is fixed to the membrane (by baking in an oven or by UV cross-linking), the membrane is placed in a hybridization solution containing a radioactively labeled DNA probe designed to anneal specifically to the sequence of the desired target fragment. Following incubation with probe, the membrane is washed to remove any probe bound nonspecifically to the membrane and X-ray film is placed over the membrane to identify those clones annealing to probe.

Annealing of probe to complementary sequence follows association/dissociation kinetics and is influenced by probe length, G/C content, temperature, salt concentration, and the amount of formamide in the hybridization solution. These factors affect annealing in the following way: Lower G/C content, increased probe length, decreased temperature, and higher salt concentrations all favor probe association to target. Formamide is used in a hybridization solution as a solvent. It acts to discourage intrastrand annealing.

The stability of the interaction between an annealing probe and the target sequence as influenced by these variables is best measured by their effect on the probe's melting temperature (T_m), the temperature at which half of the probe has dissociated from its complementary sequence. Estimation of a probe's melting temperature is important to optimizing the hybridization reaction. Davis et al. (1986) recommend that the temperature at which hybridization is performed (T_i) should be 15°C below T_m:

$$T_i = T_m - 15°\text{C}$$

T_m is calculated using the equation

$$T_m = 16.6 \log [M] + 0.41[P_{\text{GC}}] + 81.5 - P_m - B/L - 0.65[P_f]$$

where M is the molar concentration of Na^+, to a maximum of 0.5 M (1X SSC contains 0.165 M Na^+); P_{GC} is the percent of G and C bases in the oligonucleotide probe (between 30% and 70%); P_m is the percent of mismatched bases, if known (each percent of mismatch will lower T_m by approximately 1°C); P_f is the percent formamide; B is 675 (for synthetic probes up to 100 nts in length); and L = probe length in nts.

For probes longer than 100 bases, the following formula can be used to determine T_m:

$$T_m = 81.5°\text{C} + 16.6(\log[\text{Na}^+]) + 0.41(\%\,\text{GC}) - 0.63(\%\text{ formamide}) - 600/L$$

For these longer probes, hybridization is performed at 10 to 15°C below T_m.

Problem 10.13 A labeled probe has the following sequence:

5′- GAGGCTTACGCGCATTGCCGCGATTTGCCC
ATCGCAAGTACGCAATTAGCAC-3′

It will be used to detect a positive clone from plates of λEMBL3 recombinant plaques transferred to a nitrocellu lose membrane filter. Hybridization will take place in 1X SSC (0.15 M NaCl, 0.015 M Na_3Citrate•$2H_2O$) containing 50% formamide. At what temperature should hybridization be performed? (Assume no mismatches between the probe and the target sequence.)

Solution 10.13

The probe is 52 nts in length. It contains a total of 29 G and C residues. Its P_{GC} value, therefore, is

$$P_{GC} = \frac{29 \text{ nucleotides}}{52 \text{ nucleotides}} \times 100 = 56\%$$

Both sodium chloride (0.15 *M*) and sodium citrate (0.015 *M*) in the hybridization buffer contribute to the overall sodium ion concentration. The total sodium ion concentration is

$$0.15\,M + 0.015\,M = 0.165\,M$$

Using these values, the equation for T_m of the 52-mer probe is calculated as follows:

$$T_m = 16.6 \log[0.165] + 0.41[56] + 81.5° - 0 - 675/52 - 0.65[50]$$

$$T_m = 16.6[-0.783] + 22.96 + 81.5° - 12.98 - 32.5$$

$$T_m = 46°\text{C}$$

The hybridization temperature (T_i) is calculated as follows:

$$T_i = 46°\text{C} - 15°\text{C} = 31°\text{C}$$

Therefore, the hybridization should be performed at 31°C.

10.7.1 Oligonucleotide probes

Either synthetic oligonucleotides or amplified DNA fragments can be used as probes against membrane-bound DNA from a recombinant library. If the exact gene sequence of the target fragment is known, a synthetic oligonucleotide can be made having a perfect match to the sought-after clone.

The number of sites on a genome to which an oligonucleotide probe will anneal depends on the genome's complexity and the number of mismatches between the probe and the target. A mismatch is a position within the probe/target hybrid occupied by noncomplementary bases. Complexity (*C*) is the size of the genome (in bp) represented by unique (single copy) sequence and one copy of each repetitive sequence region. The genomes of most bacteria and single-celled organisms do not have many repetitive sequences. The complexity of such organisms, therefore, is similar to genome size. Mammalian genomes, however, from mice to men, contain repetitive sequences in abundance. Although most mammalian genomes are roughly 3×10^9 bp in size, Laird (1971) has estimated their complexity to be approximately equivalent to 1.8×10^9 bp.

The number of positions on a genome having a certain complexity to which an oligonucleotide probe will hybridize is given by the equation

$$P_0 = \left(\frac{1}{4}\right)^L \times 2C$$

where P_0 is the number of independent perfect matches, L is the length of the oligonucleotide probe, and C is the target genome's complexity (this value is multiplied by two to represent the two complementary strands of DNA, either of which could potentially hybridize to the probe) (Maniatis et al., 1982).

Problem 10.14 An oligonucleotide probe is 12 bases in length. To how many positions on human genomic DNA could this probe be expected to hybridize?

Solution 10.14

Using the preceding equation, L is equal to 12 and C is equivalent to 1.8×10^9. Placing these values into the equation gives the following result:

$$P_0 = \left(\frac{1}{4}\right)^{12} \times 2(1.8 \times 10^9) = (6 \times 10^{-8}) \times (3.6 \times 10^9) = 216$$

Therefore, by chance, an oligonucleotide 12 bases long should hybridize to 216 places on the human genome.

Problem 10.15 What is the minimum length an oligonucleotide should be to hybridize to only one site on the human genome?

Solution 10.15

In Problem 10.14, it was assumed that the term $2C$ for the human genome is equivalent to 3.6×10^9. We want P_0 to equal 1. Our equation then becomes

$$1 = \left(\frac{1}{4}\right)^L \times 3.6 \times 10^9$$

$\frac{1}{3.6 \times 10^9} = (0.25)^L$ Convert the fraction 1/4 into a decimal and divide each side of the equation by 3.6×10^9.

$2.8 \times 10^{-10} = (0.25)^L$

$\log 2.8 \times 10^{-10} = \log 0.25^L$ Take the common logarithm of both sides of the equation.

$\log 2.8 \times 10^{-10} = L(\log 0.25)$ Use the Power Rule for Logarithms (for any positive numbers M and a (where a is not equal to 1) and any real number p, the logarithm of a power of M is the exponent times the logarithm of M: $\log_a M^p = p \log_a M$).

$\frac{\log 2.8 \times 10^{-10}}{\log 0.25} = L$ Divide each side of the equation by log 0.25.

$\frac{-9.6}{-0.6} = 16$ Take the log values.

Therefore, an oligonucleotide 16 bases in length would be expected to hybridize to the human genome at only one position.

10.7.2 Hybridization conditions

Determination of an oligonucleotide probe's melting temperature (T_m) can help the experimenter decide on a temperature at which to perform a hybridization step. Hybridization is typically performed at 2–5°C below the oligonucleotide's T_m. However, the length of time that the probe is allowed to anneal to target sequence is another aspect of the reaction that must be considered. For an oligonucleotide probe, the length of time allowed for the hybridization reaction to achieve half-completion is given by the equation described by Wallace and Miyada (1987):

$$t_{1/2} = \frac{\ln 2}{kC}$$

where k is a first-order rate constant and C is the molar concentration of the oligonucleotide probe (in moles of nucleotide per liter). This equation should be used only as an approximation. The actual rate of hybridization can be three to four times slower than the calculated rate (Wallace et al., 1979).

The rate constant, k, represents the rate of hybridization of an oligonucleotide probe to an immobilized target nucleic acid in 1 M sodium ion. It is given by the following equation:

$$k = \frac{3 \times 10^5\ L^{0.5}\ \text{L/mol/s}}{N}$$

where k is calculated in liter/mole of nt per second, L is the length of the oligonucleotide probe in nucleotides, and N is the probe's complexity.

Complexity, as it relates to an oligonucleotide, is calculated as the number of different possible oligonucleotides in a mixture. The following oligonucleotide, an 18-mer, since it has only one defined sequence (no redundancies), has a complexity of one:

```
5′-GGACCTATAGCCGTTGCG-3′
```

When oligonucleotides are constructed by reverse translation of a protein sequence, as might be the case when probing for a gene for which there exists only protein sequence information, a degenerate oligonucleotide might be synthesized having several possible different sequences at the wobble base position. For example, reverse translation of the following protein sequence would require the synthesis of an oligonucleotide with three possible sequences:

```
MetTrpMetIleTrpTrp
ATGTGGATGATATGGTGG
           C
           T
```

Methionine (Met) and tryptophan (Trp) are each encoded by a single triplet. Isoleucine (Ile), however, can be encoded for by three possible triplets: ATA, ATC, and ATT. The oligonucleotide is synthesized as a mixed probe containing either an A, a C, or a T at the 12th position. Since there are three possible sequences for the oligonucleotide, it has a complexity of 3.

When designing a probe based on protein sequence, the experimenter should choose an area to reverse translate that has the least amount of codon redundancy. Nevertheless, since most amino acids are encoded for by two or more triplets (see Table 10.1), most mixed probes will contain a number of redundancies. The total number of different oligonucleotides in a mixture (its complexity) is calculated by multiplying the number of possible nts at all positions. For example, here is a protein sequence reverse translated using all possible codons:

```
ArgProLysPheTrpIleCysAla
AGACCAAAATTCTGGATATGCGCA
C C  C  G  T     C  T  C
  G  G           T     G
  T  T                 T
```

To determine how many different sequences are possible in such a mixture, the number of possible nts at each position are multiplied together. In this example, the total number of different sequences is calculated as

$$2 \times 1 \times 4 \times 1 \times 1 \times 4 \times 1 \times 1 \times 2 \times 1 \times 1 \times 2 \times 1 \times 1 \\ \times 1 \times 1 \times 1 \times 3 \times 1 \times 1 \times 2 \times 1 \times 1 \times 4 = 3072$$

Therefore, there are 3072 possible different oligonucleotides in this mixture.

Table 10.1 The genetic code. Codons are shown as DNA sequence rather than RNA sequence (Ts are used instead of Us) to facilitate their conversion to synthetic oligonucleotide probes.

Amino acid	Abbreviation	Codons
Alanine	Ala	GCA GCC GCG GCT
Arginine	Arg	AGA AGG
		CGA CGC CGG CGT
Asparagine	Asn	AAC AAT
Aspartic acid	Asp	GAC GAT
Cysteine	Cys	TGC TGT
Glutamine	Gln	CAA CAG
Glutamic acid	Glu	GAA GAG
Glycine	Gly	CGA GGC GGG GGT
Histidine	His	CAC CAT
Isoleucine	Ile	ATA ATC ATT
Leucine	Leu	CTA CTC CTG CTT
		TTA TTG
Lysine	Lys	AAA AAG
Methionine	Met	ATG
Phenylalanine	Phe	TTC TTT
Proline	Pro	CCA CCC CCG CCT
Serine	Ser	TCA TCC TCG TCT
		AGC AGT
Threonine	Thr	ACA ACC ACG ACT
Tryptophan	Trp	TGG
Tyrosine	Tyr	TAC TAT
Valine	Val	GTA GTC GTG GTT
		TAA TAG TGA

Problem 10.16 An oligonucleotide probe 21 nts in length is used for hybridization to a colony filter containing recombinant clones. The oligonucleotide probe is a perfect match to the target gene to be identified. The hybridization is be performed in 1 *M* sodium chloride at 2°C below the probe's T_m. The hybridization solution contains oligonucleotide probe at a concentration of 0.02 μg/mL.

a) When will hybridization be half complete?

b) How long should hybridization be allowed to proceed?

Solution 10.16(a)

We will first calculate the value for *k*, the rate constant, using the following expression:

$$k = \frac{3 \times 10^5 L^{0.5} \text{ L/mol/s}}{N}$$

For this problem, *L*, the oligonucleotide length, is 21. Since the probe is an exact match, *N*, the complexity of the oligonucleotide, is equal to 1. Placing these values into the equation for *k* gives

$$k = \frac{(3 \times 10^5)(21)^{0.5} \text{ L/mol/s}}{1} = (3 \times 10^5)(4.58) = 1.37 \times 10^6 \text{ L/mol/s}$$

So that this *k* value can be used in the equation for calculating the half-complete hybridization reaction, a value for *C*, the molar concentration of the oligonucleotide probe, must also be calculated. In this problem, the probe has a concentration of 0.02 μg/mL. To convert this to a molar concentration (moles/liter), we must first determine the molecular weight (grams/mole) of the probe. Since the oligonucleotide probe is 21 nt in length, we can determine its molecular weight by multiplying the length in nts by the molecular weight of a single nt:

$$21 \text{ nt} \times \frac{330 \text{ g/mol}}{1 \text{ nt}} = 6930 \text{ g/mol}$$

Therefore, the molecular weight of a 21-mer is approximately 6930 g/mol. This value can then be used to calculate the molarity of the probe in the hybridization solution by the use of several conversion factors, as shown in the following equation.

$$\frac{0.02\ \mu\text{g}}{\text{mL}} \times \frac{1000 \text{ mL}}{\text{L}} \times \frac{1 \text{mol}}{6930 \text{ g}} \times \frac{1 \text{g}}{1 \times 10^6\ \mu\text{g}} = C \text{ mol/L}$$

$$\frac{20\text{ mol}}{6.93 \times 10^9\text{ L}} = C\text{ mol/L}$$

$$C = 2.89 \times 10^{-9}\text{ mol/L}$$

The values calculated for k and C can now be used in the expression for half-complete hybridization:

$$t_{1/2} = \frac{\ln 2}{kC}$$

$$t_{1/2} = \frac{\ln 2}{(1.37 \times 10^6\text{ L/mol/s})(2.89 \times 10^{-9}\text{ mol/L})}$$

$$t_{1/2} = \frac{0.693}{\frac{3.96 \times 10^{-3}}{1\text{ s}}} = 175\text{ s}$$

Therefore, hybridization is half complete in 175 sec.

Solution 10.16(b)

Since the actual rate of hybridization can be approximately four times slower than the calculated value (Wallace et al., 1979), the value determined in Solution 10.16(a) can be multiplied by four:

$$175\text{ sec} \times 4 = 700\text{ sec}$$

Converting this to minutes gives the following result:

$$700\text{ sec} \times \frac{1\text{ min}}{60\text{ sec}} = 11.7\text{ min}$$

Therefore, the hybridization reaction described in this problem is half complete in about 11 min, 42 sec and should be allowed to proceed for 23 min, 21 sec.

Problem 10.17 A small amount of a novel protein is purified and sequenced. A region of the protein is chosen from which an oligonucleotide probe is to be designed. The sequence of that region is as follows:

```
TrpTyrMetHisGlnLysPheAsnTrp
```

The mixed probe synthesized from this protein sequence will be added at a concentration of 0.02 μg/mL to a hybridization solution containing 1 *M* sodium chloride for the purpose of identifying a recombinant clone within a library. Hybridization will be performed at a temperature several degrees below the average T_m of the mixed probe. When will the hybridization be half complete?

Solution 10.17

The first step in solving this problem is to determine the complexity, *N*, of the reverse-translated mixed probe that would be extrapolated from the protein sequence. Using Table 10.1, the protein fragment is reverse translated as follows:

```
TrpTyrMetHisGlnLysPheAsnTrp
TGGTACATGCACCAAAAATTCAACTGG
     T     T  G  G  T  T
```

The number of possible oligonucleotides in the mixed oligo pool is calculated by multiplying all possible codons at each amino acid position (we will include here only those positions having greater than one possibility). This gives

$$2 \times 2 \times 2 \times 2 \times 2 \times 2 = 64$$

Therefore, the mixed oligonucleotide synthesized as a probe has a complexity, *N*, equal to 64. The oligonucleotide probe, no matter which base is at any position, has a length, *L*, of 27 nt.

The value for *k*, the hybridization rate, for this oligonucleotide is calculated as follows:

$$k = \frac{3 \times 10^5 \, L^{0.5} \text{ L/mol/s}}{N}$$

Substituting the given values for *N* and *L* into the equation yields the following result:

$$k = \frac{(3 \times 10^5)(27)^{0.5} \text{ L/mol/s}}{64}$$

$$k = \frac{(3 \times 10^5)(5.2) \text{ L/mol/s}}{64} = \frac{1.6 \times 10^6 \text{ L/mol/s}}{64} = 2.5 \times 10^4 \text{ L/mol/s}$$

The probe has a molecular weight, calculated as follows:

$$27 \text{ nucleotides} \times \frac{330 \text{ g/mol}}{\text{nucleotide}} = 8910 \text{ g/mol}$$

The molarity of the probe, when at a concentration of 0.02 μg/mL in the hybridization solution, is calculated as follows:

$$\frac{0.02\ \mu\text{g probe}}{\text{mL}} \times \frac{1000 \text{ mL}}{\text{L}} \times \frac{1 \text{ mol}}{8910 \text{ g}} \times \frac{1 \text{ g}}{1 \times 10^6\ \mu\text{g}} = C \text{ mol/L}$$

$$\frac{20 \text{ mol probe}}{8.91 \times 10^9 \text{ L}} = C \text{ mol/L}$$

$$C = 2.24 \times 10^{-9} \text{ mol/L}$$

Placing the calculated values for k and C into the equation for half-complete hybridization gives the following result:

$$t_{1/2} = \frac{\ln 2}{(2.5 \times 10^4 \text{ L/mol/s})(2.24 \times 10^{-9} \text{ mol/L})}$$

$$t_{1/2} = \frac{0.693}{\frac{5.6 \times 10^{-5}}{\text{s}}} = 12\,375 \text{ s}$$

Since, as discussed previously, the actual half-complete hybridization reaction can be as many as four times longer than the calculated value, we will allow for maximum chances for hybridization between probe and target to proceed by multiplying our value by four:

$$12\,375 \text{ sec} \times 4 = 49\,500 \text{ sec}$$

Converting this answer into hr gives the following result:

$$49\,500 \text{ sec} \times \frac{1 \text{ min}}{60 \text{ sec}} \times \frac{1 \text{ hr}}{60 \text{ min}} = 13.75 \text{ hr}$$

Therefore, using this mixed probe and the appropriate hybridization temperature, the hybridization reaction should be half complete in 13.75 hr.

10.7.3 Hybridization using double-stranded DNA (dsDNA) probes

Hybridization can also be performed using a dsDNA probe prepared by nick translation. When using a hybridization temperature of 68°C in aqueous solution or 42°C in 50% formamide, Maniatis et al. (1982) recommend the use of the following equation to estimate the amount of time to achieve half-complete hybridization:

$$t_{1/2} = \frac{1}{X} \times \frac{Y}{5} \times \frac{Z}{10} \times 2$$

where X is the amount of probe added to the hybridization reaction (in μg); Y is probe complexity, which for most probes is proportional to the length of the probe in kb; and Z is the volume of the hybridization reaction (in mL).

Nearly complete hybridization is achieved after three times the $t_{1/2}$ period.

Problem 10.18 Filters prepared for plaque hybridization are placed in a plastic bag with 15 mL of an aqueous hybridization buffer and 0.5 μg of a nick-translated probe 7500 bp in length. Hybridization is allowed to proceed at 68°C. How long should the hybridization reaction continue for nearly complete hybridization of probe to target?

Solution 10.18

Using the preceding equation, X is 0.5 μg, Y is 7.5 kb (7500 bp = 7.5 kb), and Z is 15 mL. Placing these values into the equation for half-complete hybridization gives

$$t_{1/2} = \frac{1}{0.5} \times \frac{7.5}{5} \times \frac{15}{10} \times 2 = \frac{225}{25} = 9$$

Therefore, the hybridization reaction is half complete in 9 hr. The reaction can be allowed to proceed for three times as long (27 hr) to ensure nearly complete hybridization.

10.8 SIZING DNA FRAGMENTS BY GEL ELECTROPHORESIS

Once a candidate recombinant clone has been identified by either colony hybridization to an allele-specific probe or by expression of a desired protein, it should be adequately characterized to ensure it contains the expected genetic material. One of the most straightforward and essential characterizations of a nucleic acid is the determination of its length. Fragment sizing is usually performed after a PCR amplification or in the screening of recombinant clone inserts that have been excised by restriction digestion.

Fragment sizing is accomplished by electrophoresing the DNA fragments in question on an agarose or acrylamide gel in which one lane is used for the separation of DNA fragments of known size (molecular weight or size markers). Following gel staining, the migration distance of each band of the size marker is measured. These values are plotted on a semilog graph. The rate at which a DNA fragment travels during electrophoresis is inversely proportional to the $\log_{10}$ of its length in bp. The graph generated for the size markers should contain a region represented by a straight line where this relationship is most pronounced. Larger DNA fragments show this relationship to a lesser degree, depending on the concentration of the gel matrix. Agarose is an effective gel matrix to use for separation of DNA fragments ranging in length between 50 and 30 000 bp. The experimenter, however, should use a percentage of agarose suitable for the fragments to be resolved. Low-molecular-weight fragments should be run on higher-percentage gels (1 to 4% agarose). High-molecular-weight fragments are best resolved on low-percentage gels (0.3 to 1% agarose). The best resolution of low-molecular-weight fragments can be achieved on a polyacrylamide gel. No matter what type of gel matrix is used, the method for determining fragment size is the same.

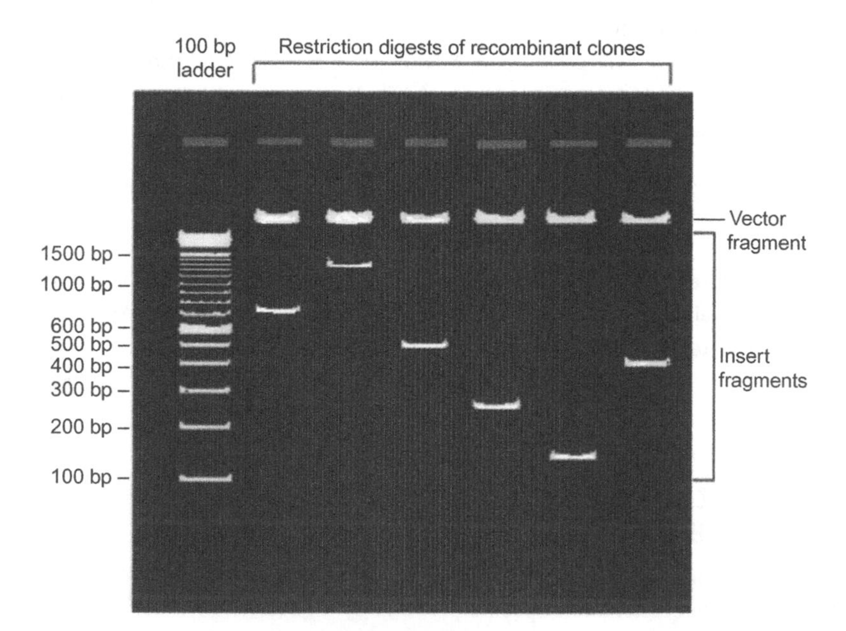

■ **FIGURE 10.1** Representation of an agarose gel used to size cDNA insert fragments. DNA of cDNA clones is digested with restriction endonucleases at unique sites on both sides of the insert. The digests are electrophoresed to separate the vector fragment from the insert fragment. The 100 bp ladder loaded in lane 1 is a size marker in which each fragment differs in size by 100 bp from the band on either side of it. For this size marker, the 600 and 1000 bp bands stain more intensely than the others.

Figure 10.1 shows a depiction of an agarose gel run to determine the sizes of cDNA fragments inserted into a plasmid vector. On this gel, a 100 bp ladder is run in the first lane as a size marker.

In Problem 10.19, the fragment size of a cDNA insert will be determined by two methods. In the first approach, a semilog graph and a standard curve will be used to extrapolate the fragment size. In the second approach, linear regression performed in Microsoft Excel will be used to derive a best-fit line from a standard curve and the line's equation will be used to calculate the fragment's size.

To determine fragment sizes on an agarose gel, a ruler is placed on the gel photograph along the size markers such that the 0 cm mark is aligned with the bottom of the well (Figure 10.2). The distance from the bottom of the well to the bottom of each band is then measured and recorded (Table 10.2). The distance from the well to each band is then plotted on a semilog graph and a straight line is drawn on the plot, connecting as many points as possible (Figure 10.3).

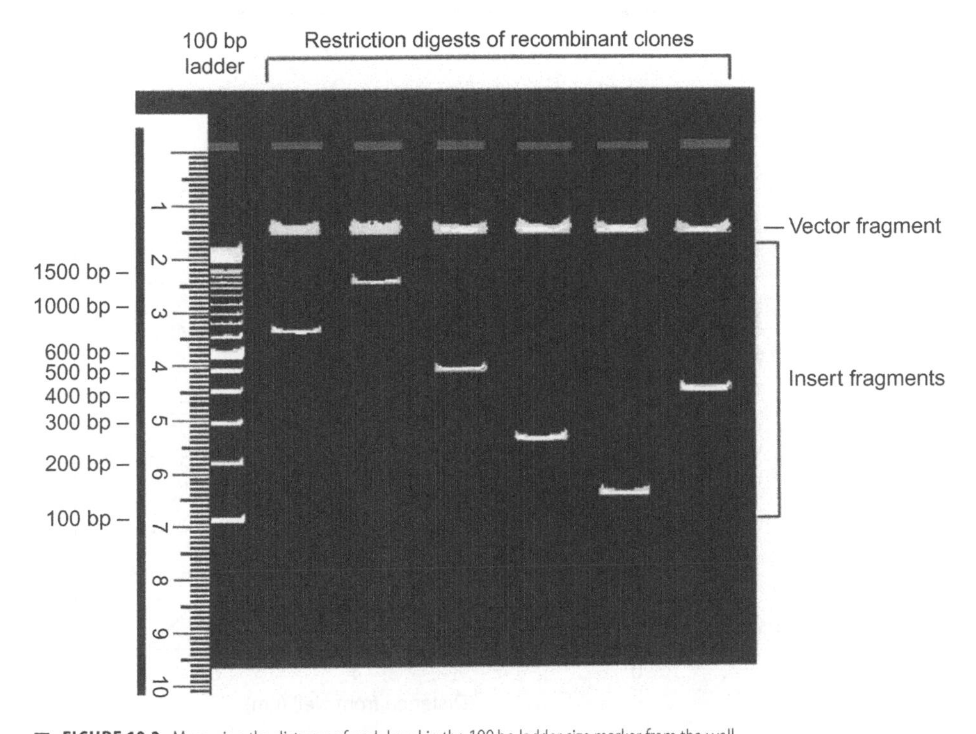

■ **FIGURE 10.2** Measuring the distance of each band in the 100 bp ladder size marker from the well.

Table 10.2 Migration distances for fragments of the 100 bp ladder shown in Figure 10.2.

Size marker band (bp)	Distance from well (cm)
1500	2.3
1400	2.35
1300	2.45
1200	2.55
1100	2.7
1000	2.85
900	3.0
800	3.2
700	3.5
600	3.9
500	4.15
400	4.55
300	5.1
200	5.85
100	7.0

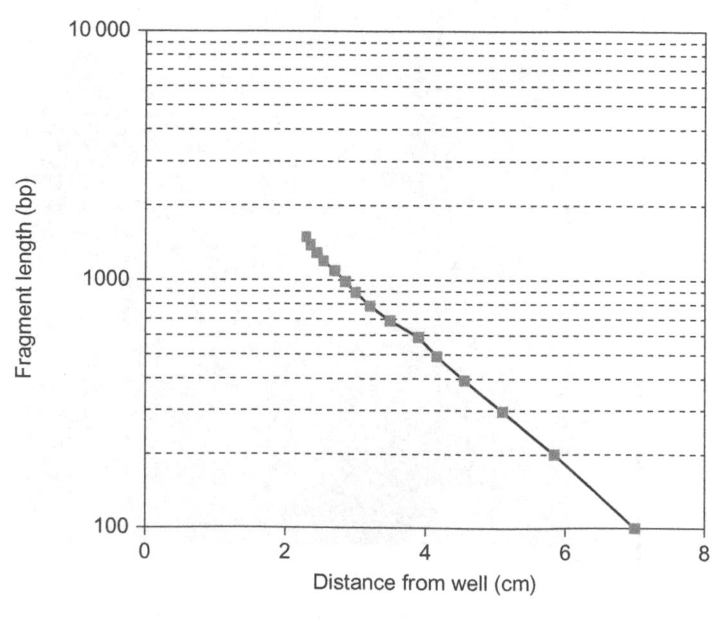

■ **FIGURE 10.3** Plotting the distance of each band of the 100 bp ladder from the well on a semilog graph.

Problem 10.19 What is the size of the insert shown in lane four of the gel in Figure 10.1? Determine the fragment's size,

a) using a semilog plot.
b) by linear regression using Microsoft Excel.

Solution 10.19(a)

Place a ruler on lane four of the photograph and measure the distance from the well to the bottom of the band representing the insert fragment (Figure 10.4).

The band measures 4.2 cm from the well. On the semilog plot of the standard curve (Figure 10.5), 4.2 cm on the *x* axis corresponds to 470 bp on the *y* axis.

Solution 10.19(b)

The Microsoft Office software package is one of the most popular computer applications for home and business use. Included in this software

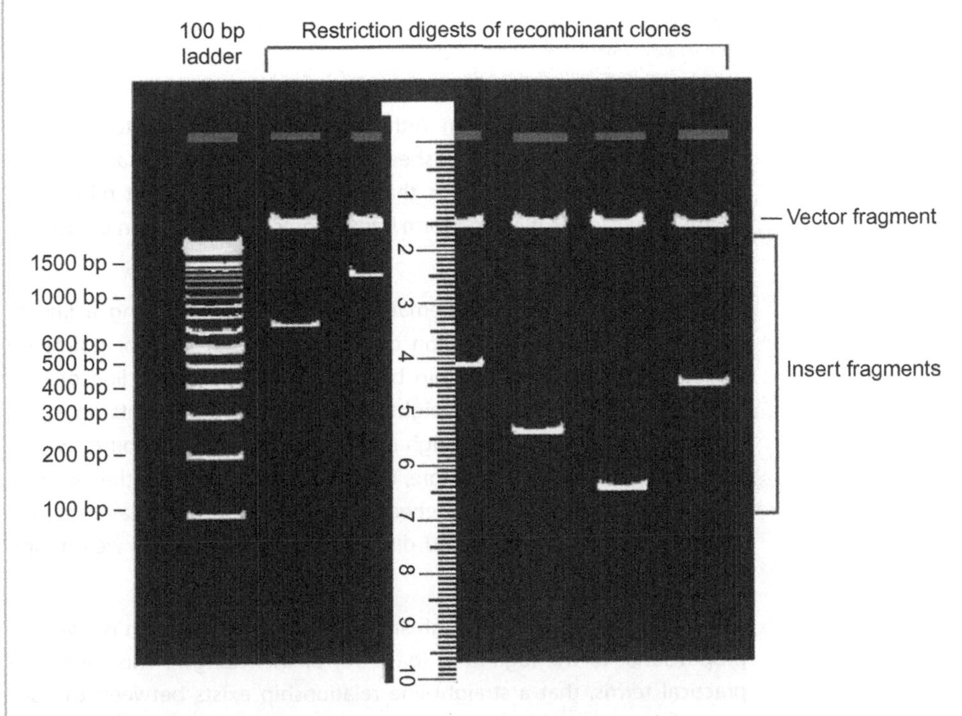

■ **FIGURE 10.4** Measuring the distance of the insert band from the well for lane 4.

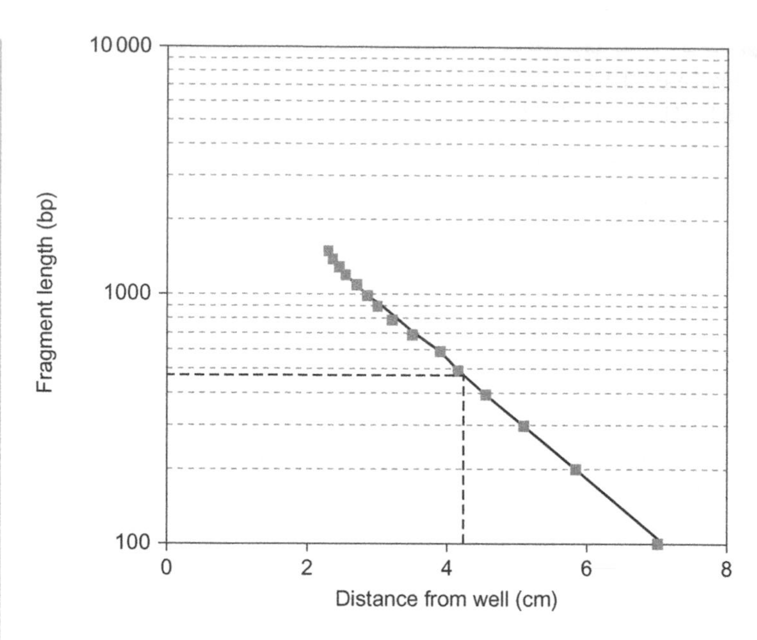

■ **FIGURE 10.5** Extrapolating the size of the DNA fragment in lane 4 of Figure 10.1. A vertical line is drawn from 4.2 cm on the x axis to the curve, and then a horizontal line is drawn to the y axis to determine the fragment's size in base pairs. A band 4.2 cm from the well corresponds to a size of approximately 470 bp.

package is the Excel program. Although designed for inventory maintenance and for creating spreadsheets to track financial transactions, it is equipped with math functions that the molecular biologist might find useful. In particular, it can perform linear regression, which can be used to calculate fragment sizes.

Linear regression is a mathematical technique for deriving a line of best fit, as well as the equation that describes that line, for points on a plot showing the relationship between two variables. A line of best fit, or **regression line**, is a straight line drawn through points on a plot such that the line passes through (or as close to) as many points as possible. Linear regression, therefore, is used to determine whether a direct and linear relationship exists between two variables. In this case, the two variables are fragment size and distance migrated from the well during electrophoresis.

DNA fragments migrate through an agarose gel at a rate that is inversely proportional to the logarithm (base 10) of their length. This means, in practical terms, that a straight-line relationship exists between the distance a fragment has migrated on a gel and the log of the fragment's size (in bp). This can be seen graphically if a plot of migration distance vs. bp size is drawn on a semilog graph (as shown in Solution 10.19(a)) or if migration distance vs. log of fragment size is graphed on a standard plot. By either technique, a straight line describes the relationship between these two variables.

A straight line is defined by the general equation

$$y = mx + b$$

For use of this equation to calculate fragment sizes by linear regression, y is the log of the fragment size (plotted on the vertical (y) axis); m is the slope of the regression line; x is the distance (in cm) from the well (plotted on the horizontal (x) axis); and b is the point where the regression line intercepts the y axis.

A measure of how closely the points on a plot fall along a regression line is given by the correlation coefficient (represented by the symbol R^2). R^2 will have a positive value if the slope of the regression line is positive (i.e., increases from left to right) and a negative value if the slope of the regression line is negative (i.e., decreases from left to right). The correlation coefficient will always have a value between -1 and $+1$. The closer the R^2 value is to -1 or $+1$, the better is the correlation between the two variables being graphically compared. If all the points of a plot fall exactly on a regression line having a positive slope, R^2 will have a value of $+1$. If all the points of a plot fall exactly on a regression line having a negative slope, R^2 will have a value of -1. If there is no correlation between the two variables (i.e., if the points on a graph appear random), R^2 will be close to 0.

In the following procedure, Microsoft Excel will be used to convert the lengths of the 100 bp size marker fragments into log values. These values will be graphed on the y axis against their migration distance on the x axis. Using linear regression, a line of best fit will be drawn through these points and an equation describing that line will be derived. The unknown fragment's migration distance will be placed into the equation for the regression line to determine a log value for the fragment's size. Taking the antilog of this value will yield the unknown fragment's size in bp.

The protocol for using Microsoft Excel to graph the data for this problem is found in Appendix A. It yields the result shown in Figure 10.6.

The equation for the regression line is $y = -0.2416x + 3.7009$. The line has an R^2 value of 0.998. All the points on the graph, therefore, lie

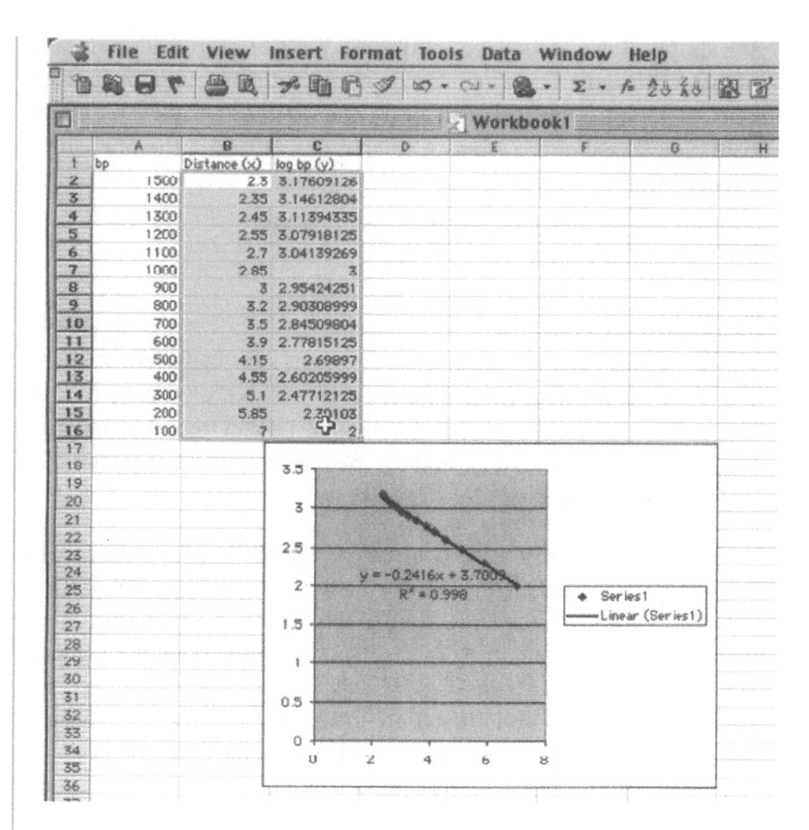

	A	B	C
1	bp	Distance (x)	log bp (y)
2	1500	2.3	3.17609126
3	1400	2.35	3.14612804
4	1300	2.45	3.11394335
5	1200	2.55	3.07918125
6	1100	2.7	3.04139269
7	1000	2.85	3
8	900	3	2.95424251
9	800	3.2	2.90308999
10	700	3.5	2.84509804
11	600	3.9	2.77815125
12	500	4.15	2.69897
13	400	4.55	2.60205999
14	300	5.1	2.47712125
15	200	5.85	2.30103
16	100	7	2

■ **FIGURE 10.6** The regression line for Solution 10.19(b) as determined using Microsoft Excel. The equation for the regression line is $y = -0.2416x + 3.7009$.

very close to or exactly on the regression line; there is a strong linear relationship between the distance from the well and the log of fragment size.

The unknown fragment is measured to be 4.2 cm from the well. This represents the x variable in the regression equation. The equation for calculating y, the log of fragment size, is then

$$y = -0.2416(4.2) + 3.7009$$

$$y = 2.6862$$

Therefore, the y value for the regression equation is 2.6862. This value is the logarithm of the fragment size (log bp). The size of the unknown fragment in bp is equal to the antilog of this value.

$$\text{Antilog } 2.6862 = 485.5$$

Therefore, the fragment is 485.5 bp in length. Note that linear regression gave a size of 485.5 bp whereas graphing on a semilog plot gave a size of 470 bp. Given the possible error involved in measuring bands on a photograph and possible migration anomalies due to salt or base sequence effects, neither answer is necessarily more accurate.

10.9 GENERATING NESTED DELETIONS USING NUCLEASE BAL 31

Once a recombinant clone carrying the sought-after gene has been identified, it is often desirable to create deletions of the insert fragment. This can facilitate at least three objectives:

1. Creation of a deletion set of the insert fragment can assist in the construction of a restriction map.
2. DNA sequencing of the entire insert fragment can be achieved by preparing a series of deletions that stretch further and further into the insert.
3. The beginning of the inserted gene (if working from a genomic source) can be brought closer to the control elements of the vector.

BAL 31 is an exonuclease that degrades both 3′ and 5′ termini of duplex DNA. The enzyme also has endonuclease activity and cleaves at nicks, gaps, and single-stranded sections of duplex DNA and RNA. Digestion of linear DNA with BAL 31 produces truncated molecules having blunt ends or ends having short overhanging 5′ termini.

The most frequently used method for generating deletion mutants using BAL 31 in a recombinant plasmid is first to cut it at a unique restriction site and then to digest nts away from the ends. The enzyme's activity is dependent on the presence of calcium. The addition of ethylene glycol tetraacetic acid (EGTA) will chelate the calcium and stop the reaction. The extent of digestion, therefore, can be controlled by adding the chelating agent after defined intervals of time.

The incubation time required to produce the desired deletion can be estimated using the equation of Legersky et al. (1978):

$$M_t = M_0 - \frac{2M_n V_{max} t}{[K_m + (S)_0]}$$

where M_t is the molecular weight of the duplex DNA after t minutes of incubation; M_0 is the original molecular weight of the duplex DNA; M_n is the average molecular weight of a mononucleotide (taken as 330 Da); V_{max} is the maximum reaction velocity (in moles of nt removed/L/min); t is the length of time of incubation, in min; K_m is the Michaelis–Menten constant (in moles of dstDNA termini/L); and S_0 is moles of dsDNA termini/L at $t = 0$ min.

Maniatis et al. (1982) recommend using a V_{max} value of 2.4×10^{-5} moles of nt removed/L/min when the reaction contains 40 units of BAL 31/mL. They also make the assumption that the value of V_{max} is proportional to the amount of enzyme in the reaction. For example, if 0.8 units are used in a 100 μL reaction, the concentration, in units/mL, becomes

$$\frac{0.8 \text{ units}}{100\ \mu\text{L}} \times \frac{1000\ \mu\text{L}}{\text{mL}} = 8 \text{ units/mL}$$

The V_{max} value is proportional to this amount, as follows:

$$\frac{8 \text{ units/mL}}{40 \text{ units/mL}} = \frac{x \text{ mol/L/min}}{2.4 \times 10^{-5} \text{ mol/L/min}}$$

Solving for x yields

$$\frac{(8 \text{ units/mL})(2.4 \times 10^{-5} \text{ mol/L/min})}{40 \text{ units/mL}} = x$$

$$\frac{1.92 \times 10^{-4} \text{ mol/L/min}}{40} = x = 4.8 \times 10^{-6} \text{ mol/L/min}$$

Therefore, if using a BAL 31 concentration of 8 units/mL in the reaction, the new V_{max} value you should used is 4.8×10^{-6} mol/L/min.

Maniatis et al. (1982) also recommend using a K_m value of 4.9×10^{-9} mol of dsDNA termini/L.

Problem 10.20 Into a 100 μL reaction you have placed 1 unit of BAL 31 nuclease and 5 μg of a 15 000 bp fragment of DNA. You wish to digest 5000 bp from the termini. How many minutes should you allow the reaction to proceed before stopping it with EGTA?

Solution 10.20

We need to determine values for M_0 (the initial molecular weight of the 15 000 bp fragment); M_t (the molecular weight of the 10 000 bp fragment made by removing 5000 bp from the 15 000 bp original fragment); V_{max} (the new value for this enzyme concentration); and S_0 (the number of double-stranded termini present in 5 μg of a 15 000 bp fragment).

The molecular weight of a 15 000 bp fragment of duplex DNA can be calculated as follows (assuming an average molecular weight of 660 g/mol/bp):

$$15\,000\text{ bp} \times \frac{660\text{ g/mol}}{\text{bp}} = 9.9 \times 10^6\text{ g/mol}$$

Therefore, a 15 000 bp fragment has a molecular weight of 9.9×10^6 g/mol. This is equivalent to 9.9×10^6 Da. This is the M_0 value.

The M_t value, the molecular weight of the 10 000 bp desired product, is then

$$10\,000\text{ bp} \times \frac{660\text{ g/mol}}{\text{bp}} = 6.6 \times 10^6\text{ g/mol}$$

Therefore, the desired 10 000 bp product has a molecular weight of 6.6×10^6 g/mol, which is equivalent to 6.6×10^6 Da.

To calculate the new V_{max} value for the enzyme amount used in this problem, we first need to determine the concentration of BAL 31 in the reaction in units/mL. We have 1 unit in a 100 μL reaction. This converts to units/mL as follows:

$$\frac{1\text{ unit}}{100\,\mu\text{L}} \times \frac{1000\,\mu\text{L}}{\text{mL}} = \frac{10\text{ units}}{\text{mL}}$$

The new V_{max} value can then be calculated by a relationship of ratios:

$$\frac{10\text{ units/mL}}{40\text{ units/mL}} = \frac{x\text{ mol/L/min}}{2.4 \times 10^{-5}\text{ mol/L/min}}$$

Solving for x gives

$$x\text{ mol/L/min} = \frac{(10\text{ units/mL})(2.4 \times 10^{-5}\text{ mol/L/min})}{40\text{ units/mL}}$$
$$= 6 \times 10^{-6}\text{ mol/L/min}$$

Therefore, the V_{max} value for this problem is 6×10^{-6} mol/L/min.

The S_0 value (the number of double-stranded termini present in 5 μg of a 15 000 bp fragment) is calculated in several steps. We have determined that a 15 000 bp fragment has a molecular weight of 9.9×10^6 g/mol. We can now determine how many moles of this 15 000 bp fragment are in 5 μg (the amount of DNA added to our reaction):

$$5\,\mu\text{g} \times \frac{1\text{ g}}{1 \times 10^6\,\mu\text{g}} \times \frac{1\text{ mol}}{9.9 \times 10^6\text{ g}} = \frac{5\text{ mol}}{9.9 \times 10^{12}} = 5.1 \times 10^{-13}\text{ mol}$$

Since we are using a 15 000 bp linear fragment, there are two termini per fragment, or 1×10^{-12} mol of dsDNA termini:

$$2 \times (5.1 \times 10^{-13} \text{ mol}) = 1 \times 10^{-12} \text{ mol termini}$$

To obtain the S_0 value, we need to determine how many moles of termini there are per liter. Since our reaction has a volume of 100 μL, this converts to moles per liter as follows:

$$\frac{1 \times 10^{-12} \text{ mol termini}}{100\,\mu\text{L}} \times \frac{1 \times 10^{6}\,\mu\text{L}}{\text{L}} = \frac{1 \times 10^{-8} \text{ mol termini}}{\text{L}}$$

Therefore, the S_0 value is 1×10^{-8} moles termini/L. We now have all the values needed to set up the equation. The original expression is

$$M_t = M_0 - \frac{2M_nV_{\max}t}{[K_m + (S)_0]}$$

Substituting the given and calculated values yields the following equation:

$$6.6 \times 10^{6}\text{ Da} = 9.9 \times 10^{6}\text{ Da} - \frac{2(330\text{ Da})(6 \times 10^{-6}\text{ mol/L/min})t\text{ min}}{\left(\frac{4.9 \times 10^{-9}\text{ mol termini}}{\text{L}}\right) + \left(\frac{1 \times 10^{-8}\text{ mol termini}}{\text{L}}\right)}$$

Simplifying the equation gives the following result:

$$6.6 \times 10^{6}\text{ Da} = 9.9 \times 10^{6}\text{ Da} - \frac{(3.96 \times 10^{-3}\text{ Da})t}{1.49 \times 10^{-8}}$$

$$6.6 \times 10^{6}\text{ Da} = 9.9 \times 10^{6}\text{ Da} - (2.66 \times 10^{5}\text{ Da})t$$

Subtracting 9.9×10^6 from each side of the equation yields

$$-3.3 \times 10^{6}\text{ Da} = (-2.66 \times 10^{5}\text{ Da})t$$

Dividing each side of the equation by -2.66×10^5 yields the following result:

$$\frac{-3.3 \times 10^{6}\text{ Da}}{-2.66 \times 10^{5}\text{ Da}} = t$$

$$t = 12.4\text{ min}$$

Therefore, to remove 5000 bp from a 15 000 bp linear fragment in a 100 μL reaction containing 1 unit of BAL 31 and 5 μg of 15 000 bp fragment, allow the reaction to proceed for 12.4 min before adding EGTA.

CHAPTER SUMMARY

Recombinant DNA is the method of joining two or more DNA molecules to create a hybrid. The technology is made possible by two types of enzymes, restriction endonucleases and ligase. A restriction endonuclease recognizes a specific sequence of DNA and cuts within, or close to, that sequence. By chance, a restriction enzyme's recognition sequence will occur every $(1/4)^n$ bases along a random DNA chain.

The amount of fragment ends (in moles) generated by cutting DNA with a restriction enzyme is given by the equation

$$\text{moles of DNA ends} = \frac{2 \times (\text{grams of DNA})}{(\text{number of bp}) \times \left(\frac{660\ \text{g/mol}}{\text{bp}}\right)}$$

The amount of ends generated by a restriction enzyme digest of circular DNA is given by the equation

$$\text{moles ends} = 2 \times (\text{moles DNA}) \times (\text{number of restriction sites})$$

When a linear molecule is digested with a restriction endonuclease, the amount of ends generated is calculated by the following equation.

$$\text{moles ends} = [2 \times (\text{moles DNA}) \times (\text{number of restriction sites})] + [2 \times (\text{moles DNA})]$$

DNA fragments generated by digestion with a restriction endonuclease can be joined together again by the enzyme ligase. The likelihood that two DNA molecules will ligate to each other is dependent on the concentration of their ends; the higher the concentration of compatible ends, the greater the likelihood that two termini will meet and be ligated. This parameter is designated by the term i and is defined as the *total* concentration of complementary ends in the ligation reaction. For a linear fragment of duplex DNA with cohesive ends, i is given by the formula

$$i = 2N_0M \times 10^{-3}\ \text{ends/mL}$$

where N_0 is Avogadro's number (6.023×10^{23}) and M is the molar concentration of the DNA.

Ligation of DNA molecules can result in their circularization. The amount of circularization is dependent on the parameter j, the concentration of same-molecule ends in close enough proximity to each other that they can effectively interact. For any DNA fragment, j is a constant value dependent on the fragment's length. It can be calculated as

$$j = \left(\frac{3}{2\pi lb}\right)^{3/2} \text{ ends/mL}$$

where l is the length of the DNA fragment, b is the minimal length of DNA that can bend around to form a circle, and π is the number pi. For bacteriophage λ DNA, j has a value of 3.22×10^{11} ends/mL. The j value for any DNA molecule can be calculated in relation to j_λ by the equation

$$j = j_\lambda \left(\frac{\text{MW}_\lambda}{\text{MW}}\right)^{1.5} \text{ ends/mL}$$

where j_λ is equal to 3.22×10^{11} ends/mL and MW represents molecular weight. Under circumstances in which j is equal to i ($j = i$ or $j/i = 1$), the end of any particular DNA molecule is just as likely to join with another molecule as it is to interact with its own opposite end. If j is greater than i ($j > i$), intramolecular ligation events predominate and circles are the primary product. If i is greater than j ($i > j$), intermolecular ligation events are favored and hybrid linear structures predominate.

Ligation of fragments to plasmid vectors may be most efficient when i is greater than j by two- to three-fold – a ratio that will favor intermolecular ligation but will still allow for circularization of the recombinant molecule. In addition, the concentration of the termini of the insert (i_{insert}) should be approximately twice the concentration of the termini of the linearized plasmid vector ($i_{\text{insert}} = 2i_{\text{vector}}$).

Transformation efficiency is a measure of how many bacteria are able to take in recombinant plasmids. It is expressed as transformants/μg DNA.

The likelihood of finding a particular clone within a randomly generated recombinant library can be estimated by the following equation:

$$N = \frac{\ln(1 - P)}{\ln\left[1 - \left(\frac{I}{G}\right)\right]}$$

where I is the size of the average cloned insert, in bp; G is the size of the target genome, in bp; N is the number of independent clones; and P is the probability of isolating a specific DNA segment.

The number of clones that need to be screened to obtain the recombinant of a rare mRNA in a cDNA library at a certain probability is given by the equation

$$N = \frac{\ln(1-P)}{\ln\left[1-\left(\frac{n}{T}\right)\right]}$$

where N is the number of cDNA clones in the library; P is the probability that each mRNA type will be represented in the library at least once (P is usually set to 0.99; a 99% chance of finding the rare mRNA represented in the cDNA library); n is the number of molecules of the rarest mRNA in a cell; and T is the total number of mRNA molecules in a cell.

When constructing an expression library, the probability of obtaining a recombinant clone with the insert fragment positioned correctly within the vector is given by the equation

$$P = \frac{\text{coding sequence size (in kb)}}{\text{genome size (in kb)} \times 6}$$

Hybridization of a probe to a recombinant library should be carried out at a temperature (T_i) 15°C below the probe's T_m:

$$T_i = T_m - 15°\text{C}$$

T_m is calculated using the equation

$$T_m = 16.6 \log[M] + 0.41[P_{GC}] + 81.5 - P_m - B/L - 0.65[P_f]$$

where M is the molar concentration of Na^+, P_{GC} is the percent of G and C bases in the oligonucleotide probe, P_m is the percent of mismatched bases, P_f is the percent formamide, B is equal to 675 (for probes up to 100 nt in length), and L is the probe length in nt. The T_m of probes longer than 100 bases can be calculated using the following formula

$$T_m = 81.5°\text{C} + 16.6(\log[\text{Na}^+]) + 0.41(\%\,\text{GC}) \\ - 0.63(\%\text{ formamide}) - 600/L$$

where M is the molar concentration of Na^+, to a maximum of 0.5 M (1XSSC contains 0.165 M Na^+); P_{GC} is the percent of G and C bases in the oligonucleotide probe (between 30 and 70%); P_m is the percent of mismatched bases, if known (each percent of mismatch will lower T_m by approximately 1°C); P_f is the percent formamide; B is 675 (for synthetic probes up to 100 nt in length); and L is the probe length in nts.

The number of positions on a genome having a certain complexity to which an oligonucleotide probe will hybridize is given by the equation

$$P_0 = \left(\frac{1}{4}\right)^L \times 2C$$

where P_0 is the number of independent perfect matches, L is the length of the oligonucleotide probe, and C is the target genome's complexity.

Hybridization of an oligonucleotide probe to the DNA of a recombinant library is typically performed at 2–5°C below the oligonucleotide's T_m. The approximate length of time allowed for the hybridization reaction to achieve half completion is given by the equation

$$t_{1/2} = \frac{\ln 2}{kC}$$

where k is a first-order rate constant and C is the molar concentration of the oligonucleotide probe (in moles of nt per L).

The rate constant, k, represents the rate of hybridization of an oligonucleotide probe to an immobilized target nucleic acid in 1 M sodium ion and is given by the equation

$$k = \frac{3 \times 10^5\ L^{0.5}\ \text{L/mol/s}}{N}$$

where k is calculated in L/M of nt per second, L is the length of the oligonucleotide probe in nts, and N is the probe's complexity.

Complexity, as it relates to an oligonucleotide, is calculated as the number of different possible oligonucleotides in a mixture. The total number of different oligonucleotides in a mixture, its complexity, is calculated by multiplying the number of possible nts at all positions.

Hybridization to recombinant clone DNA can be performed using a dsDNA probe prepared by nick translation. If using a hybridization temperature of 68°C in aqueous solution or 42°C in 50% formamide, the following equation to estimate the amount of time to achieve half-complete hybridization can be used:

$$t_{1/2} = \frac{1}{X} \times \frac{Y}{5} \times \frac{Z}{10} \times 2$$

where X is the amount of probe added to the hybridization reaction (in μg); Y is the probe complexity, which for most probes is proportional to the length of the probe in kb; and Z is the volume of the hybridization reaction (in mL).

Nearly complete hybridization is achieved after three times the $t_{1/2}$ period.

Recombinant clones can be identified by a restriction digest that removes the insert (or a characterized piece of it). The generated fragments are sized by electrophoresis on a gel also carrying a size ladder. The ladder is used to generate a standard curve and regression line equation that can be used to determine the size of the fragments from the recombinant clones.

Deletions of dsDNA can be prepared by the BAL 31 nuclease. The incubation time required to produce the desired deletion can be estimated using the equation

$$M_t = M_0 - \frac{2M_n V_{\max} t}{[K_m + (S)_0]}$$

where M_t is the molecular weight of the duplex DNA after t minutes of incubation; M_0 is the original molecular weight of the duplex DNA; M_n is the average molecular weight of a mononucleotide (taken as 330 Da); $V_{\max}$ is the maximum reaction velocity (in moles of nt removed/L/min); t is the length of time of incubation, in min; K_m is the Michaelis–Menten constant (in moles of dsDNA termini/L); and S_0 moles of dsDNA termini/L at $t = 0$ min.

REFERENCES

Clark, L., and J. Carbon (1976). A colony bank containing synthetic ColE1 hybrid plasmids representative of the entire *E. coli* genome. *Cell* 9:91.

Davis, L.G., M.D. Dibner, and J.F. Battey (1986). *Basic Methods in Molecular Biology*. Elsevier Science, New York.

Dugaiczyk, A., H.W. Boyer, and H.M. Goodman (1975). Ligation of *Eco*R I endonuclease-generated DNA fragments into linear and circular structures. *J. Mol. Biol.* 96:174–184.

Laird, C.D. (1971). Chromatid structure: relationship between DNA content and nucleotide sequence diversity. *Chromosoma* 32:378.

Legersky, R.J., J.L. Hodnett, and H.B. Gray Jr. (1978). Extracellular nucleases of pseudomonad *Bal*31 III. Use of the double-strand deoxyriboexonuclease activity as the basis of a convenient method for mapping fragments of DNA produced by cleavage with restriction enzymes. *Nucl. Acids Res.* 5:1445.

Maniatis, T., E.F. Fritsch, and J. Sambrook (1982). *Molecular Cloning: A Laboratory Manual*. Cold Spring Harbor Laboratory, Cold Spring Harbor, NY.

Wallace, B.R., and C.G. Miyada (1987). Oligonucleotide probes for the screening of recombinant DNA libraries. *Meth. Enzymol.* 152:432.

Wallace, B.R., J. Shaffer, R.C. Murphy, J. Bonner, T. Hirose, and K. Itakura (1979). Hybridization of synthetic oligodeoxyribonucleotides to ϕX174 DNA: the effect of single base pair mismatch. *Nucl. Acids Res.* 6:3543.

Chapter 11

Protein

INTRODUCTION

There are a number of reasons for creating recombinant DNA clones. One is to test the activity of a foreign promoter or enhancer control element. This is accomplished by placing the control element in close proximity to a reporter gene encoding an easily assayed enzyme. The most popular genes for this purpose are the *lacZ* gene encoding β-galactosidase, the *cat* gene encoding chloramphenicol acetyltransferase (CAT), and the *luc* gene encoding the light-emitting protein luciferase.

The amount of enzyme produced in a reporter gene system is usually expressed in terms of units. For most proteins, a **unit** is defined as the amount of enzyme activity per period of time divided by the quantity of protein needed to give that activity under defined reaction conditions in a specified volume. A unit of the restriction endonuclease *Eco*R I, for example, is defined as the amount of enzyme required to completely digest 1 μg of double-stranded substrate DNA in 60 min at 37°C in a 50 μL reaction volume. Calculation of units of protein, therefore, requires the assay of both protein activity and protein quantity.

This chapter addresses the calculation needed to quantify protein and, for the cases of several examples, to assess protein activity.

11.1 CALCULATING A PROTEIN'S MOLECULAR WEIGHT FROM ITS SEQUENCE

Knowing the molecular weight of a protein can assist in its characterization, assay, or purification. If you know the amino acid sequence of a protein or if you know the DNA sequence of the gene that encodes that protein and can thereby extrapolate its amino acid sequence, you can determine the protein's molecular weight by adding up the molecular weights of the individual amino acids. Since the amino acids are bound within the protein chain all linked by peptide bonds between them, and since peptide bonds are formed by the removal of water, when doing this calculation, you should use the molecular weight value of each amino acid minus that contributed by H_2O (18.02 Da),

Calculations for Molecular Biology and Biotechnology. DOI: 10.1016/B978-0-12-375690-9.00011-5
© 2010 Elsevier Inc. All rights reserved.

as shown in Table 11.1. You should *not* use the molecular weights of the free amino acids as this would overestimate the protein's true molecular weight value. Once all the individual amino acid molecular weights have been summed, you need to add an additional 18.02 Da to account for the extra H on the amino-terminal amino acid and the extra OH on the amino acid at the carboxyl end. The general expression for this operation is

$$MW_{protein} = \sum_{Ala}^{Val}(n_{aa} \times MW_{aa}) + 18.02$$

where $MW_{protein}$ is the protein's molecular weight, Σ is the sum of the molecular weights in the protein, n_{aa} is the number of times that amino acid appears in the protein, and MW_{aa} is the molecular weight of each amino acid appearing in the protein.

Table 11.1 The 20 amino acids, their symbols, and their molecular weights. The molecular weights are calculated as the nonionized species minus water.

Amino acid	Three-letter abbreviation	Single-letter abbreviation	Molecular weight
Alanine	Ala	A	71.09
Arginine	Arg	R	156.19
Asparagine	Asn	N	114.11
Aspartic Acid	Asp	D	115.09
Cysteine	Cys	C	103.15
Glutamic Acid	Glu	E	129.12
Glutamine	Gln	Q	128.14
Glycine	Gly	G	57.05
Histidine	His	H	137.14
Isoleucine	Ile	I	113.16
Leucine	Leu	L	113.16
Lysine	Lys	K	128.17
Methionine	Met	M	131.19
Phenylalanine	Phe	F	147.18
Proline	Pro	P	97.12
Serine	Ser	S	87.08
Threonine	Thr	T	101.11
Tryptophan	Trp	W	186.21
Tyrosine	Tyr	Y	163.18
Valine	Val	V	99.14
Weighted Average			119.40

Problem 11.1 A protein fragment has the following amino acid sequence. What is its molecular weight?

MGLKPCERVWFIIQHDDCYAARP

Solution 11.1

We find the sum of all the individual amino acid molecular weights (from Table 11.1) and add the molecular weight of one water molecule (18.02):

$$\begin{aligned}\text{Molecular weight} &= 131.19 + 57.05 + 113.16 + 128.17 + 97.12 + 102.15 \\ &\quad + 129.12 + 156.19 + 99.14 + 186.21 + 147.18 \\ &\quad + 113.16 + 113.16 + 128.14 + 137.14 + 115.09 \\ &\quad + 115.09 + 103.15 + 163.18 + 71.09 + 71.09 \\ &\quad + 156.19 + 97.12 + 18.02 \\ &= 2749.3\end{aligned}$$

Therefore, this protein has a molecular weight of 2749.3 Da.

Problem 11.2 What is a quick estimate of the molecular weight of the protein described in Problem 11.1?

Solution 11.2

A simple and quick calculation of the protein's molecular weight is made by multiplying the number of amino acids in the protein fragment (in this case, 23 residues) by the average molecular weight of an amino acid (119.4 Da; Table 11.1):

$$23 \times 119.4 = 2746.2 \text{ Da}$$

To this value, we add the molecular weight of one water molecule:

$$2746.2 + 18.02 = 2764.22 \text{ Da}$$

Therefore, as a rough estimate, the protein fragment described in Problem 11.1 has a molecular weight of 2764 Da.

As calculating the molecular weight of large proteins can be cumbersome, a search of the internet will provide links to a number of sites that will calculate a protein's molecular weight automatically (such as the site maintained by the University of Delaware and the Georgetown University Medical Center; Figures 11.1 and 11.2).

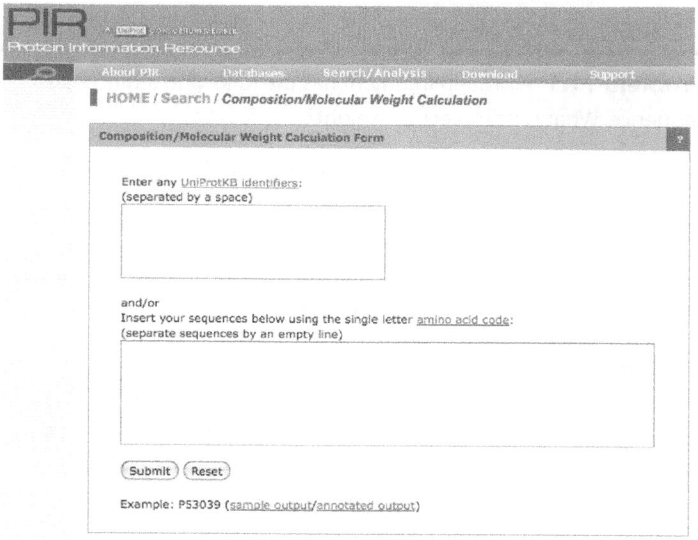

PIR Protein Information Resource

About PIR Databases Search/Analysis Download Support

HOME / Search / *Composition/Molecular Weight Calculation*

Composition/Molecular Weight Calculation Form

Enter any UniProtKB identifiers:
(separated by a space)

and/or
Insert your sequences below using the single letter amino acid code:
(separate sequences by an empty line)

Submit Reset

Example: P53039 (sample output/annotated output)

■ **FIGURE 11.1** Molecular weight calculators are available on the internet. On this site, found at http://pir.georgetown.edu/pirwww/search/comp_ms.shtml, the experimenter enters the name of the protein (in UniProtKB format) or the protein's single letter amino acid code and a molecular weight is calculated automatically.

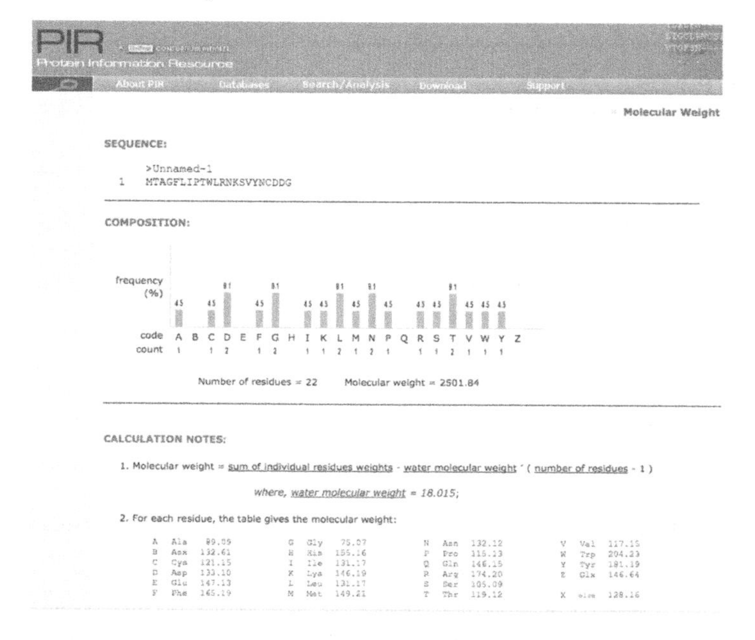

■ **FIGURE 11.2** The molecular weight of a protein is automatically calculated when its amino acid sequence is entered into the field shown in Figure 11.1. The protein's molecular weight is calculated as the sum of each amino acid's molecular weight minus the contribution from the molecular weight of water, which has been multiplied by one less than the number of amino acids in the chain.

11.2 PROTEIN QUANTIFICATION BY MEASURING ABSORBANCE AT 280 nm

There are a number of methods for estimating protein concentration. One of the simplest is to measure absorbance at 280 nm. Proteins absorb UV light at 280 nm primarily because of the presence of the ringed (aromatic) amino acids tyrosine and tryptophan and because of disulfide bonds between cysteine residues (phenylalanine, also an aromatic amino acid, absorbs at 260 nm much more efficiently than at 280 nm). Since different proteins have different amounts of the amino acids with aromatic side chains, there will be variability in the degree to which different proteins absorb light at 280 nm. In addition, conditions that alter a protein's tertiary structure (buffer type, pH, and reducing agents) can affect its absorbance. Despite the variability, reading absorbance at 280 nm is often used because few other chemicals also absorb at this wavelength.

As a very rough approximation, one absorbance unit at 280 nm is equal to 1 mg protein/mL. The protein solution being assayed should be diluted if its absorbance at 280 nm is greater than 2.0.

Problem 11.3 A solution of a purified protein is diluted 0.2 mL into a total volume of 1.0 mL in a cuvette having a 1 cm light path. A spectrophotometer reading at 280 nm gives a value of 0.75. What is a rough estimate of the protein concentration?

Solution 11.3

This problem can be solved by setting up a relationship of ratios as follows:

$$\frac{1\ \text{mg/mL}}{1\ \text{absorbance unit}} = \frac{x\ \text{mg/mL}}{0.75\ \text{absorbance units}}$$

Solving for x yields the following result:

$$\frac{(1\ \text{mg/mL}) \times (0.75)}{1} = x = 0.75\ \text{mg protein/mL}$$

Since the protein sample is diluted 0.2 mL/1.0 mL, to determine the protein concentration of the original sample this result must be multiplied by the inverse of the dilution factor:

$$\frac{0.75\ \text{mg}}{\text{mL}} \times \frac{1.0\ \text{mL}}{0.2\ \text{mL}} = 3.75\ \text{mg protein/mL}$$

Therefore, the purified protein solution has an approximate concentration of 3.75 mg/mL.

11.3 USING ABSORBANCE COEFFICIENTS AND EXTINCTION COEFFICIENTS TO ESTIMATE PROTEIN CONCENTRATION

The **absorbance coefficient** of a protein is its absorbance for a certain concentration at a given wavelength in a 1 cm cuvette. Absorbance coefficients are usually determined for convenient concentrations, such as 1 mg/mL, 1%, or 1 molar. In the *CRC Handbook of Biochemistry and Molecular Biology* (CRC Press, Boca Raton, FL), absorbance coefficients are given for 1% solutions ($A^{1\%}$). The absorbance coefficient of a 1% protein solution is equivalent to the absorbance of a 1% solution of a given protein at a given wavelength in a 1 cm cuvette. The absorbance coefficient of a 1 mg/mL protein solution (written $A^{1\text{mg/mL}}$) is equivalent to the absorbance of a 1 mg/mL solution of a given protein at a given wavelength in a 1 cm cuvette. Since absorbance coefficients are dependent on pH and ionic strength, it is important when using the published values to duplicate the protein's environment under which the coefficients were determined.

The **molar extinction coefficient** (E_{M}; also called **molar absorptivity** and designated with the symbol ε) of a protein is equivalent to the absorbance of a 1 *M* solution of a given protein at a given wavelength in a 1 cm cuvette. It has units of $\text{M}^{-1}\text{cm}^{-1}$. However, since it is standard practice to use a cuvette with a 1 cm width, the units are usually just dropped. Although most extinction coefficients are determined at 280 nm, this is not always the case and so close attention should be paid to the wavelength of light used to make the measurement.

Using the coefficients, protein concentration can be determined using the following relationships (Stoscheck, 1990):

$$\text{Protein concentration (in mg/mL)} = \frac{\text{Absorbance}}{A_{1\text{ cm}}^{1\text{ mg/mL}}}$$

$$\text{Protein concentration (in \%)} = \frac{\text{Absorbance}}{A_{1\text{ cm}}^{1\%}}$$

$$\text{Protein concentration (in molarity)} = \frac{\text{Absorbance}}{E_{\text{M}}}$$

A 1% solution used to determine the percent solution extinction coefficient, $\text{A}^{1\%}$, is equivalent to 1 g/100 mL. Since 1 g/100 mL is equivalent to 10 mg/mL, if the concentration of the protein is to be reported as a mg/mL concentration, the absorbance/$\text{A}^{1\%}$ value must be multiplied by a factor of 10:

$$\text{Protein concentration (in mg/mL)} = \frac{\text{Absorbance}}{A_{1\text{ cm}}^{1\%}} \times 10$$

Problem 11.4 Acetylcholinesterase from *E. electricus* has an absorbance coefficient ($A^{1\%}$) of 22.9 in 0.1 *M* NaCl, 0.03 *M* sodium phosphate, pH 7.0 at 280 nm. A solution of this protein gives a reading at 280 nm of 0.34 in a 1 cm cuvette. What is its percent concentration?

Solution 11.4

The following equation will be used:

$$\text{Protein concentration} = \frac{\text{Absorbance}}{A^{1\%}_{1\,\text{cm}}}$$

Substituting the values provided gives the following result:

$$\%\ \text{protein concentration} = \frac{0.34}{22.9}$$
$$= 0.015\%$$

Therefore, a solution of acetylcholinesterase having an absorbance of 0.34 has a concentration of 0.015%.

Problem 11.5 What is the concentration of acetylcholinesterase from Problem 11.4 in mg/mL?

Solution 11.5

The following equation will be used:

$$\text{Protein concentration (in mg/mL)} = \frac{\text{Absorbance}}{A^{1\%}_{1\,\text{cm}}} \times 10$$

Inserting the values from Problem 11.2, we have

$$\text{Protein concentration} = \frac{0.34}{22.9} \times 10$$
$$= 0.15\,\text{mg/mL}$$

Therefore, the acetylcholinesterase solution from Problem 11.4 has a concentration of 0.15 mg/mL.

Problem 11.6 Acetylcholinesterase has a molar extinction coefficient (E_M) of 5.27×10^5. In a 1 cm cuvette, a solution of the protein has an A_{280} of 0.22. What is its molar concentration?

Solution 11.6

The following equation will be used:

$$\text{Protein concentration (in molarity)} = \frac{\text{Absorbance}}{E_M}$$

Substituting the values provided gives the following result:

$$M = \frac{0.22}{5.27 \times 10^5} = 4.2 \times 10^{-7}\ M$$

Therefore, the solution of acetylcholinesterase has a concentration of $4.2 \times 10^{-7} M$.

Software tools for calculating protein concentration based on its absorbance and extinction coefficient can be found on the web. An example is the Protein Concentration Calculator maintained by the University of Oxford (Figure 11.3).

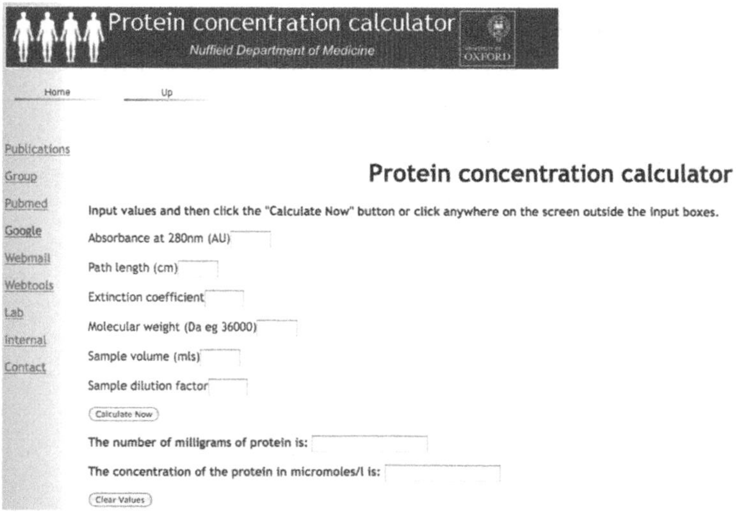

Protein concentration calculator
Nuffield Department of Medicine
OXFORD
Home
Up
Publications
Group
Pubmed
Google
Webmail
Webtools
Lab
Internal
Contact
Protein concentration calculator
Input values and then click the "Calculate Now" button or click anywhere on the screen outside the input boxes.
Absorbance at 280nm (AU)
Path length (cm)
Extinction coefficient
Molecular weight (Da eg 36000)
Sample volume (mls)
Sample dilution factor
Calculate Now
The number of milligrams of protein is:
The concentration of the protein in micromoles/l is:
Clear Values

FIGURE 11.3 A software tool maintained by the University of Oxford for determining a protein's concentration based on its molar extinction coefficient and absorbance: http://www.ccmp.ox.ac.uk/ocallaghan/webtools/protein_-concentration_calculator.htm.

11.3.1 Relating absorbance coefficient to molar extinction coefficient

The absorbance coefficient for a 1% protein solution ($A_{1cm}{}^{1\%}$) is related to the molar extinction coefficient (E_M) in the following way (Kirschenbaum, 1976):

$$E_M = \frac{(A_{1\text{ cm}}^{1\%})(\text{molecular weight})}{10}$$

A 0.1% protein solution is equivalent to 1 mg protein/mL. Its absorbance can be written as $A^{0.1\%}$. It can be calculated by dividing the molar extinction coefficient by the molecular weight:

$$A_{1\text{ cm}}^{0.1\%} = \frac{E_M}{Molecular\ Weight}$$

Problem 11.7 β-galactosidase from *E. coli* has a molecular weight of 750 000. Its absorbance coefficient for a 1% solution is 19.1. What is its molar extinction coefficient?

Solution 11.7

Substituting the values provided into the preceding equation gives the following result:

$$E_M = \frac{(19.1)(750\,000)}{10} = \frac{1.4 \times 10^7}{10} = 1.4 \times 10^6$$

Therefore, β-galactosidase has a molar extinction coefficient of 1.4×10^6.

Problem 11.8 The molar extinction coefficient of a protein you are studying is 6.0×10^4. On an SDS PAGE gel, the protein shows a molecular weight of 75 000. What absorbance should be expected for a 1 mg/mL solution of this protein?

Solution 11.8

We will use the following relationship to solve this problem:

$$A_{1\text{ cm}}^{0.1\%} = \frac{E_M}{Molecular\ Weight}$$

Substituting in our values yields

$$A_{1\,cm}^{0.1\%} = \frac{6.0 \times 10^4}{75\,000} = 0.8$$

Therefore, a 1 mg/mL solution of this protein should have an absorbance of 0.8.

11.3.2 Determining a protein's extinction coefficient

A protein's extinction coefficient at 205 nm can be determined using the following formula (Scopes, 1974):

$$E_{205\ nm}^{1\ mg/mL} = 27 + 120\left(\frac{A_{280}}{A_{205}}\right)$$

This equation can be used when the protein has an unknown concentration. Once a concentration has been determined, the extinction coefficient at 280 nm can be determined using the following equation:

$$E_{280\ nm} = \frac{\text{mg protein/mL}}{A_{280}}$$

Note: This equation will not give an accurate result for proteins having an unusual phenylalanine content.

Problem 11.9 A solution of protein having an unknown concentration is diluted 30 μL into a final volume of 1 mL. At 205 nm, the diluted solution gives an absorbance of 0.648. Absorbance at 280 nm is 0.012. What is the molar extinction coefficient at 205 nm and at 280 nm?

Solution 11.9

The given values are substituted into the equation for determining a protein's extinction coefficient:

$$E_{205\,nm}^{1\,mg/mL} = 27 + 120\left(\frac{A_{280}}{A_{205}}\right)$$

$$E_{205\ nm} = 27 + 120\left(\frac{0.012}{0.648}\right) = 27 + 120(0.019) = 27 + 2.28 = 29.28$$

Therefore, the extinction coefficient at 205 nm is 29.28 for the diluted sample. To obtain the extinction coefficient for the undiluted sample, this value is multiplied by the dilution factor:

$$29.28 \times \frac{1000\,\mu L}{30\,\mu L} = 976$$

Therefore, the molar extinction coefficient at 205 nm is 976.

The protein concentration can now be determined using the earlier equation relating molarity, absorbance, and the molar extinction coefficient:

$$\text{Molarity} = \frac{A_{205}}{E_{205}}$$

Since the molar extinction coefficient, E_{205nm}, was calculated for the sample (not the diluted sample), the A_{205} value must be multiplied by the dilution factor so that all terms are determined from equivalent data:

$$A_{205} = 0.648 \times \frac{1000\,\mu L}{30\,\mu L} = 21.6$$

The equation for calculating molarity is then written as follows:

$$\text{Molarity} = \frac{21.6}{976} = 0.22\,M$$

The molar extinction coefficient at 280 nm (E_{280nm}) is now calculated using the following relationship:

$$E_{280\,nm} = \frac{\text{Molarity}}{A_{280}}$$

To use this equation for this problem, the A_{280} value must first be multiplied by the dilution factor so that all terms are treated equivalently:

$$A_{280} = 0.012 \times \frac{1000\,\mu L}{30\,\mu L} = 0.4$$

The molar extinction coefficient at 280 nm can now be calculated:

$$E_{280\,nm} = \frac{0.022\,M}{0.4} = 0.055$$

Therefore, the molar extinction coefficient for this protein at 205 nm is 976 and at 280 nm is 0.4.

A protein's absorbance at 280 nm is dictated by its content of tryptophan, tyrosine, and cystine (the molecule formed by two cysteines linked by a disulfide bond); Pace et al. (1995) have shown that a protein's molar extinction coefficient at 280 nm can be estimated by adding the number of times each of these occurs in the protein times their own extinction coefficients ($5500\,M^{-1}cm^{-1}$ for tryptophan (W), $1490\,M^{-1}cm^{-1}$ for tyrosine (Y), and $125\,M^{-1}cm^{-1}$ for cystine (C)) according to the following relationship:

$$E_M = (nW \times 5500) + (nY \times 1490) + (nC \times 125)$$

where n is the number of times that residue occurs in the protein.

Problem 11.10 A protein has four tryptophan residues, six tyrosine residues, and three cysteines. What is its estimated molar extinction coefficient?

Solution 11.10

We will use the following relationship:

$$E_M = (nW \times 5500) + (nY \times 1490) + (nC \times 125)$$

Substituting in our values, we have

$$E_M = (4 \times 5500) + (6 \times 1490) + (3 \times 125)$$
$$E_M = (22000) + (8940) + (375)$$
$$E_M = 31315$$

Therefore, this protein has a predicted molar extinction coefficient of $31\,315\,M^{-1}cm^{-1}$.

11.4 RELATING CONCENTRATION IN MILLIGRAMS PER MILLILITER TO MOLARITY

Protein concentration in milligrams per milliliter is related to molarity in the following way:

$$\text{Concentration (in mg/mL)} = \text{Molarity} \times \text{Protein molecular weight}$$

This relationship can be used to convert a concentration in milligrams per milliliter to molarity, and vice versa. Note that this equation will actually give a result in grams per liter. However, this is the same as milligrams per milliliter. For example, 4 g/L is the same as 4 milligrams per milliliter as shown by the following relationship:

$$\frac{4\text{ g}}{\text{L}} \times \frac{1\text{ L}}{1000\text{ mL}} \times \frac{1000\text{ mg}}{\text{g}} = \frac{4\text{ mg}}{\text{mL}}$$

Problem 11.11 A protein has a molecular weight of 135 000 g/mol. What is the concentration of a $2 \times 10^{-4} M$ solution in mg/mL?

Solution 11.11

Using the relationship between concentration in mg/mL and molarity we have

$$\begin{aligned}\text{mg/mL} &= \frac{2 \times 10^{-4}\text{ mol}}{\text{L}} \times \frac{135\,000\text{ g}}{\text{mol}} \times \frac{1000\text{ mg}}{\text{g}} \times \frac{1\text{ L}}{1000\text{ mL}} \\ &= 27\text{ mg/mL}\end{aligned}$$

Therefore, a $2 \times 10^{-4} M$ solution of a protein having a molecular weight of 135 000 has a concentration of 27 mg/mL.

Problem 11.12 A protein has a molecular weight of 167 000. A solution of this protein has a concentration of 2 mg/mL. What is the molarity of the solution?

Solution 11.12

To solve for molarity, the equation for converting molarity and molecular weight to mg/mL must be rearranged as follows:

$$\text{Molarity} = \frac{\text{Concentration (in mg/mL)}}{\text{Protein molecular weight}}$$

Substituting the given values into this equation gives the following result:

$$\text{Molarity} = \frac{2\text{ mg/mL}}{167\,000\text{ g/mol}} = 1.2 \times 10^{-5}\ M$$

Therefore, the protein solution has a concentration of $1.2 \times 10^{-5} M$.

11.5 PROTEIN QUANTITATION USING A_{280} WHEN CONTAMINATING NUCLEIC ACIDS ARE PRESENT

Although nucleic acids maximally absorb UV light at 260 nm, they do show strong absorbance at 280 nm. Measuring protein concentration of crude extracts also containing nucleic acids (RNA and DNA) will give artificially high protein estimates. The reading at 280 nm must therefore be corrected by subtracting the contribution from nucleic acids. This relationship is given by the following equation (Layne, 1957):

$$\text{Protein concentration (in mg/mL)} = (1.55 \times A_{280}) - (0.76 \times A_{260})$$

Problem 11.13 An experimenter is to determine the concentration of protein in a crude extract. She zeros the spectrophotometer with buffer at 260 nm, dilutes 10 μL of the extract into a total volume of 1.0 mL in a 1 cm quartz cuvette, and then takes a reading of the crude extract. She obtains an A_{260} reading of 0.075. She zeros the spectrophotometer with buffer at 280 nm and reads the diluted extract at that wavelength. She obtains an A_{280} value of 0.25. What is the approximate concentration of protein in the crude extract?

Solution 11.13

Using the preceding equation and substituting in the values the experimenter obtained on the spectrophotometer gives the following result:

$$\begin{aligned} \text{mg/mL} &= (1.55 \times 0.25) - (0.76 \times 0.075) \\ &= 0.39 - 0.06 = 0.33\ \text{mg/mL} \end{aligned}$$

Therefore, the protein in the 1 mL cuvette has an approximate concentration of 0.33 mg/mL. The experimenter, however, dilutes her sample 10 μL/1000 μL to take a measurement. To determine the protein concentration of the undiluted sample, this value must be multiplied by the inverse of the dilution factor:

$$0.33\ \text{mg/mL} \times \frac{1000\ \mu\text{L}}{10\ \mu\text{L}} = 33\ \text{mg/mL}$$

Therefore, the original, undiluted sample has an approximate protein concentration of 33 mg/mL.

11.6 PROTEIN QUANTIFICATION AT 205 nm

Proteins also absorb light at 205 nm and with greater sensitivity than at 280 nm. Absorbance at this wavelength is due primarily to the peptide bonds between amino acids. Since all proteins have peptide bonds, there is less variability between proteins at the 205 nm wavelength than at the 280 nm wavelength. Whereas quantification using 280 nm works best for protein in the range of 0.02 to 3 mg in a 1 mL cuvette, absorbance at 205 nm is more sensitive and works for 1 to 100 μg of protein per milliliter.

Protein concentration at 205 nm can be estimated using the equation from Stoscheck (1990):

$$\text{Protein concentration (in mg/mL)} = 31 \times A_{205}$$

Problem 11.14 A sample of purified protein is diluted 2 μL into a final volume of 1.0 mL in a quartz cuvette. A reading at 205 nm gives a value of 0.0023. What is the approximate protein concentration of the original sample?

Solution 11.14
Using the preceding equation with these values gives

$$\text{mg/mL} = 31 \times 0.146 = 4.53\,\text{mg/mL}$$

Therefore, the concentration of protein in the cuvette is 0.07 mg/mL. However, this represents the concentration of the diluted sample. To obtain the protein concentration of the original, undiluted sample, we must multiply this value by the inverse of the dilution factor:

$$4.53\,\text{mg/mL} \times \frac{1000\,\mu\text{L}}{2\,\mu\text{L}} = 2265\,\text{mg/mL}$$

Therefore, the undiluted sample has a concentration of 2265 mg protein/mL.

11.7 PROTEIN QUANTITATION AT 205 nm WHEN CONTAMINATING NUCLEIC ACIDS ARE PRESENT

Nucleic acids contribute very little to an absorbance reading at 205 nm. However, they do contribute more substantially at the other wavelength at which proteins absorb: 280 nm. To correct for the contribution from RNA and DNA to protein estimation in crude extracts, the following equation is used (Peterson, 1983):

$$\text{Protein concentration (in mg/mL)} = \frac{A_{205}}{\left(27 + (120)\frac{A_{280}}{A_{205}}\right)}$$

Problem 11.15 A crude extract is diluted 2.5 μL into a final volume of 1000 μL in a 1 cm cuvette. Absorbency readings at 205 and 280 nm are taken, with the following results. What is the concentration of protein in the crude extract?

$$A_{205} = 0.648$$

$$A_{280} = 0.086$$

Solution 11.15

Using the preceding formula and substituting the values provided gives the following result:

$$\text{mg/mL} = \frac{0.648}{\left[27 + (120)\frac{0.086}{0.648}\right]}$$

$$= \frac{0.648}{27 + (120)0.13}$$

$$= \frac{0.648}{27 + 15.6}$$

$$= \frac{0.648}{42.6} = 0.015\ \text{mg/mL}$$

Therefore, the concentration of protein in the cuvette is 0.015 mg/mL. However, since this is a diluted sample, to obtain the protein concentration in the original, undiluted sample, this value must be multiplied by the inverse of the dilution factor:

$$0.015\ \text{mg/mL} \times \frac{1000\ \mu\text{L}}{2.5\ \mu\text{L}} = 6\ \text{mg/mL}$$

Therefore, the crude extract has a protein concentration of approximately 6 mg/mL.

11.8 MEASURING PROTEIN CONCENTRATION BY COLORIMETRIC ASSAY – THE BRADFORD ASSAY

There are several published methods for measuring protein concentration by colorimetric assay. Each of these relies on an interaction between the protein and a chemical reagent dye. This interaction leads to a color change detectable by a spectrophotometer. The absorbance of an unknown protein sample is compared to a standard curve generated by a protein diluted in a series of known concentrations. The concentration of the unknown sample is interpolated from the standard curve.

One of the most popular methods for quantifying proteins by a dye reagent is called the Bradford Assay (after its developer, Marion Bradford (1976)). The method uses Coomassie Brilliant Blue G-250 as a protein-binding dye. When protein binds to the dye, an increase in absorbance at 595 nm occurs. The dye-binding reaction takes only a couple of minutes and the color change is stable for up to an hour. The assay is sensitive down to approximately 200 μg protein/mL.

Note: Replicates (at least two, preferably three) of each standard should be used to construct a standard curve. Only one of each concentration is used here to simplify the math.

Problem 11.16 An experimenter prepares dilutions of bovine serum albumin (BSA) in amounts ranging from 10 μg to 100 μg in a volume of 0.1 mL (adjusted to volume with buffer where necessary). Five mL of dye reagent (Bradford, 1976) is added to each tube, and the mixture is vortexed and allowed to sit for at least 2 min at room temperature. The absorbance at 595 nm is then read on a spectrophotometer. The results shown in Table 11.2 are obtained. A purified protein of unknown concentration is diluted 2 μL into a total volume of 100 μL and reacted with the Bradford dye reagent in the same manner as the protein standard. Its absorbance at 595 nm is 0.44. What is the concentration of protein in the original, undiluted sample, in mg/mL?

Solution 11.16

The data is plotted as a standard curve (Figure 11.4) using Microsoft Excel (see Appendix A). Linear regression analysis for the standard curve yields the following equation:

$$y = 0.0087x + 0.0396$$

Table 11.2 Absorbance at 595 nm for samples of bovine serum albumin (BSA) having known protein amounts.

μg BSA	A_{595}
10	0.11
20	0.21
30	0.30
40	0.39
50	0.48
60	0.58
75	0.71
100	0.88

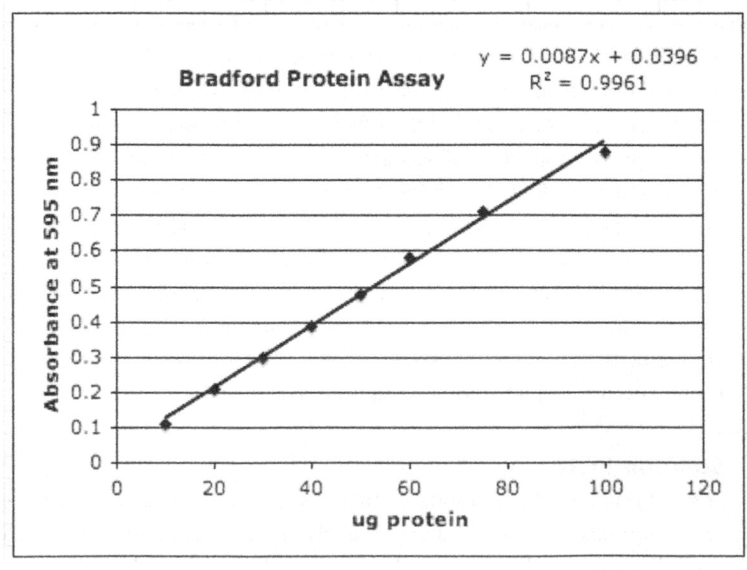

■ **FIGURE 11.4** Data plot for Problem 11.16.

In this equation, y represents the absorbance at 595 nm and x represents the μg amount of protein. Since we have the y value (0.44), we can solve for x:

$$0.44 = 0.0087x + 0.0396$$

$$0.44 - 0.0396 = 0.0087x$$

$$0.4004 = 0.0087x$$

$$\frac{0.4004}{0.0087} = x = 46$$

Therefore, in the assay reaction tube, there are 46 μg of protein, or, thought of another way, the 2 μL taken from the protein prep and placed in the assay tube contains 46 μg of protein. This represents a concentration of 23 mg/mL in the original sample, as shown in the following conversion:

$$\frac{46\,\mu g}{2\,\mu L} \times \frac{1000\,\mu L}{1\,mL} \times \frac{1\,mg}{1000\,\mu g} = \frac{23\,mg}{mL}$$

Therefore, the original, undiluted protein preparation has a concentration of 23 mg protein/mL.

11.9 USING β-GALACTOSIDASE TO MONITOR PROMOTER ACTIVITY AND GENE EXPRESSION

The sugar lactose is a disaccharide of galactose and glucose joined by a β-galactoside bond. The enzyme β-galactosidase, the product of the *lacZ* gene, breaks the β-galactoside bond to release the galactose and glucose monosaccharides. A compound known as ONPG (*o*-nitrophenol-β-D-galactopyranoside) has a structure that mimics lactose and can be cleaved by β-galactosidase (Figure 11.5). *O*-nitrophenol-β-D-galactopyranoside, however, carries an *o*-nitrophenol (ONP) ring group rather than glucose. When ONPG, a colorless compound, is cleaved, ONP, a yellow compound, is released. *O*-nitrophenol absorbs light at 420 nm. This provides a convenient assay for β-galactosidase activity. In the presence of excess ONPG, the amount of yellow color (i.e., the amount of ONP released) is directly proportional to the amount of β-galactosidase in the assay and to the time allowed for the reaction. As β-galactosidase is so easily assayed, the *lacZ* gene can be placed downstream from a promoter to determine its activity under different conditions. The *lacZ* gene can also be fused in reading frame to the gene of another protein under the control of its natural promoter to monitor expression of that promoter.

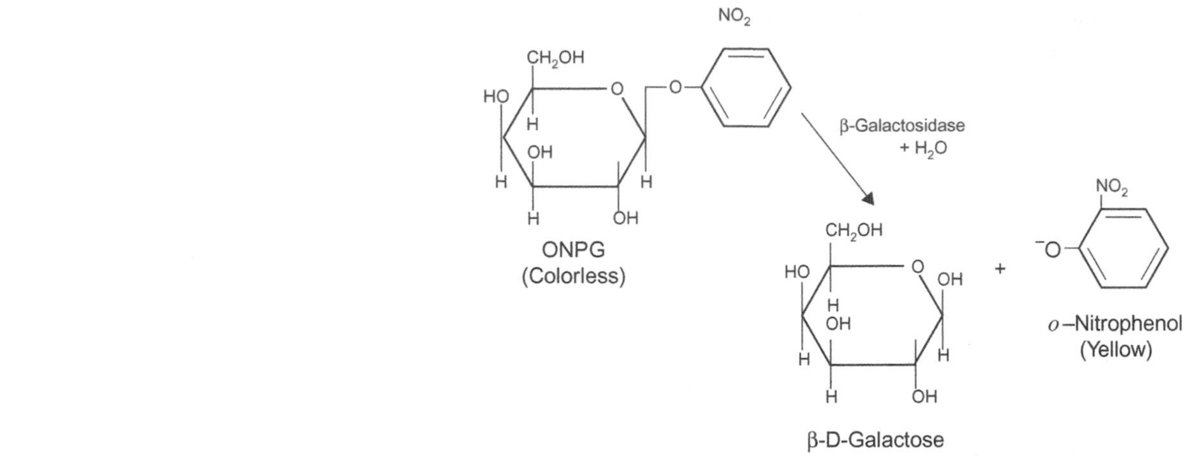

■ **FIGURE 11.5** β-galactosidase splits *o*-nitrophenol-β-D-galactopyranoside (ONPG), a colorless compound, into galactose and *o*-nitrophenol (ONP). The ONP released is yellow and absorbs light at 420 nm.

11.9.1 Assaying β-galactosidase in cell culture

β-galactosidase is typically assayed from *E. coli* cells growing in culture. Cells are grown to log phase ($2 - 5 \times 10^8$ cells/mL; OD_{600} of 0.28 to 0.70), the culture is cooled on ice for 20 min to prevent further growth, and 1 mL is withdrawn for an OD_{600} reading. An aliquot of the culture is added to a phosphate buffer (Z buffer) containing KCl, magnesium sulfate, and β-mercaptoethanol such that the final volume of the assay reaction is 1 mL. To make the cells permeable to ONPG, a drop of toluene is added to the reaction tube and it is then vortexed for 10 sec. The toluene is allowed to evaporate and the assay tube is then placed in a 28°C waterbath for 5 min. The reaction is initiated by adding ONPG. The reaction time is monitored, and when sufficient yellow color has developed, the reaction is stopped by the addition of sodium carbonate to raise the pH to a level at which β-galactosidase activity is inactivated. The reaction should take between 15 min and 6 hr to produce sufficient yellow color. The OD is then measured at 420 and 550 nm. The optimal 420 nm reading should be between 0.6 and 0.9. If it does not fall within this range, the sample should be diluted appropriately.

Both the ONP and the light scatter from cell debris contribute to the absorbance at 420 nm. Light scatter is corrected for by taking an absorbance measurement at 550 nm, a wavelength at which there is no contribution from ONP. The amount of light scattering at 420 nm is proportional to the amount at 520 nm. This relationship is described by the equation

$$OD_{420} \text{ light scattering} = 1.75 \times OD_{550}$$

The absorbance of only the ONP component at 420 nm can be computed by using this relationship as a correction factor and subtracting the OD_{550} multiplied by 1.75 from the 420 nm absorbance reading. Units of β-galactosidase are then calculated using the formula of Miller (1972):

$$\text{Units} = 1000 \times \frac{OD_{420} - 1.75 \times OD_{550}}{t \times v \times OD_{600}}$$

where OD_{420} and OD_{550} are read from the reaction mixture, OD_{600} is the absorbance of the cell culture just before assay, t is equal to the reaction time (in min), and v is the volume (in mL) of aliquoted culture used in the assay.

Problem 11.17 The *lacZ* gene (β-galactosidase) has been cloned adjacent to an inducible promoter. Recombinant cells are grown to log phase and a sample is withdrawn to measure β-galactosidase levels during the uninduced state. Inducer is then added and the cells are allowed to grow for another 10 min before being chilled on ice. A 1 mL aliquot of the culture has an OD_{600} of 0.4. A 0.2 mL aliquot of the culture is added to 0.8 mL of Z buffer, toluene, and ONPG. After 42 min, Na_2CO_3 is added to stop the reaction. A measurement at 420 nm gives a reading of 0.8. Measurement at 550 nm gives a reading of 0.02. How many units of β-galactosidase are present in the 0.2 mL of induced culture?

Solution 11.17

Using the equation

$$\text{Units} = 1000 \times \frac{OD_{420} - 1.75 \times OD_{550}}{t \times v \times OD_{600}}$$

and substituting our values gives the following result:

$$\begin{aligned}\text{Units} &= 1000 \times \frac{(0.8) - (1.75 \times 0.02)}{42 \times 0.2 \times 0.4} \\ &= \frac{765}{3.36} \\ &= 228 \text{ units enzyme}\end{aligned}$$

Therefore, there are 228 units of β-galactosidase in 0.2 mL of culture, or 228 units/0.2 mL = 1140 units β-galactosidase/mL.

11.9.2 Specific activity

It may sometimes be desirable to calculate the specific activity of β-galactosidase in a reaction. **Specific activity** is defined as the number of units of enzyme per milligram of protein. Protein quantification was covered earlier in this chapter. However, even without directly quantifying the amount of protein in an extract, an amount can be estimated by assuming that there are 150 μg of protein per 10^9 cells and that, for *E. coli*, an OD_{600} of 1.4 is equivalent to 10^9 cells/mL.

Problem 11.18 If it is assumed that 1 mL of cells described in Problem 11.17 contains 43 μg of protein, what is the specific activity of the β-galactosidase for that problem?

Solution 11.18

It was calculated in Problem 11.17 that 1 mL of culture contains 1140 units of β-galactosidase/mL. The specific activity can now be calculated as follows:

$$\frac{1140 \text{ units}}{43\,\mu\text{g protein}} \times \frac{1000\,\mu\text{g}}{\text{mg}} = 26512 \text{ units/mg protein}$$

Therefore, the β-galactosidase in the cell culture has a specific activity of 26512 units/mg protein.

11.9.3 Assaying β-galactosidase from purified cell extracts

For some applications, β-galactosidase might be assayed from a more purified source than that described in Problem 11.12. For example, if the cells are removed by centrifugation prior to assay or if the protein (or fusion protein) is being purified by biochemical techniques, then the OD_{600} value that is part of the calculation for determining β-galactosidase units is irrelevant.

In a protocol described by Robertson et al. (1997), β-galactosidase activity (Z_a) is calculated as the number of micromoles of ONPG converted to ONP per unit time and does not require a cell density measurement at 600 nm. It is given by the equation

$$Z_a = \frac{\left(\frac{A_{410}}{E_{410}}\right)V}{t}$$

where A_{410} is the absorbance at 410 nm; E_{410} is the molar extinction coefficient of ONP at 410 nm (3.5 mL/μmol), assuming the use of a cuvette having a 1 cm light path; V is reaction volume; and t is time, in minutes, allowed for the development of yellow color.

To perform the β-galactosidase assay, 2.7 mL of 0.1 M sodium phosphate, pH 7.3; 0.1 mL of 3.36 M 2-mercaptoethanol; 0.1 mL of 30 mM $MgCl_2$; and 0.1 mL of 34 mM ONPG are combined in a spectrophotometer cuvette and mixed. The reaction mix is zeroed in the spectrophotometer at 410 nm. Ten μL of the sample to be assayed for β-galactosidase activity are added to the 3 mL of reaction mix, and the increase in absorbance at 410 nm is monitored. Monitoring is discontinued when A_{410} has reached between 0.2 and 1.2.

Problem 11.19

a) A gene having an in-frame *lacZ* fusion is placed under control of the SV40 early promoter on a recombinant plasmid that is transfected into mammalian cells in tissue culture. After 48 hr, 1 mL of cells is sonicated and centrifuged to remove cellular debris. After blanking the spectrophotometer against reagent, 10 μL of sample are added to 3 mL of reagent mix prepared as described earlier and the increase in absorbance at 410 nm is monitored. After 12 min, an absorbance of 0.4 is achieved. What is the β-galactosidase enzyme activity, in micromoles per minute, for the reaction sample?

b) A 20 μL sample of cleared cell lysate is used in the Bradford protein assay to measure protein quantity. It is discovered that this aliquot contains 180 μg of protein. What is the specific activity of the sample?

Solution 11.19(a)

The assay gives the following results:

$A_{410} = 0.4$,

$t = 12$ min, and

$V = 3.01$ mL (3 mL of reaction mix + 10 μL sample).

These values are placed into the equation for calculating β-galactosidase activity:

$$Z_a = \frac{\left(\frac{A_{410}}{E_{410}}\right)V}{t}$$

$$Z_a = \frac{\left(\frac{0.4}{3.5\ \text{mL}/\mu\text{mol}}\right)3.01\ \text{mL}}{12\ \text{min}}$$

$$Z_a = \frac{(0.114\,\mu\text{mol})(3.01)}{12\,\text{min}} = \frac{0.343\,\mu\text{mol}}{12\,\text{min}} = 0.03\,\mu\text{mol/min}$$

Therefore, the 3.01 mL reaction produces 0.03 μmol of ONP per minute.

Solution 11.19(b)

The Bradford assay for protein content gives 180 μg of protein in the 20 μL sample. This is converted to micrograms per milliliter as follows:

$$\frac{180\,\mu\text{g protein}}{20\,\mu\text{L}} = 9\,\mu\text{g}/\mu\text{L}$$

The amount of protein in the 10 μL sample used in the β-galactosidase assay is calculated as follows:

$$10\,\mu\text{L} \times \frac{9\,\mu\text{g protein}}{\mu\text{L}} = 90\,\mu\text{g protein}$$

The specific activity of the 10 mL sample in the enzyme assay is the activity divided by the amount of protein in that sample:

$$\text{Specific activity} = \frac{0.03\,\mu\text{mol/min}}{90\,\mu\text{g}} = \frac{0.0003\,\mu\text{mol/min}}{\mu\text{g}}$$

Therefore, the sample has a specific activity of $3 \times 10^{-4}\,\mu\text{mol/min}/\mu\text{g}$.

11.10 THIN LAYER CHROMATOGRAPHY (TLC) AND THE RETENTION FACTOR (R_f)

Thin layer chromatography (TLC) is a method of separating compounds by their rate of movement through a thin layer of silica gel (the stationary phase) coated on a glass plate. A protein sample is applied to a spot at the bottom of the plate (about 1 cm from the bottom edge) along a line marked in pencil perpendicular to the long axis of the plate. This spot is designated as the origin. The TLC plate, spotted with sample, is then placed upright in a sealed developing chamber containing a small volume of solvent (such as ethyl acetate or an alcohol). The solvent (known as the stationary phase) should not be of a volume that will submerge the origin. When the TLC plate is in the solvent, capillary action draws the solvent mobile phase up the plate. The sample compounds, as they dissolve in the ascending solvent, also move up the plate and separate at a rate that is based on their polarity, their solubility, and their interaction with the stationary phase.

When the solvent has almost reached the top, the plate is removed from the developing chamber and the solvent front is marked with pencil. The plate can then be developed with a reagent (iodine or ninhydrin) that aids in the visualization of the separated molecules, which produce roughly circular spots on the plate. The spots are outlined in pencil and their distances from the origin are measured with a ruler. The distance from the origin to the geometric center of a spot divided by the distance from the origin to the solvent front is called the retention factor (R_f). The larger an R_f value, the further the compound will have traveled from the origin.

$$R_f = \frac{\text{Distance compound traveled}}{\text{Distance solvent traveled}}$$

The R_f value is a physical constant for an organic molecule and can be used to corroborate the identity of a molecule. If two molecules have different R_f values under identical run conditions – if they are run on the same TLC plate – they are different compounds. If two molecules have the same R_f value, they *may* be the same compound.

Since it is difficult to exactly match the chromatography conditions (solvent concentration, temperature, silica gel thickness, amount spotted, etc.) between runs, **relative R_f** (R_{rel}) values are often used and reported. A R_{rel} value is calculated as a relationship to the R_f of a known standard run on the same plate at the same time. The R_{rel} values in a given solvent should remain a constant.

$$R_{rel} = \frac{R_f \text{ of unknown}}{R_f \text{ of standard}}$$

Problem 11.20 A sample of purified protein discovered in the laboratory freezer has lost its label. It is run on a TLC plate to see if it can be quickly identified. When the TLC plate is taken from the developing chamber, the solvent front is marked and the plate is treated with ninhydrin spray. Measurements reveal that the solvent front traveled 7.3 cm from the origin, the unidentified sample ran 4.1 cm, and a standard ran 5.4 cm (Figure 11.6). What is the R_f value of the unknown sample and what is its R_{rel} compared with the standard?

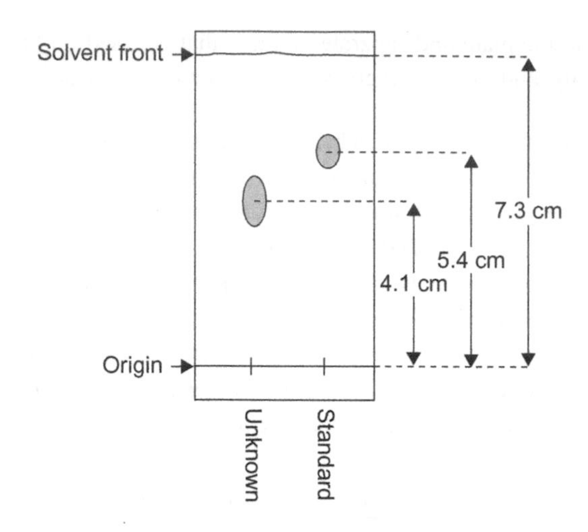

FIGURE 11.6 The thin layer chromatography (TLC) experiment described in Problem 11.20.

Solution 11.20

The R_f of the unknown sample is calculated as the distance it traveled from the origin (4.1 cm) divided by the distance traveled by the solvent front (7.3 cm). This yields a R_f value of 0.56:

$$R_f = \frac{4.1\,\text{cm}}{7.3\,\text{cm}} = 0.56$$

The standard ran 5.4 cm. It has an R_f value, therefore, of 0.74 (5.4 cm divided by 7.3 cm). The R_{rel} is the unknown's R_f divided by the standard's R_f:

$$R_{rel} = \frac{0.56}{0.74} = 0.76$$

The R_{rel}, therefore, is 0.76.

11.11 ESTIMATING A PROTEIN'S MOLECULAR WEIGHT BY GEL FILTRATION

Gel filtration chromatography (also called **size exclusion chromatography**) is a method of separating molecules on the basis of their size. By this technique, a protein sample is suspended in an aqueous solution (the mobile phase) and applied to the top of a chromatography column filled with a matrix of porous beads (the stationary phase). The smaller molecules within the sample will enter into the pores of the beads and, by gravity, will travel down the column, taking a long and circuitous route through the bead matrix. As a consequence of the complicated path they take in and out of the beads, their movement through the column is slowed. Any molecules within the sample mixture too large to enter into the pores of the beads take a more simple and direct route through the column. They will elute quickly. Molecules of intermediate size may diffuse into the beads to varying extents – smaller molecules more so than larger ones. The molecules of the sample elute from the column, therefore, in order of decreasing size – the largest molecules (those completely excluded) elute first while the smallest molecules (those able to diffuse into the pores of the beads) elute last. A UV monitor detects the different molecules as they come off the column.

The matrix selected for sample fractionation and molecular size determination depends on the size range of the molecules within the experimental sample. Most commercial suppliers offer an assortment of column matrix materials having their own unique pore sizes, recommended fractionation ranges, and **exclusion limits** – the molecular weight of the smallest molecule too large to pass through the beads' pores. Some of the more widely used column materials include Sephacryl®, Sephadex®, and Sepharose®.

No matter what beads are used, however, the column must first be calibrated. A molecule having a size smaller than that of the pores is passed through the column and the volume of liquid it takes to completely elute the molecule is measured. This volume is referred to as the 'total volume' and is designated V_t. Likewise, a molecule having a size in large excess to the gel's exclusion limit – a molecule sure to be excluded from entering the pores of the beads – is passed through the column and the volume it takes to elute the molecule is measured. This volume is referred to as void volume. It represents the amount of liquid in the column between the beads. It is designated V_0. The total volume, V_t, is usually about 90% or so of the bed volume – the volume taken up by the hydrated beads packed within the column. (The bed volume is calculated as $\pi r^2 h$ where r is the inner radius of the column and h is the height of the packed column material.) The void volume, V_0, is in the range of roughly 40% or so of the bed volume.

To determine the molecular size of a molecule, the gel filtration column is then calibrated with a set of standards of known molecular size. The volume taken to elute each molecule of the standard from the column is recorded. This volume is designated V_e and is unique to each molecule within the standard. To provide consistency in reporting, a distribution coefficient (K_D) for each molecule is determined. It is calculated as

$$K_D = \frac{(V_e - V_0)}{(V_t - V_0)}$$

Plotting the $\log_{10}$ of the molecular weights of the standards vs. the K_D values generates a straight-line standard curve. An unknown protein can then be run through the column and its molecular weight extrapolated from the standard curve as shown in Problem 11.21 below.

Note: Proteins have different shapes. They can be globular, linear, or twisted into any of an infinite array of configurations. The shape of a molecule will affect how it interacts with the pores of the beads in the column. Elution, therefore, is more strictly related to molecular size and shape than to molecular weight. The proteins used to make the standard curve should have, as close as possible, the same overall shape as the unknown protein. Nevertheless, this method gives a fair approximation of a protein's size in Da and therefore has value as a means to aid in its characterization.

Problem 11.21 Gel filtration chromatography is being used to give an estimate of a protein's molecular weight. A column having a fractionation range of 5000 to 250 000 is calibrated and gives a void volume, V_0, of 40 mL and a total volume, V_t, of 90 mL. Molecular weight standards are applied to the column and their elution volumes (V_e values) are measured with the results shown in Table 11.3. 1.1 A purified protein has an elution volume of 60 mL. What is its approximate molecular weight?

Table 11.3 Dataset for Problem 11.21, showing the molecular weights of the standards and their elution volumes.

Protein	Molecular weight	Elution volume (V_e) in mL
Standard A	12 500	82.5
Standard B	15 300	81.25
Standard C	22 000	77.5
Standard D	25 100	75
Standard E	28 700	70
Standard F	33 300	68.75
Standard G	45 200	67.5
Standard H	74 500	55
Standard I	96 800	52.5
Standard J	120 000	50
Standard K	244 000	41.25

Solution 11.21

We will first calculate the distribution coefficient, K_D, for each standard. Since the V_0 for this column has been measured to be 40 mL and the V_t has been measured to be 90 mL, the K_D is calculated as

$$K_D = \frac{(V_e - V_0)}{(V_t - V_0)}$$

$$K_D = \frac{(V_e - 40)}{(90 - 40)} = \frac{(V_e - 40)}{50}$$

For standard sample A, we have

$$K_D = \frac{(82.5 - 40)}{50} = \frac{42.5}{50} = 0.85$$

The K_D values for the other standards are calculated in a similar manner and are shown in Table 11.4.

We can now plot these values into Microsoft Excel (see Appendix A) against the log values of the molecular weight of each standard; linear regression analysis will give us the equation for the line that best fits, which, for this example, is $y = -0.6824x + 3.6742$ (Figure 11.7).

The unknown protein has a V_e value of 60 mL. Its K_D, therefore, is 0.4:

$$K_D = \frac{(60 - 40)}{50} = \frac{20}{50} = 0.4$$

Table 11.4 Dataset for Problem 11.21, showing the elution volumes for each standard and their K_D values.

Protein	Elution volume (in mL)	K_D
Standard A	82.5	0.85
Standard B	81.25	0.825
Standard C	77.5	0.75
Standard D	75	0.7
Standard E	70	0.6
Standard F	68.75	0.575
Standard G	67.5	0.55
Standard H	55	0.3
Standard I	52.5	0.25
Standard J	50	0.2
Standard K	41.25	0.025

MW	log MW	KD
12500	4.09691001	0.85
15300	4.18469143	0.825
22000	4.34242268	0.75
25100	4.39967372	0.7
28700	4.4578819	0.6
33300	4.52244423	0.575
45200	4.65513843	0.55
74500	4.87215627	0.3
96800	4.98587536	0.25
120000	5.07918125	0.2
244000	5.38738983	0.025

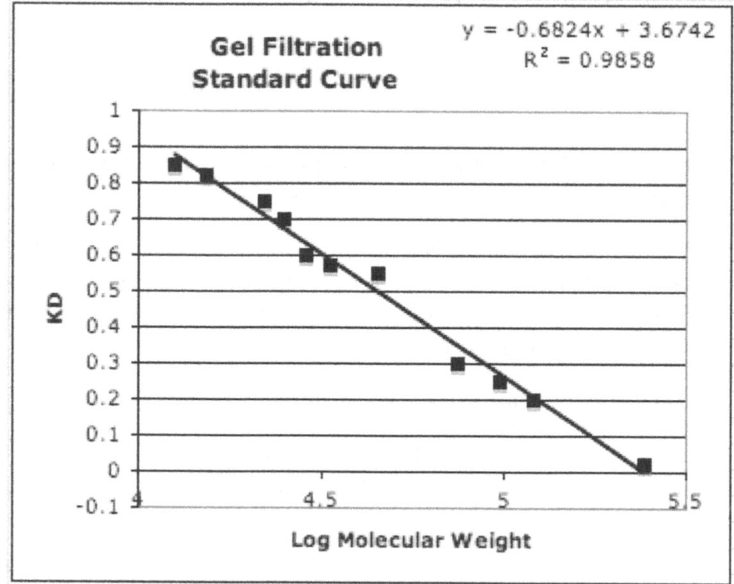

■ **FIGURE 11.7** The standard curve generated by gel filtration of proteins of known molecular weight.

This K_D value represents the y value in the equation for the linear regression line. Using this equation and substituting our y value, we can solve for the log of the protein's molecular weight:

$$\begin{aligned} y &= -0.6824x + 3.6742 \\ 0.4 &= -0.6824x + 3.6742 \\ 0.4 - 3.6742 &= -0.6824x \\ -3.2742 &= -0.6824x \\ \frac{-3.2742}{-0.6824} &= x = 4.798 \end{aligned}$$

Therefore, the log of the molecular weight of the unknown protein is 4.798. To find its molecular weight, we must take the antilog of this value:

$$\text{antilog } 4.798 = 62\,800$$

Therefore, the unknown protein has an approximate molecular weight of 62 800 Da.

11.12 THE CHLORAMPHENICOL ACETYLTRANSFERASE (CAT) ASSAY

The *cat* gene, encoding chloramphenicol acetyltransferase (CAT), confers resistance to the antibiotic chloramphenicol to those bacteria that carry and express it. The *cat* gene, having evolved in bacteria, is not found naturally in mammalian cells. However, when introduced into mammalian cells by transfection and when under the control of an appropriate promoter, the *cat* gene can be expressed. It has found popularity, therefore, as a reporter gene for monitoring transfection and for studying gene expression within transfected cells.

Chloramphenicol acetyltransferase transfers an acetyl group ($—COCH_3$) from acetyl coenzyme A to two possible positions on the chloramphenicol molecule. To assay for CAT activity, an extract from cells expressing CAT is incubated with [^{14}C]-labeled chloramphenicol. Transfer of the acetyl group from acetyl coenzyme A to chloramphenicol alters the mobility of the antibiotic on a TLC plate. The amount of acetylated chloramphenicol is directly proportional to the amount of CAT enzyme.

Protocols for running a CAT assay have been described by Gorman et al. (1982) and by Kingston (1989). Typically, 1×10^7 mammalian cells are transfected with a plasmid carrying *cat* under a eukaryotic promoter. After two to three days, the cells are resuspended in 100 μL of 250 m*M* Tris, pH 7.5. The cells are sonicated and centrifuged at 4°C to remove debris. The extract is incubated at 60°C for 10 min to inactivate endogenous acetylases and spun again at 4°C for 5 min. Twenty μL of extract is added to a reaction containing 1 μCi [^{14}C] of chloramphenicol (50 μCi/mmol), 140 m*M* Tris-HCl, pH 7.5, and 0.44 m*M* acetyl coenzyme A in a final volume of 180 μL. The reaction is allowed to proceed at 37°C for 1 hr. Four hundred μL of cold ethyl acetate are added and mixed in, and the reaction is spun for 1 min. The upper organic phase is transferred to a new tube and dried under vacuum. The dried sample is resuspended in 30 μL of ethyl acetate, and a small volume (5–10 μL) is spotted close to the edge of a silica gel TLC plate. The TLC plate is placed in a chromatography tank pre-equilibrated with 200 mL of chloroform : methanol (19 : 1). Chromatography is allowed to proceed until the solvent front is ⅔ of the way up the plate. The plate is air-dried, a pen with radioactive ink is used to mark the plate for orientation, and the plate is exposed to X-ray film. Since chloramphenicol has two potential acetylation sites, two acetylated forms, running faster than the unacetylated form, are produced by the CAT reaction (Figure 11.8). A third, higher spot, corresponding to double acetylation, can be seen under conditions of high CAT activity when single-acetylation sites have been saturated. When this third spot appears, the assay will be out of its linear range and the extract should be diluted or less time allowed for the assay. The area on the TLC plate corresponding to the chloramphenicol spot and the acetylated chloramphenicol spots are scraped off and counted in a liquid scintillation counter. Chloramphenicol acetyltransferase activity is then calculated as follows:

$$\text{CAT activity} = \%\ \text{chloramphenicol aceylated}$$

$$= \frac{\text{Counts in acetylated species}}{\begin{pmatrix}\text{Counts in}\\ \text{acetylated species}\end{pmatrix} + \begin{pmatrix}\text{Nonacetylated}\\ \text{chloramphenicol counts}\end{pmatrix}} \times 100$$

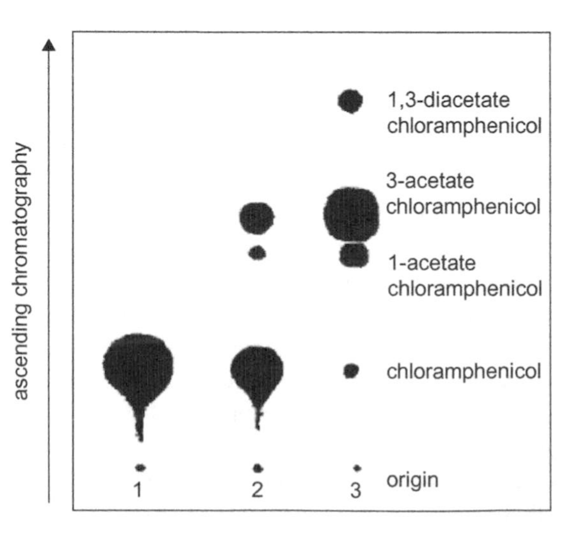

■ **FIGURE 11.8** Thin layer chromatography (TLC) plate assay of CAT activity. Spot 1 contains only labeled chloramphenicol. Spot 2 shows the two spots created by single-chloramphenicol-acetylation events. Spot 3 shows the spot toward the top of the plate representing double-acetylation events.

Problem 11.22 A transfection experiment using CAT as a reporter is conducted as just described. The acetylated species of chloramphenicol gives 3×10^5 cpm. The unacetylated chloramphenicol spot gives 1.46×10^6 cpm. What is the CAT activity?

Solution 11.22

The values obtained in the liquid scintillation counter are placed into the formula for calculating CAT activity, as follows:

$$\text{CAT activity} = \%\ \text{chloramphenicol aceylated}$$

$$= \frac{\text{Counts in acetylated species}}{\begin{pmatrix}\text{Counts in}\\ \text{acetylated species}\end{pmatrix} + \begin{pmatrix}\text{Nonacetylated}\\ \text{chloramphenicol counts}\end{pmatrix}} \times 100$$

$$\%\ \text{CAT activity} = \frac{3 \times 10^5\ \text{cpm}}{3 \times 10^5\ \text{cpm} + 1.46 \times 10^6\ \text{cpm}} \times 100 = 17\%$$

Therefore, 17% of the chloramphenicol is acetylated by CAT.

11.12.1 Calculating molecules of chloramphenicol acetyltransferase (CAT)

Purified CAT protein has a specific activity of 153 000 nmol/min/mg of protein (Shaw and Brodsky, 1968) and has a molecular weight of 25 668 g/mol (Shaw et al., 1979). These values can be used to calculate the number of molecules of CAT in an assay, as shown in the next problem.

Problem 11.23 For Problem 11.22, how many molecules of CAT are contained in the extract?

Solution 11.23

The CAT assay contains 1 μCi of [^{14}C] chloramphenicol having a specific activity of 50 μCi/mmol. The amount of chloramphenicol in the reaction is then calculated as follows:

$$1\,\mu\text{Ci} \times \frac{\text{mmol}}{50\,\mu\text{Ci}} = 0.02\,\text{mmol}$$

Seventeen percent of the [^{14}C] chloramphenicol is converted to the acetylated forms during the assay. The number of millimoles of chloramphenicol acetylated by CAT is then calculated by multiplying 0.02 mmol by 17/100 (17%):

$$0.02\text{ mmol} \times \frac{17}{100} = 0.0034\text{ mmol}$$

The reaction in the assay is incubated at 37°C for 60 min. The amount of [^{14}C] chloramphenicol acetylated per minute is calculated as follows:

$$\frac{0.0034\text{ mmol}}{60\text{ min}} = 5.7 \times 10^{-5}\text{ mmol/min}$$

Since the specific activity of CAT is given in units of nanomoles, a conversion factor must be used to convert millimoles to nanomoles. Likewise, since it also contains a 'milligram' term and since molecular weight is defined as grams/mole, another conversion factor will be used to convert milligrams to grams. Avogadro's number (6.023×10^{23} molecules per mole) will be used to calculate the final number of molecules. The expression for calculating the number of molecules of CAT in the assay can be written as follows:

$$\frac{5.7 \times 10^{-5}\text{ mmol}}{\text{min}} \times \frac{1 \times 10^{6}\text{ nmol}}{\text{mmol}} \times \frac{\text{mg}}{153000\text{ nmol/min}} \times \frac{\text{g}}{1 \times 10^{3}\text{ mg}} \times \frac{1\text{mol}}{25668\text{ g}} \times \frac{6.023 \times 10^{23}\text{ molecules}}{\text{mol}}$$

Note that the term

$$\frac{\text{mg}}{153000\text{ nmol/min}}$$

in the equation is the same as

$$\text{mg} \times \frac{\text{min}}{153000\text{ nmol}}$$

All units cancel, therefore, except 'molecules,' the one term desired. Multiplying numerator and denominator values gives the following result:

$$= \frac{3.4 \times 10^{25}\text{ molecules}}{3.9 \times 10^{12}} = 8.7 \times 10^{12}\text{ molecules}$$

Therefore, the reaction in Problem 11.15 contains 8.7×10^{12} molecules of CAT.

11.13 USE OF LUCIFERASE IN A REPORTER ASSAY

The *luc* gene encoding the enzyme luciferase from the firefly *Photinus pyralis* has become a popular reporter for assaying gene expression. Luciferase from transformed bacterial cells or transfected mammalian cells expressing the *luc* gene is assayed by placing cell extract in a reaction mix containing excess amounts of luciferin (the enzyme's substrate), ATP, $MgSO_4$, and molecular oxygen (O_2). ATP-dependent oxidation of luciferin leads to a burst of light that can be detected and quantitated (as light units (LUs)) at 562 nm in a luminometer. The amount of light generated in the reaction is proportional to luciferase concentration. The luciferase assay is more sensitive than that for either β-galactosidase or CAT.

Because the luminometer used for assaying luciferase activity provides a reading in LUs, very little math is required to evaluate the assay. In the following problem, however, the number of molecules of luciferase enzyme in a reporter gene assay will be determined.

Problem 11.24 To derive a standard curve, different amounts of purified firefly luciferase are assayed for activity in a luminometer. A consistent peak emission of 10 LU is the lower level of activity reliably detected by the instrument, and this corresponds to 0.35 pg of enzyme. How many molecules of luciferase does this represent?

Solution 11.24

Luciferase has a molecular weight of 60 746 g/mol (De Wet et al., 1987). The number of molecules of luciferase is calculated by multiplying the protein quantity by its molecular weight and Avogadro's number. A conversion factor is used to convert pg to g. The calculation is as follows:

$$0.35\ \text{pg} \times \frac{1\ \text{g}}{1 \times 10^{12}\ \text{pg}} \times \frac{1\ \text{mol}}{60746\ \text{g}} \times \frac{6.023 \times 10^{23}\ \text{molecules}}{\text{mol}}$$

$$= \frac{2.11 \times 10^{23}\ \text{molecules}}{6.075 \times 10^{16}}$$

$$= 3.47 \times 10^{6}\ \text{molecules}$$

Therefore, 0.35 pg of luciferase corresponds to 3.47×10^{6} molecules.

11.14 *IN VITRO* TRANSLATION – DETERMINING AMINO ACID INCORPORATION

In vitro translation can be used to radiolabel a protein from a specific gene. A small amount of the tagged protein can then be added as a tracer to follow the purification of the same protein from a cell extract. As for any experiment using radioactive labeling, it is important to determine the efficiency of its incorporation into the target molecule.

To prepare an *in vitro* translation reaction, the gene to be expressed should be under the control of an active promoter (*lac*UV5, *tac*, λ P_L, etc.). The DNA carrying the desired gene is placed in a reaction containing S30 or S100 extract freed of exogenous DNA and mRNA (Spirin et al., 1988). The reaction should also contain triphosphates, tRNAs, ATP, GTP, and a mixture of the amino acids in which one of them (typically Met) is added as a radioactive compound.

To determine the amount of incorporation of the radioactive amino acid, an aliquot of the reaction is precipitated with cold TCA and casamino acids (as carrier). The precipitate is collected on a glass fiber filter, rinsed with TCA several times, and then rinsed a final time with acetone to remove unincorporated [^{35}S] Met. The filter is dried and counted in a liquid scintillation counter.

Problem 11.25 A gene is transcribed and translated in a 100 μL reaction. Into that reaction is added 2 μL of 20 mCi/mL [^{35}S] Met (2000 Ci/mmol). After incubation at 37°C for 60 minutes, a 4 μL aliquot is TCA-precipitated to determine incorporation of radioactive label. In the liquid scintillation counter, the filter gives off 1.53×10^6 cpm. A 2 μL aliquot of the reaction is counted directly in the scintillation counter to determine total counts in the reaction. This sample gives 2.4×10^7 cpm. What percent of the [^{35}S] Met has been incorporated into polypeptide?

Solution 11.25

To solve this problem, the total number of counts in the reaction must be determined. The number of counts in the aliquot from the reaction precipitated by TCA will then be calculated and the number of counts expected in the entire translation reaction will be determined. The percent [^{35}S] Met incorporated will then be calculated as a final step.

Two μL of the reaction is placed directly into scintillation fluid and counted. Since 2 μL gives 2.4×10^7 cpm and there is a total of 100 μL in the reaction, the total number of counts in the reaction is calculated as follows:

$$\frac{2.4 \times 10^7 \text{ cpm}}{2\,\mu\text{L}} \times 100\,\mu\text{L} = 1.2 \times 10^9 \text{ cpm}$$

Therefore, a total of 1.2×10^9 cpm is in the 100 μL reaction.

Four μL are treated with TCA. On the filter, 1.53×10^6 cpm are precipitated. The total number of counts expected in the 100 μL reaction is then determined as follows:

$$\frac{1.53 \times 10^6 \text{ cpm}}{4\,\mu\text{L}} \times 100\,\mu\text{L} = 3.83 \times 10^7 \text{ cpm}$$

Therefore, a total of 3.83×10^7 cpm would be expected to be incorporated in the reaction.

The percent total radioactivity incorporated into protein is then calculated as the total number of counts incorporated divided by the total number of counts in the reaction, multiplied by 100:

$$\frac{3.83 \times 10^7 \text{ cpm}}{1.2 \times 10^9 \text{ cpm}} \times 100 = 3.2\%$$

Therefore, by this TCA-precipitation assay, 3.2% of the [^{35}S] Met is incorporated into polypeptide.

11.15 THE ISOELECTRIC POINT (pI) OF A PROTEIN

In the method of isoelectric focusing, proteins are separated on a polyacrylamide gel having a pH gradient. A protein will stop migrating at the point in the gradient at which it has no net charge. The pH at which this occurs is called the protein's **isoelectric point (pI)**. At that point, the protein has neither a positive nor negative charge – its anion and cation charges are balanced.

A protein is made from a string of many amino acids, each one able to give or accept protons. Each individual amino acid has both a carboxyl group (—COOH), which can be ionized to the unprotonated form (—COO$^-$), and an amino group (—NH_2), which can be ionized to the protonated —NH_3^+ form. The readiness with which these groups become ionized depends on the local pH. Since, however, in a protein chain of amino acids, the amino and carboxyl groups are engaged in the formation of peptide bonds between the amino acids along the chain (except those on the two ends), a protein's overall level of ionization – its charge – depends on other molecular groups capable of being ionized. These are the R groups found on the amino acids arginine (Arg, R), aspartic acid (Asp, D), cysteine (Cys, C), glutamic acid (Glu, E), histidine (His, H), lysine (Lys, K), and tyrosine (Tyr, Y). Each protein, therefore, will have its own unique pI depending on its amino acid content. Basic proteins (those having a preponderance of basic R groups) will have a high pI while acidic proteins (those having a preponderance of acidic R groups) will have a low pI.

The pK_a of a molecule describes the readiness with which it can become protonated or deprotonated. A protein's pI, therefore, depends on the pK_a values of the amino acids it contains; more specifically and primarily, on the pK_a values of the seven amino acids described above whose R groups can be ionized. This, in turn, depends on the pH. In a buffer having a pH below the protein's pI (if the buffer is more acidic), the protein carries a net positive charge and, during electrophoresis, will migrate to the cathode (the negative pole). In a buffer having a pH above the pI, the protein carries a net negative charge and will migrate to the anode (positive pole). At a pH equivalent to the protein's pI, the protein will not migrate at all. Separating proteins by isoelectric focusing is the first step in two-dimensional gel electrophoresis – the second step taking advantage of the mass differences between the proteins being separated.

The readiness with which the specified R groups become ionized is influenced not only by the pH but also by the local topography – what amino acids are in the vicinity when the protein is folded into its three-dimensional structure. As we do not yet have a complete understanding of protein folding, the pK_a values of the amino acids a protein contains can, at best, be only estimated. Further, as most proteins are hundreds of amino acids in length, the calculation of pI is best left to a computer program. Several can be found on the web. The one maintained by the European Molecular Biology Laboratory (EMBL) is an example (Figure 11.9). Using this program, the protein's amino sequence is entered (by its single letter code) and its pI as determined by a method that 'zeroes in' on the pI value by calculating the protein's net charge over a wide pH range (Figure 11.10).

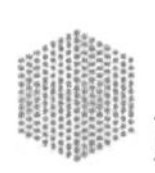

EMBL WWW Gateway to Isoelectric Point Service

Isoelectric Point determination

Reference: Lehninger (1979) Biochimie

Enter your sequence

Example:

QWERTYYTREWQRTYEYTRQWE

Please NOTE: This tool understands ONLY UPPERCASE LETTERS as input. Lowercase letters are simply NOT taken into account at all when calculating the PI. The AA contributing to the isoelectric point (in this calculation) are: C, D, E, H, K, R and Y (and the Amino- and Carboxy-Terminus of the peptide)

Determine the Isoelectric Point | Start over

■ **FIGURE 11.9** The website maintained by EMBL that can be used to calculate a protein's isoelectric point (pI) value.

EMBL WWW Gateway to Isoelectric Point Service

Sequence

QWERTYYTREWQRTYEYTRQWE

The Net Charge had been determined using the values available from the Lehninger's Biochemistry book. At the time this service being developed, they were also available on the WWW

pH	Charge
1	4.95404104510604
1.5	4.86660991080583
2	4.66393232787135
2.5	4.33903584614563
3	3.96654332613935
3.5	3.46078635650845
4	2.58166317480794
4.5	1.44659162041615
5	0.606040638811295
5.5	0.213460661548707
6	0.0695014097411031
6.5	0.0205222630320048
7	0.00107286215655567
7.5	-0.0168196965319369
8	-0.0590586750133673
8.5	-0.183541169701062
9	-0.534711898572718
9.5	-1.35835803661716
10	-2.68709458830983
10.5	-3.9546059737823
11	-4.74281788534035
11.5	-5.30939719177408
12	-6.0945110472756
12.5	-7.21509085425769
13	-8.19247441416308

Isoelectric Point: 7.03499999999987 (using a ph increment of 0.005 for the calculation of the lowest net charge. Charge at that pH is 1.53122640416115e-05)

■ **FIGURE 11.10** The EMBL isoelectric point (pI) calculator program zeroes in on a protein's pI by determining the protein's net charge incrementally over a wide pH range.

■ CHAPTER SUMMARY

Protein concentration within a solution can be estimated by measuring its absorbance at 280 nm. An absorbance of 1.0 at 280 nm is equivalent to 1 mg protein/mL. Protein concentration can also be estimated by using its absorbance coefficient – its absorbance at a certain concentration at a given wavelength in a 1 cm light path. The concentration of a protein in mg/mL is equivalent to the absorbance divided by it absorbance coefficient for a 1 mg/mL solution:

$$\text{Protein concentration (in mg/mL)} = \frac{\text{Absorbance}}{A_{1\text{ cm}}^{1\text{ mg/mL}}}$$

Protein concentration in molarity can be estimated using absorbance and the protein's molar extinction coefficient (E_M) (the absorbance of a 1 *M* solution of a given protein at a given wavelength in a 1 cm cuvette):

$$\text{Protein concentration (in molarity)} = \frac{\text{Absorbance}}{E_M}$$

The absorbance coefficient for a 1% protein solution ($A_{1\text{cm}}{}^{1\%}$) is related to the molar extinction coefficient (E_M) as follows:

$$E_M = \frac{(A_{1\text{ cm}}^{1\%})(\text{molecular weight})}{10}$$

A protein's extinction coefficient at 205 nm can be determined using the following formula:

$$E_{205\text{ nm}}^{1\text{ mg/mL}} = 27 + 120\left(\frac{A_{280}}{A_{205}}\right)$$

This equation can be used when the protein has an unknown concentration. Once a concentration has been determined, the extinction coefficient at 280 nm can be determined using the following equation (except for those proteins having a high phenylalanine content):

$$E_{280\text{ nm}} = \frac{\text{mg protein/mL}}{A_{280}}$$

A protein's molar extinction coefficient can be estimated from its content of typtophan, tyrosine, and cystine residues using the following relationship (where *n* is the number of each residue type in the protein):

$$E_M = (n\text{W} \times 5500) + (n\text{Y} \times 1490) + (n\text{C} \times 125)$$

Protein concentration in mg/mL is related to molarity in the following way:

$$\text{Concentration (in mg/mL)} = \text{Molarity} \times \text{Protein molecular weight}$$

This relationship can be used to convert a concentration in mg/mL to molarity, and vice versa.

When contaminating nucleic acids are present in the protein solution, you must correct the reading at 280 nm by subtracting the contribution from DNA or RNA. This relationship is given by the following equation:

$$\text{Protein concentration (in mg/mL)} = (1.55 \times A_{280}) - (0.76 \times A_{260})$$

When quantifying proteins at lower concentrations (1 to 100 μg protein/mL), measuring their absorbance at 205 nm can give more accurate results than those estimated from a reading at 280 nm. Protein concentration at 205 nm can be estimated using the following equation:

$$\text{Protein concentration (in mg/mL)} = 31 \times A_{205}$$

If estimating protein concentration by its absorbance at 205 nm, contamination from nucleic acids can be corrected using the following equation:

$$\text{Protein concentration (in mg/mL)} = \frac{A_{205}}{\left(27 + (120)\dfrac{A_{280}}{A_{205}}\right)}$$

Protein concentration can be determined by reaction with a reagent such as Coomassie Brilliant Blue G-250, as employed in the Bradford Assay, whereby a standard curve is used to extrapolate the concentration of an unknown sample.

The characterization of genetic elements involved in transcription (promoters, repressors, silencers, etc.) can be monitored in recombinant constructs by positioning them adjacent to the β-galactosidase gene. β-galactosidase protein, which converts the colorless chemical ONPG into the yellowish compound ONP, can be quantified by the amount of yellow color generated in the presence of ONPG. Units of β-galactosidase are calculated as

$$\text{Units} = 1000 \times \frac{\text{OD}_{420} - 1.75 \times \text{OD}_{550}}{t \times v \times \text{OD}_{600}}$$

where OD_{420} and OD_{550} are read from the reaction mixture, OD_{600} is the absorbance of the cell culture just before assay, t is equal to the reaction time (in min), and v is the volume (in mL) of aliquoted culture used in the assay. The specific activity of β-galactosidase is defined as the number of units of enzyme per milligram of protein.

β-galactosidase activity (Z_a) can also be calculated as the number of micromoles of ONPG converted to ONP per unit time, as shown in the following relationship:

$$Z_a = \frac{\left(\frac{A_{410}}{E_{410}}\right)V}{t}$$

where A_{410} is the absorbance at 410 nm; E_{410} is the molar extinction coefficient of ONP at 410 nm (3.5 mL/μM), assuming the use of a cuvette having a 1 cm light path; V is the reaction volume; and t is the time, in min, allowed for the development of yellow color.

Thin layer chromatography (TLC) can be used to separate compounds based on their solubility in a solvent phase and their mobility dictated by their affinity for a silica gel stationary phase layered on a glass plate. The compound's R_f is a physical characteristic of a specific compound and can be used to help in its identification. It is calculated as

$$R_f = \frac{\text{Distance compound traveled}}{\text{Distance solvent traveled}}$$

A protein's molecular weight can be estimated by gel filtration chromatography from a standard curve.

The *cat* gene, encoding CAT, has also been used as a reporter for gene expression in mammalian cells. To assay for CAT activity, an extract from cells expressing CAT is incubated with [^{14}C]-labeled chloramphenicol. Transfer of the acetyl group from acetyl coenzyme A to chloramphenicol alters the mobility of the antibiotic on a TLC plate. The area on the TLC plate corresponding to the chloramphenicol spot and the acetylated chloramphenicol spots are scraped off and counted in a liquid scintillation counter. The amount of acetylated chloramphenicol is directly proportional to the amount of CAT enzyme and CAT activity is then calculated as

$$\begin{aligned}\text{CAT activity} &= \text{\% chloramphenicol aceylated} \\ &= \frac{\text{Counts in acetylated species}}{\left(\begin{array}{l}\text{Counts in} \\ \text{acetylated species}\end{array}\right) + \left(\begin{array}{l}\text{Nonacetylated} \\ \text{chloramphenicol counts}\end{array}\right)} \times 100\end{aligned}$$

The *luc* gene encoding the enzyme luciferase from the firefly *Photinus pyralis* has become a popular reporter for assaying gene expression. Luciferase from transformed bacterial cells or transfected mammalian cells expressing the *luc* gene is assayed by placing cell extract in a reaction mix containing excess amounts of luciferin (the enzyme's substrate), ATP, $MgSO_4$, and molecular oxygen (O_2). ATP-dependent oxidation of luciferin leads to a burst of light that can be detected and quantitated (as LUs) at 562 nm in a luminometer. The amount of light generated in the reaction is proportional to luciferase concentration.

Translation *in vitro* can be monitored by following the amount of radioactive amino acids incorporated into protein in the form of TCA-precipitable counts collected on a glass fiber filter.

REFERENCES

Bradford, M.M. (1976). A rapid and sensitive method for the quantitation of microgram quantities of protein utilizing the principle of protein-dye binding. *Analyt. Biochem.* 72:248.

De Wet, J.R., K.V. Wood, M. DeLuca, D.R. Helinski, and S. Subramani (1987). Firefly luciferase gene: structure and expression in mammalian cells. *Mol. Cell Biol.* 7:725–737.

Gorman, C.M., L.F. Moffat, and B.H. Howard (1982). Recombinant genomes which express chloramphenicol acetyltransferase in mammalian cells. *Molec. Cell. Biol.* 2:1044–1051.

Kennedy, J.F., Z.S. Rivera, L.L. Lloyd, F.P. Warner, and M.P.C. da Silva (1995). Determination of the molecular weight distribution of hydroxyethylcellulose by gel filtration chromatography. *Carbohydrate Polymers* 26:31–34.

Kingston, R.E. (1989). Uses of fusion genes in mammalian transfection: harvest and assay for chloramphenicol acetyltransferase. In *Short Protocols in Molecular Biology: A Compendium of Methods from Current Protocols in Molecular Biology* (F.M. Ausubel, R. Brent, R.E. Kingston, D.D. Moore, J.G. Seidman, J.A. Smith, K. Struhl, P. Wang-Iverson, and S.G. Bonitz, eds.). Wiley, New York, pp. 268–270.

Kirschenbaum, D.M. (1976). In *CRC Handbook of Biochemistry and Molecular Biology* (G.D. Fasman, ed.), 3rd ed, Vol. 2. CRC Press, Boca Raton, FL, p. 383.

Layne, E. (1957). Spectrophotometric and turbidimetric methods for measuring proteins. *Meth. Enzymol.* 3:447–454.

Miller, J.H. (1972). *Experiments in Molecular Genetics*. Cold Spring Harbor Laboratory, Cold Spring Harbor, NY, pp. 352–355.

Pace, C.N., F. Vajdos, L. Fee, G. Grimsley, and T. Gray (1995). How to measure and predict the molar absorption coefficient of a protein. *Protein Sci.* 4:2411–2423.

Peterson, G.L. (1983). Determination of total protein. *Meth. Enzymol.* 91:95–119.

Robertson, D., S. Shore, and D.M. Miller (1997). *Manipulation and Expression of Recombinant DNA: A Laboratory Manual*. Academic Press, San Diego, CA, pp. 133–138.

Scopes, R.K. (1974). Measurement of protein by spectrophotometry at 205 nm. *Analyt. Biochem.* 59:277–282.

Shaw, W.V., and R.F. Brodsky (1968). Characterization of chloramphenicol acetyltransferase from chloramphenicol-resistant *Staphylococcus aureus*. *J. Bacteriol.* 95:28–36.

Shaw, W.V., L.C. Packman, B.D. Burleigh, A. Dell, H.R. Morris, and B.S. Hartley (1979). Primary structure of a chloramphenicol acetyltransferase specified by R plasmids. *Nature* 282:870–872.

Spirin, A.S., V.I. Baranov, L.A. Ryabova, S.Y. Ovodov, and Y.B. Alakhov (1988). A continuous cell-free translation system capable of producing polypeptides in high yield. *Science* 242:1162.

Stoscheck, C.M. (1990). Quantitation of protein. *Meth. Enzymol.* 182:50–68.

FURTHER READING

Andrews, P. (1962). Estimation of molecular weights of proteins by gel filtration. *Nature* 196:36–39.

Ezzeddine, S., and U. Al-Khalidi (1982). A short note on the molecular weight determination by gel filtration in minutes. *Molec. Cell. Biochem.* 45:127–128.

Le Maire, M.A., A. Ghazi, J.V. Moller, and L.P. Aggerbeck (1987). The use of gel chromatography for the determination of sizes and relative molecular masses of proteins. Interpretation of calibration curves in terms of gel-pore-size distribution. *Biochem. J.* 243:399–404.

Margolis, S. (1967). Separation and size determination of human serum lipoproteins by agarose gel filtration. *J. Lipid Research* 8:501–507.

Steere, R.L., and G.K. Ackers (1962). Restricted-diffusion chromatography through calibrated columns of granular agar gel: a simple method for particle-size determination. *Nature* 196:475.

Whitaker, J.R. (1963). Determination of molecular weights of proteins by gel filtration on Sephadex. *Anal. Chem.* 35:1950–1953.

Chapter 12

Centrifugation

INTRODUCTION

Centrifugation is a method of separating molecules having different densities by spinning them in solution around an axis (in a centrifuge rotor) at high speed. It is one of the most useful and frequently employed techniques in the molecular biology laboratory. Centrifugation is used to collect cells, to precipitate DNA, to purify virus particles, and to distinguish subtle differences in the conformation of molecules. Most laboratories undertaking active research will have more than one type of centrifuge, each capable of using a variety of rotors. Small tabletop centrifuges can be used to pellet cells or to collect strands of DNA during ethanol precipitation. Ultracentrifuges can be used to band plasmid DNA in a cesium chloride gradient or to differentiate various structures of replicating DNA in a sucrose gradient.

12.1 RELATIVE CENTRIFUGAL FORCE (RCF) (*g* FORCE)

During centrifugation, protein or DNA molecules in suspension are forced to the furthest point from the center of rotation. The rate at which this occurs depends on the speed of the rotor (measured as **rpm** (**revolutions per minute**)). The force generated by the spinning rotor is described as the **relative centrifugal force** (**RCF**), also called the ***g*** force, and is proportional to the square of the rotor speed and the radial distance (the distance of the molecule being separated from the axis of rotation). The RCF is calculated using either of the following equations (they are equivalent):

$$RCF = 11.18(r_{cm})\left(\frac{rpm}{1000}\right)^2$$

$$RCF = 1.118 \times 10^{-5}\, r_{cm}(rpm)^2$$

where r is equal to the radius from the centerline of the rotor to the point in the centrifuge tube at which the RCF value is needed (in cm) and rpm is equal to the rotor speed in rpm.

Calculations for Molecular Biology and Biotechnology. DOI: 10.1016/B978-0-12-375690-9.00012-7
© 2010 Elsevier Inc. All rights reserved.

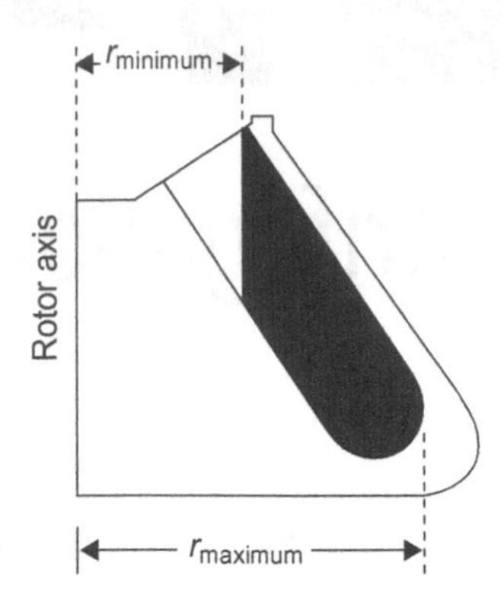

■ **FIGURE 12.1** Cross section of a typical centrifuge rotor showing the minimum and maximum radii as measured from the rotor axis.

As the RCF increases with the square of the rotor speed, doubling the rpm will increase the RCF four-fold.

The RCF is usually expressed in units of gravitational force written as '$\times$ *g*' (times the force of gravity). One *g* is equivalent to the force of gravity. Two *g* is twice the force of gravity. An RCF of 12 500 is 12 500 times the force of gravity and is written as 12 500 $\times$ *g*.

As the distance from the center of the rotor increases, so does the centrifugal force. Most rotor specifications from the manufacturer will give the rotor's radius (***r***) in three measurements. The **minimum radius** ($\boldsymbol{r_{min}}$) is the distance from the rotor's centerline to the top of the liquid in the centrifuge tube when placed in the rotor at its particular angle (see Figure 12.1). The **maximum radius** ($\boldsymbol{r_{max}}$) is the distance from the rotor centerline to the furthest point in the centrifuge tube away from the axis of rotation (Figure 12.1). The **average radius** ($\boldsymbol{r_{avg}}$) is equal to the minimum radius plus the maximum radius divided by two:

$$r_{ave} = \frac{r_{min} + r_{max}}{2}$$

Since the minimum radius and the average radius will vary depending on the volume of liquid being centrifuged, RCF is usually calculated using the maximum radius value.

Problem 12.1 *E. coli* cells are to be pelleted in an SS-34 rotor (maximum radius of 10.7 cm) by centrifugation at 7000 rpm. What is the RCF (*g* force)?

Solution 12.1

We will use the following equation:

$$RCF = 11.17(r_{cm})\left(\frac{rpm}{1000}\right)^2$$

Substituting the values provided gives the following result:

$$\text{RCF} = 11.17(10.7)\left(\frac{7000}{1000}\right)^2$$

$$\text{RCF} = (119.52)(7)^2 = (119.52)(49) = 5856$$

Therefore, 7000 rpm in an SS-34 rotor is equivalent to 5856 $\times$ *g*.

12.1.1 Converting *g* force to revolutions per minute (rpm)

The majority of protocols written in the 'Materials and Methods' sections of journal articles will give a description of centrifugation in terms of the *g* force. This allows other researchers to reproduce the laboratory method even if they have a different type of centrifuge and rotor. To solve for rpm when the *g* force (RCF) is given, the equation

$$RCF = 11.18(r_{cm})\left(\frac{rpm}{1000}\right)^2$$

becomes

$$rpm = 1000\sqrt{\frac{RCF}{11.18(r_{cm})}}$$

which can also be written as

$$rpm = \sqrt{\frac{RCF}{(1.118 \times 10^{-5})(r_{cm})}}$$

Problem 12.2 A protocol calls for centrifugation at $6000 \times g$. What rpm should be used with an SS-34 rotor (maximum radius of 10.7 cm) to attain this *g* force?

Solution 12.2

Placing our values into the preceding equation gives

$$\text{rpm} = 1000\sqrt{\frac{6000}{11.17(10.7)}}$$

$$\text{rpm} = 1000\sqrt{\frac{6000}{119.52}} = 1000\sqrt{50.2} = 1000(7.1) = 7100$$

Therefore, an SS-34 rotor should be spun at 7100 rpm to obtain $6000 \times g$.

A search of the internet will turn up a number of RCF calculators, such as the one maintained by DJB Labcare, Buckingham, UK (Figure 12.2), by which RCF or rpm are calculated automatically.

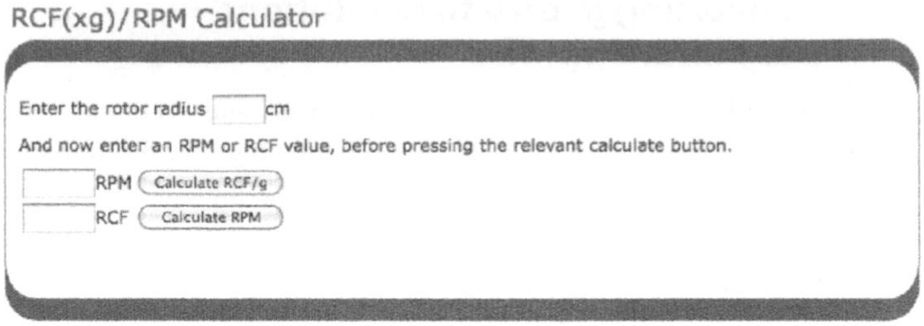

RCF(xg)/RPM Calculator

Enter the rotor radius [ ] cm

And now enter an RPM or RCF value, before pressing the relevant calculate button.

[ ] RPM (Calculate RCF/g)

[ ] RCF (Calculate RPM)

■ **FIGURE 12.2** A relative centrifugal force (RCF) calculator found at centrifuge.org.uk/page.p?p=rcfrpmcalc.

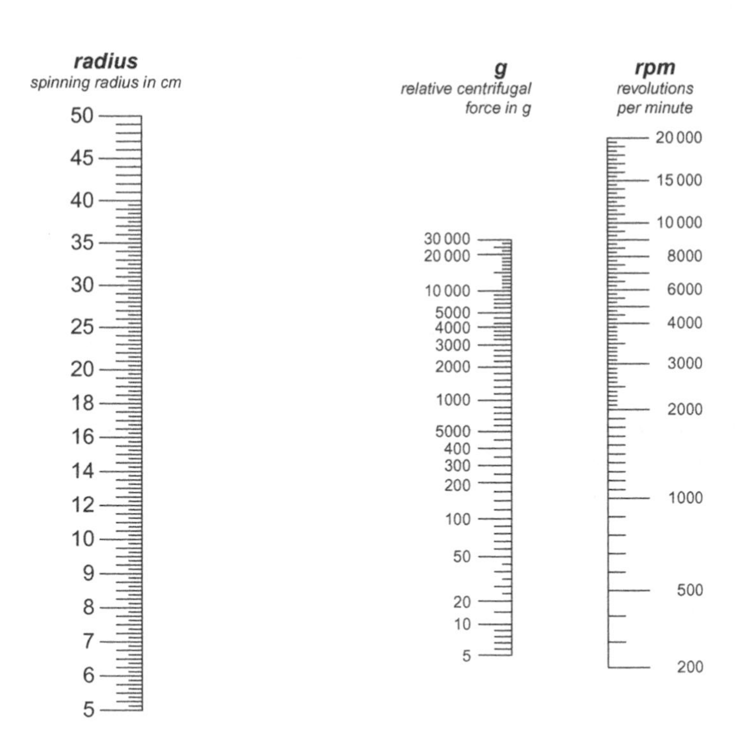

■ **FIGURE 12.3** Nomogram for determining centrifugation parameters.

12.1.2 Determining *g* force and revolutions per minute (rpm) by use of a nomogram

We have determined *g* force and rpm values using the following equation:

$$g = 11.18(r_{cm})\left(\frac{rpm}{1000}\right)^2$$

If a large number of *g* values are calculated using a large number of different values for rpm and spinning radii, a chart approximating Figure 12.3

is generated. This chart, called a **nomogram,** is a graphical representation of the relationship between rpm, *g* force, and a rotor's radius. If any two values are known, the third can be determined by use of a straightedge. The straightedge is lined up such that it intersects the two known values on their respective scales. The third value is given by the intersection of the line on the third scale.

Problem 12.3 A rotor with a 7.5 cm spinning radius is used to pellet a sample of cells. The cells are to be spun at 700 $\times$ *g*. What rpm should be used?

Solution 12.3

Using the nomogram in Figure 12.3, a straight line is drawn connecting the three scales such that it passes through both the 7.5 cm mark on the 'radius' scale and the 700 mark on the '*g*' scale. This line will intersect the 'rpm' scale at 3000 (Figure 12.4). Therefore, the rotor should be spun at 3000 rpm.

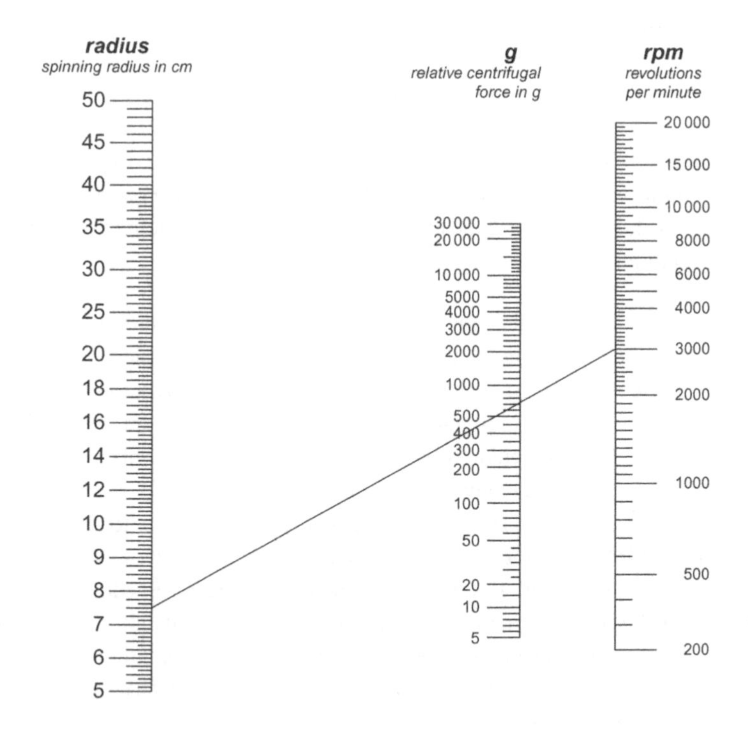

■ **FIGURE 12.4** Solution to Problem 12.3.

12.2 CALCULATING SEDIMENTATION TIMES

The time (in hours) required to pellet a particle in water at 20°C through the maximum rotor path length (the distance between r_{min} and r_{max}) is given by the equation

$$t = \frac{k}{s}$$

where t is the sedimentation time in hours, k is the clearing factor for the rotor, and s is the sedimentation coefficient of the particle being pelleted in water at 20°C as expressed in Svedberg units (see below).

The clearing factor, k, is defined by the equation

$$k = \frac{(253\,000)\left[\ln\left(\frac{r_{max}}{r_{min}}\right)\right]}{\left(\frac{\text{rpm}}{1000}\right)^2}$$

The k factor is a measure of the pelleting efficiency of the rotor. Its value decreases with increasing rotor speed. The lower the k factor, the more efficient is the rotor's ability to pellet the desired molecule. The k factor for any rotor, as provided by the manufacturer, assumes that the fluid being centrifuged has the density of water at 20°C. The k factor will increase if the medium has a higher density or viscosity than water. For example, at the same rotor speed, the k factor is higher for a sucrose gradient than for a water medium.

Several factors will affect the rate at which a particular biological molecule will pellet during centrifugation. Its size, shape, and buoyancy all influence the rate at which a molecule passes through the medium. A molecule having a high molecular weight will sediment faster than one having a low molecular weight. A **Svedberg unit** describes the sedimentation rate of a molecule and is related to its size. It is expressed as a dimension of time such that 1 Svedberg unit is equal to 10^{-13} sec. For most biological molecules, the Svedberg unit value lies between 1 S (1 Svedberg unit; 1×10^{-13} sec) and 200 S (200 Svedberg units; 200×10^{-13} sec). The names of the ribosomal subunits 30S and 50S from *E. coli,* for example, derive from their sedimentation rates as expressed in Svedberg units. The same is true for the bacterial ribosomal RNAs 5S, 16S, and 23S.

Problem 12.4 An SS-34 rotor has an r_{min} of 3.27 cm and an r_{max} of 10.70 cm for a sealed tube containing the maximum allowable volume. If a 120 S molecule is to be spun in an aqueous medium at 20 000 rpm at 20°C, how long will it take to pellet the molecule?

Solution 12.4

The *k* factor must first be calculated. Substituting the given values into the given equation yields the following result:

$$k = \frac{(253\,000)\left[\ln\left(\frac{10.70}{3.27}\right)\right]}{\left(\frac{20\,000}{1000}\right)^2}$$

$$k = \frac{(253\,000)[\ln(3.27)]}{(20)^2}$$

$$= \frac{(253\,000)(1.18)}{400}$$

$$= \frac{298\,540}{400}$$

$$= 746$$

Therefore, an SS-34 rotor spinning at 20 000 rpm has a *k* factor value of 746. Placing this value into the equation for calculating sedimentation time gives

$$t = \frac{749}{120\text{ S}}$$

$$= 6.24\text{ hr}$$

$$= 6\text{ hr}, 14\text{ min}$$

Therefore, to pellet the 120 S molecule, it must be spun in an SS-34 rotor at 20 000 rpm for 6 hr and 14 min.

Software can be found on the internet that will automatically calculate *k* values. An example is that constructed by the Hu Laboratory at Texas A&M University, College Station, TX, USA (Figure 12.5).

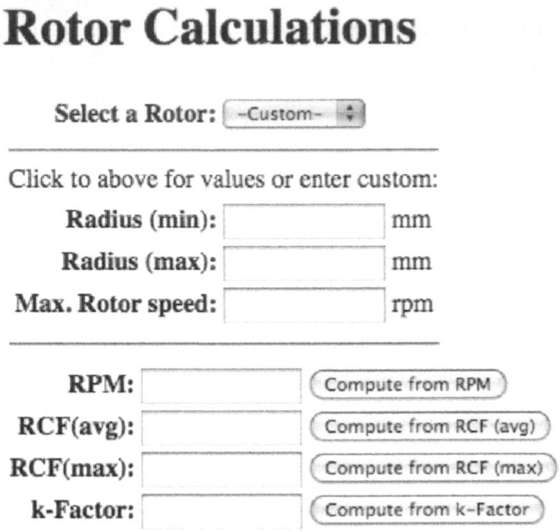

Rotor Calculations

Select a Rotor: -Custom-

Click to above for values or enter custom:

Radius (min): mm

Radius (max): mm

Max. Rotor speed: rpm

RPM: Compute from RPM

RCF(avg): Compute from RCF (avg)

RCF(max): Compute from RCF (max)

k-Factor: Compute from k-Factor

■ **FIGURE 12.5** A relative centrifugal force (RCF), revolutions per minute (rpm), and *k* calculator found at hulab.tamu.edu/rcfcalc.html.

■ CHAPTER SUMMARY

Relative centrifugal force (RCF), or *g* force, is proportional to the square of the rotor speed (in rpm) and the radial distance (r):

$$RCF = 11.18(r_{cm})\left(\frac{rpm}{1000}\right)^2$$

A nomogram can be used to determine *g* force or rpm for any rotor radius.

The time (t) required to pellet a particle in water at 20°C through the maximum rotor path length is given by the relationship

$$t = \frac{k}{s}$$

where k is the rotor's clearing factor and s (Svedberg unit) is the particle's sedimentation coefficient. The value for k is calculated as

$$k = \frac{(253000)\left[\ln\left(\frac{r_{\max}}{r_{\min}}\right)\right]}{\left(\frac{rpm}{1000}\right)^2}$$

REFERENCES

Torri, Al F., and P.T. Englund (1992). Purification of a mitochondrial DNA polymerase from Crithidia fasciculate. *J. Biol. Chem.* 267:4786–4792.

FURTHER READING

Graham, J. (2001). *Biological Centrifugation: The Basics from Background to Bench.* BIOS Scientific Publishers, Ltd., Oxford, UK.

Rickwood, D. (ed.) (1993). *Preparative Centrifugation: A Practical Approach.* Oxford University Press, Oxford, UK.

Rickwood, D., T.C. Ford, and J. Steensgaard (1994). *Centrifugation: Essential Data.* John Wiley & Sons in association with BIOS Scientific Publishers, Ltd., Oxford, UK.

Chapter 13

Forensics and paternity

INTRODUCTION

In almost all crimes of violence, whether murder, rape, or assault, there is usually some type of biological material left at the scene or on the victim. This material may be hair, saliva, sperm, skin, or blood. All of these contain DNA – DNA that may belong to the victim or to the perpetrator. The application of PCR to these samples has revolutionized law enforcement's capabilities in providing evidence to the courtroom when a suspect's guilt or innocence is disputed.

In forensics, PCR is used for the amplification of **polymorphic** sites, those regions on DNA that are variable among people. Some polymorphisms result from specific point mutations in the DNA, others from the addition or deletion of repeat units. A number of polymorphic sites have been identified and exploited for human identification. There are also a number of different techniques by which PCR products can be assayed and polymorphisms identified.

If the suspect's DNA profile and that of the DNA recovered from the crime scene do not match, then it can be said that the suspect is excluded as a contributor of the sample. If there is a match, then the suspect is included as a possible contributor of DNA evidence and it becomes important to determine the significance of the match. A lawyer involved in the case will want to know the probability that someone other than the suspect left the DNA evidence at the scene. The question is one of population statistics, and that is the primary focus of this chapter.

Concomitantly with its development as a tool in forensics, PCR also found application in paternity testing. In fact, the same polymorphic sites used by criminalists to explore the involvement of a suspect in a criminal act are also used by testing laboratories to identify the father of a child whose parentage is in dispute. The recovery and isolation of DNA samples as well as their amplification are essentially the same in forensics and paternity testing. The math used to calculate the significance of a match, however, is handled slightly differently between the two applications.

Calculations for Molecular Biology and Biotechnology. DOI: 10.1016/B978-0-12-375690-9.00013-9
© 2010 Elsevier Inc. All rights reserved.

13.1 ALLELES AND GENOTYPES

An **allele** is one of two or more alternate forms of a genetic site (or **locus**). Some sites may have many different alleles. This is particularly true for sites carrying tandem repeats, where the DNA segment may be repeated a number of times, each different repeat number representing a different allele designation.

A **genotype** is an organism's particular combination of alleles. For any one polymorphic site, an individual will carry two possible alleles – one obtained from the mother, the other from the father. For an entire population, however, any possible combination of alleles is possible for the individuals within that population. For a locus having three possible alleles, *A*, *B*, and *C*, there are six possible genotypes. An individual may have any of the following allele combinations: *AA*, *AB*, *AC*, *BB*, *BC*, and *CC*.

The number of possible genotypes can be calculated in two ways. It can be calculated by adding the number of possible alleles to each positive integer below that number. For example, for a locus having three possible alleles, the number of possible genotypes is

$$3 + 2 + 1 = 6$$

Likewise, for a five-allele system, there are 15 possible genotypes:

$$5 + 4 + 3 + 2 + 1 = 15$$

The number of genotypes can also be calculated using the equation

$$\text{number of genotypes} = \frac{n(n+1)}{2}$$

where n is equal to the number of possible alleles.

The first method is easier to remember but becomes cumbersome for systems having a large number of alleles.

Problem 13.1 A VNTR (variable number of tandem repeats) locus has 14 different alleles. How many genotypes are possible in a population for this VNTR?

Solution 13.1

Using the preceding equation, we have

$$\text{number of genotypes} = \frac{14(14+1)}{2} = \frac{(14)(15)}{2} = \frac{210}{2} = 105$$

Therefore, for a locus having 14 different alleles, 105 different genotypes are possible.

13.1.1 **Calculating genotype frequencies**

The **genotype frequency** is the proportion of the total number of people represented by a single genotype. For example, if the genotype *AA* (for a locus having three different alleles) is found to be present in six people out of 200 sampled, the genotype frequency is 6/200 = 0.03. The sum of all genotype frequencies for a single locus should be equal to 1.0. If it is not, the frequency calculations should be checked again.

Problem 13.2 A locus has three possible alleles, A, B, and C. The genotypes of 200 people chosen at random are determined for the different allele combinations, with the results shown in the following table. What are the genotype frequencies?

Genotype	Number of individuals
AA	6
AB	34
AC	46
BB	12
BC	60
CC	42
Total	200

Solution 13.2

Genotype frequencies are obtained by dividing the number of individuals having each genotype by the total number of people sampled.

Genotype frequencies for each genotype are given in the right column of the following table.

Genotype	Number of individuals	Ratio of individuals to total	Genotype frequency
AA	6	6/200 =	0.03
AB	34	34/200 =	0.17
AC	46	46/200 =	0.23
BB	12	12/200 =	0.06
BC	60	60/200 =	0.30
CC	42	42/200 =	0.21
Total	**200**		**1.00**

Note that the genotype frequencies add up to 1.00.

13.1.2 Calculating allele frequencies

The **allele frequency** is the proportion of the total number of alleles in a population represented by a particular allele. For any polymorphism other than those present on the X or Y chromosome, each individual carries two alleles per locus, one inherited from the mother, the other from the father. Within a population, therefore, there are twice as many total alleles as there are people. When calculating the number of a particular allele, homozygous individuals each contribute two of that allele to the total number of that particular allele. Heterozygous individuals each contribute one of a particular allele to the total number of that allele. For example, if there are six individuals with the *AA* genotype, they contribute 12 *A* alleles. Thirty-four *AB* heterozygous individuals contribute a total of 34 *A* alleles and 34 *B* alleles to the total. The frequency of the *A* allele in a three-allele system is calculated as

$$A\ allele\ frequency = \frac{[2(\#\ of\ AAs) + (\#\ of\ ABs) + (\#\ of\ ACs)]}{2n}$$

where n is the number of people in the survey. The frequencies for the other alleles are calculated in a similar manner. The sum of all allele frequencies for a specific locus should equal 1.0.

Problem 13.3 What are the frequencies for the *A*, *B*, and *C* alleles described in Problem 13.2?

Solution 13.3

We will assign a frequency of p to the *A* allele, q to the *B* allele, and r to the *C* allele. We have the following genotype data.

Genotype	Number of individuals
AA	6
AB	34
AC	46
BB	12
BC	60
CC	42
Total	200

The allele frequencies are then calculated as follows:

$$pA = \frac{[2(6) + 34 + 46]}{2(200)} = \frac{92}{400} = 0.230$$

$$qB = \frac{[2(12) + 34 + 60]}{2(200)} = \frac{118}{400} = 0.295$$

$$rC = \frac{[2(42) + 46 + 60]}{2(200)} = \frac{190}{400} = 0.475$$

Therefore, the allele frequency of *A* is 0.230, the allele frequency of *B* is 0.295, and the allele frequency of *C* is 0.475. Note that the allele frequencies add up to 1.000 (0.230 + 0.295 + 0.475 = 1.000).

13.2 THE HARDY–WEINBERG EQUATION AND CALCULATING EXPECTED GENOTYPE FREQUENCIES

Gregor Mendel demonstrated by crossing pea plants with different characteristics that gametes combine randomly. He used a **Punnett square** to predict the outcome of any genetic cross. For example, if he crossed two plants both heterozygous for height, where *T* represents a dominant tall phenotype and *t* represents the recessive short phenotype, the Punnett square would have the following appearance:

		Pollen	
		T	*t*
Ovules	*T*	*TT*	*Tt*
	t	*Tt*	*tt*

From the Punnett square, Mendel predicted that the offspring of the cross would have a phenotypic ratio of tall to short plants of 3 : 1.

G.H. Hardy, a British mathematician, and W. Weinberg, a German physician, realized that they could apply a similar approach to predicting the outcome of random mating, not just for an individual cross but for crosses occurring within an entire population. After all, random combination of gametes, as studied by Mendel for individual crosses, is quite similar in concept to random mating of genotypes. In an individual cross, it is a matter of chance which sperm will combine with which egg. In an infinitely large, randomly mating population, it is a matter of chance which genotypes will combine. Determining the distribution of genotype frequencies in a randomly mating population can also be accomplished using a Punnett square. However, rather than a single cross between two parents, Hardy and Weinberg examined crosses between all mothers and all fathers in a population.

For a locus having two alleles, A and B, a Punnett square can be constructed such that an allele frequency of p is assigned to the A allele and an allele frequency of q is assigned to the B allele. The allele frequencies are multiplied as shown in the following Punnett square:

		Fathers	
		pA	qB
Mothers	pA	$pApA$	$pAqB$
	qB	$pAqB$	$qBqB$

If AA represents the homozygous condition, then its frequency in the population is p^2 ($p \times p = p^2$). The frequency of the BB homozygous individuals is q^2. Since heterozygous individuals can arise in two ways, by receiving alternate alleles from either parent, the frequency of the AB genotype in the population is $2pq$ since $(p \times q) + (p \times q) = 2pq$.

The sum of genotype frequencies is given by the expression

$$p^2 + 2pq + q^2 = 1.0$$

This relationship is often referred to as the **Hardy–Weinberg equation**. It is used to determine expected genotype frequencies from allele frequencies.

Punnett squares can be used to examine any number of alleles. For example, a system having three alleles A, B, and C, with allele frequencies p, q, and r, respectively, will have the following Punnett square:

		Fathers		
		pA	qB	rC
	pA	$pApA$	$pAqB$	$pArC$
Mothers	qB	$pAqB$	$qBqB$	$qBrC$
	rC	$pArC$	$qBrC$	$rCrC$

The sum of genotype frequencies for this three-allele system is

$$p^2 + q^2 + r^2 + 2pq + 2pr + 2qr = 1.0$$

This demonstrates, that no matter how many alleles are being examined, the frequency of the homozygous condition is always the square of the allele frequency (p^2), and the frequency of the heterozygous condition is always two times the allele frequency of one allele times the frequency of the other allele ($2pq$).

The Hardy–Weinberg equation was derived for a strict set of conditions. It assumes that the population is in **equilibrium** – that it experiences no net change in allele frequencies over time. To reach equilibrium, the following conditions must be met:

1. Random mating,
2. An infinitely large population,
3. No mutation,
4. No migration into or out of the population, and
5. No selection for genotypes (all genotypes are equally viable and equally fertile).

In reality, of course, there are no populations that meet these requirements. Nevertheless, most reasonably large populations approximate these conditions to the extent that the Hardy–Weinberg equation can be applied to estimate genotype frequencies.

Problem 13.4 What are the expected genotype frequencies for the alleles described in Problem 13.3? (Assume the population is in Hardy–Weinberg equilibrium.)

Solution 13.4

From Problem 13.3, we have the following allele frequencies:

$$pA = 0.230$$

$$qB = 0.295$$

$$rC = 0.475$$

The Hardy–Weinberg equation gives the following values:

Genotype	Multiplication	Mathematics	Expected genotype frequency
AA	(freq. of A)^2	$(0.230)^2$	0.053
AB	2(freq. of A)(freq. of B)	$2(0.230)(0.295) =$	0.136
AC	2(freq. of A)(freq. of C)	$2(0.230)(0.475) =$	0.218
BB	(freq. of B)^2	$(0.295)^2 =$	0.087
BC	2(freq. of B)(freq. of C)	$2(0.295)(0.475) =$	0.280
CC	(freq. of C)^2	$(0.475)^2 =$	0.226

Note that the total of all genotype frequencies is equal to 1.000, as would be expected if the mathematics were performed correctly.

13.3 THE CHI-SQUARE TEST – COMPARING OBSERVED TO EXPECTED VALUES

In Problem 13.4, genotype frequencies were calculated from allele frequencies using the Hardy–Weinberg equation. These genotype frequencies represent what should be expected in the sampled population if that population is in Hardy–Weinberg equilibrium. In Problem 13.2, the observed genotype frequencies found within that population were calculated. It can now be asked: Do the observed genotype frequencies differ significantly from the expected genotype frequencies? This question can be addressed using a statistical test called chi-square.

Chi-square is used to test whether or not some observed distributional outcome fits an expected pattern. Since it is unlikely that the observed genotype frequencies will be exactly as predicted by the Hardy–Weinberg equation, it is important to look at the nature of the differences between the observed and expected values and to make a judgment as to the 'goodness of fit' between them. In the chi-square test, the expected value is subtracted from the observed value in each category and this value is then squared. Each squared value is then weighted by dividing it by the expected value for that category. The sum of these squared and weighted values, called chi-square (denoted as χ^2), is represented by the following equation:

$$\chi^2 = \sum \frac{(\text{observed} - \text{expected})^2}{\text{expected}}$$

In the chi-square test, two hypotheses are tested. The **null hypothesis** (H_0) states that there is no difference between the two observed and expected values; they are statistically the same and any difference that may be detected is due to chance. The alternative hypothesis (H_a) states that the two sets of data, the observed and expected values, are different; the difference is statistically significant and must be due to some reason other than chance.

In populations, a normal distribution of values around a consensus frequency should be represented by a bell-shaped curve. The degree of distribution represented by a curve is associated with a value termed the **degrees of freedom** (*df*). The smaller the *df* value (the smaller the sample size), the larger is the dispersion in the distribution, and a bell-shaped curve will be more difficult to distinguish. The larger the *df* value (the larger the sample size), the closer the distribution will come to a normal, bell-shaped curve. For most problems in forensics, the degrees of freedom is equal to one less than the number of categories in the distribution. For such problems, the number of categories is equal to the number of different alleles (or the number of different genotypes, depending upon which is the object of the test). Statistically significant differences may be observed in chi-square if the sample size is small (less than 100 individuals sampled from a population) or if there are drastic deviations from the Hardy–Weinberg equilibrium.

A chi-square test is performed at a certain level of significance, usually 5% ($\alpha = 0.05$; $p = 0.95$). At a 5% significance level, we are saying that there is less than a 5% chance that the null hypothesis will be rejected even though it is true. Conversely, there is a 95% chance that the null hypothesis is correct; that there is no difference between the observed and expected values. A chi-square distribution table (see Table 13.1) will provide the expected chi-square value for a given probability and *df* value. For example, suppose a chi-square value of 2.3 is obtained for some dataset having eight degrees of freedom and we wish to test at a 5% significance level. From the chi-square distribution table, we find a value of 15.51 for a probability value of 0.95 and eight degrees of freedom. Therefore, there is a probability of 0.95 (95%) that the chi-square will be less than 15.51. Since the chi-square value of 2.3 is less than 15.51, the null hypothesis is not rejected; the data do not provide sufficient evidence to conclude that the observed data differ from the expected.

The chi-square test can be used in forensics to compare observed genotype frequencies with those expected from Hardy–Weinberg analysis and to compare local allele frequencies (observed) with those maintained in a larger or national database (expected values).

Table 13.1 The chi-square distribution table. At eight degrees of freedom, for example, there is a probability of 0.95 (95%) that the chi-square value will be less than 15.51. If the chi-square value obtained is less than 15.51, then there is insufficient evidence to conclude that the observed values differ from the expected values. Below 15.51, any differences between the observed and expected values are merely due to chance.

df	$p = 0.75$	$p = 0.90$	$p = 0.95$	$p = 0.975$	$p = 0.99$
1	1.32	2.71	3.84	5.02	6.64
2	2.77	4.60	5.99	7.37	9.21
3	4.10	6.24	7.80	9.33	11.31
4	5.38	7.77	9.48	11.14	13.27
5	6.62	9.23	11.07	12.83	15.08
6	7.84	10.64	12.59	14.44	16.81
7	9.04	12.02	14.07	16.01	18.48
8	10.22	13.36	15.51	17.54	20.09
9	11.39	14.68	16.92	19.02	21.67
10	12.5	15.9	18.3	20.5	23.2
11	13.7	17.3	19.7	21.9	24.7
12	14.8	18.6	21.0	23.3	26.2
13	16.0	19.8	22.4	24.7	27.7
14	17.1	21.1	23.7	26.1	29.1
15	18.2	22.3	25.0	27.5	30.6
16	19.4	23.5	26.3	28.8	32.0
17	20.5	24.8	27.6	30.2	33.4
18	21.6	26.0	28.9	31.5	34.8
19	22.7	27.2	30.1	32.9	36.2
20	23.8	28.4	31.4	34.2	37.6
21	24.9	29.6	32.7	35.5	38.9
22	26.0	30.8	33.9	36.8	40.3
23	27.1	32.0	35.2	38.1	41.6
24	28.2	33.2	36.4	39.4	43.0
25	29.3	34.4	37.7	40.7	44.3
30	34.8	40.3	43.8	47.0	50.9
40	45.6	51.8	55.7	59.3	63.7
50	56.3	63.2	67.5	71.4	76.2
60	67.0	74.4	79.1	83.3	88.4
70	77.6	85.5	90.5	95.0	100.4
80	88.1	96.6	101.9	106.6	112.3
90	98.7	107.6	113.2	118.1	124.1
100	109.1	118.5	124.3	129.6	135.8

Problem 13.5 In Problems 13.2 and 13.4, observed and expected genotype frequencies were calculated for a fictitious population. These are shown in the following table. Using the chi-square test at a 5% level of significance (95% probability), determine whether the observed genotype frequencies are significantly different from the expected genotype frequencies.

Genotype	Observed frequency	Expected frequency
AA	0.03	0.053
AB	0.17	0.136
AC	0.23	0.218
BB	0.06	0.087
BC	0.30	0.280
CC	0.21	0.226

Solution 13.5

A table is prepared so that the steps involved in obtaining the chi-square value can be followed. So that we are not dealing with small decimal numbers, the frequencies given in the table will each be multiplied by 100 to give a percent value.

Genotype	Observed frequency (*O*)	Expected frequency (*E*)	Difference (*O* − *E*)	Square of difference $(O - E)^2$	$(O - E)^2/E$
AA	3	5.3	−2.3	5.3	1.00
AB	17	13.6	3.4	11.6	0.85
AC	23	21.8	1.2	1.4	0.06
BB	6	8.7	−2.7	7.3	0.84
BC	30	28	2.0	4	0.14
CC	21	22.6	−1.6	2.6	0.12
Total					$\chi^2 = 3.01$

Therefore, a chi-square value of 3.01 is obtained. The degrees of freedom is equal to the number of categories (six, because there are six genotypes) minus 1:

$$df = 6 - 1 = 5$$

At five degrees of freedom and 0.95 probability, a chi-square value of 11.07 is obtained from Table 13.1. Therefore, there is a probability of 95% that χ^2 will be less than 11.07 if the null hypothesis is true, if there is not a statistically significant difference between the observed and expected genotype frequencies. Since our derived chi-square value of 3.01 is less than 11.07, we do not reject the null hypothesis; there is insufficient evidence to conclude that the observed genotype frequencies differ from the expected genotype frequencies. This suggests that our population may be in Hardy–Weinberg equilibrium.

13.3.1 Sample variance

Sample variance (s^2) is a measure of the degree to which the numbers in a list are spread out. If the numbers in a list are all close to the expected values, the variance will be small. If they are far away, the variance will be large. Sample variance is given by the equation

$$s^2 = \frac{\sum(O - E)^2}{n - 1}$$

where n is the number of categories.

A sample variance calculation is at its most powerful when it is used to compare two different sets of data, say genotype frequency distributions in New York City as compared with those in Santa Fe, New Mexico. A population that is not in Hardy–Weinberg equilibrium is more likely to have a larger variance in genotype frequencies than one that is in Hardy–Weinberg equilibrium. Populations far from Hardy–Weinberg equilibrium may have a sample variance value in the hundreds.

Problem 13.6 What is the sample variance for the data in Problem 13.5?

Solution 13.6

The sum of the $(O - E)^2$ column in Problem 13.5 is

$$5.3 + 11.6 + 1.4 + 7.3 + 4 + 2.6 = 32.2$$

The value for n in this problem is 6 since there are six possible genotypes. Placing these values into the equation for sample variance gives the following result:

$$s^2 = \frac{32.2}{6 - 1} = \frac{32.2}{5} = 6.44$$

This value is relatively small, indicating that the observed genotype frequencies do not vary much from the expected values.

13.3.2 Sample standard deviation

The **sample standard deviation** (s) is the square root of the sample variance and is also a measure of the spread from the expected values. In its simplest terms, it can be thought of as the average distance of the observed data from the expected values. It is given by the formula

$$s = \sqrt{\frac{\sum(O - E)^2}{n - 1}}$$

If there is little variation in the observed data from the expected values, the standard deviation will be small. If there is a large amount of variation in the data, then, on average, the data values will be far from the mean and the standard deviation will be large.

Problem 13.7 What is the standard deviation for the dataset in Problem 13.5?

Solution 13.7

In Problem 13.6, it was determined that the variance is 6.44. The standard deviation is the square root of this number:

$$s = \sqrt{6.44} = 2.54$$

Therefore, the standard deviation is 2.54. The average observed genotype frequency is within one standard deviation (2.54 percentage points) from each expected genotype frequency. The *AB* and *BB* genotype frequencies are greater than one standard deviation from the expected frequencies (*AB* is 3.4 percentage points away from the expected value; *BB* is 2.7 percentage points away from the expected value; see Solution 13.5).

13.4 THE POWER OF INCLUSION (P_i)

The **power of inclusion** (P_i) gives the probability that, if two individuals are chosen at random from a population, they will have the same genotype. The smaller the P_i value, the more powerful is the genotyping method for differentiating individuals. The power of inclusion is equivalent to the sum of the squares of the expected genotype frequencies. For example, a P_i value of 0.06 means that, 6% ($0.06 \times 100 = 6\%$) of the time, if two individuals are chosen at random from a population, they will have the same genotype.

Problem 13.8 What is the power of inclusion for the genotypes in Problem 13.4?

Solution 13.8

The expected genotype frequencies for the fictitious population are as follows:

Genotype	Expected frequency
AA	0.053
AB	0.136
AC	0.218
BB	0.087
BC	0.280
CC	0.226

P_i is the sum of squares of the expected genotype frequencies. These are calculated as follows:

$$P_i = (\text{exp. freq. of } AA)^2 + (\text{exp. freq. of } AB)^2 + (\text{exp. freq. of } AC)^2 + (\text{exp. freq. of } BB)^2 + (\text{exp. freq. of } BC)^2 + (\text{exp. freq. of } CC)^2$$

Placing the appropriate values into this equation gives the following result:

$$P_i = (0.053)^2 + (0.136)^2 + (0.218)^2 + (0.087)^2 + (0.280)^2 + (0.226)^2$$
$$= 0.003 + 0.018 + 0.048 + 0.008 + 0.078 + 0.051$$
$$= 0.206$$

Therefore, 20.6% of the time, if two people are chosen at random from the population, they will have the same genotype.

13.5 THE POWER OF DISCRIMINATION (P_d)

The **power of discrimination** (P_d) is equivalent to 1 minus the P_i:

$$P_d = 1 - P_i$$

P_d gives a measure of how likely it is that two individuals will have different genotypes. For example, if a DNA typing system has a P_d of 0.9997, it means that, 99.97% of the time, if two individuals are chosen at random from a population, they will have different genotypes that can be discriminated by the test.

A **combined P_d** is equivalent to 1 minus the product of P_i of all markers. It is used when multiple loci are being used in the typing assay. The P_i value for a single genetic marker is equivalent to 1 minus the P_d value for that marker. The more loci used in a DNA typing test, the higher will be the combined P_d.

Problem 13.9 What is the power of discrimination for the genotypes in Problem 13.8?

Solution 13.9

In Problem 13.8, a P_i value of 0.206 was calculated. P_d is 1 minus P_i. This yields

$$P_d = 1 - 0.206 = 0.794$$

Therefore, 79.4% of the time, if two individuals are chosen at random from a population, they will have different genotypes at that locus.

13.6 DNA TYPING AND WEIGHTED AVERAGE

Genotype frequency is derived by using the Hardy–Weinberg equation and the observed allele frequencies. Many crime laboratories obtain allele frequency values from national databases. These databases may have allele frequencies for the major population groups (Caucasian, African American, and Hispanic).

A crime occurring within a city is usually perpetrated by one of its inhabitants. For this reason, most major metropolitan areas will develop an allele frequency database from a random sample of its own population. Even without this data, however, a city can still derive a fairly good estimate of frequencies for the genotypes within its boundaries. This can be done using census data. For example, the proportion of Caucasians within the city's census area can be multiplied by a national database genotype frequency to give a weighted average for that population group. The weighted averages for each population group are then added to give an overall estimate of the frequency of a particular genotype within the census area.

Problem 13.10 A blood spot is typed for a single STR (short tandem repeat) locus. It is found that the DNA contains five repeats and six repeats (designated 5, 6; read 'five comma six') for this particular STR. The FBI database shows that this heterozygous genotype has a frequency of 0.003 in Caucasians, 0.002 in African Americans, and 0.014 in Hispanic populations. The city in which this evidence is collected and typed has a population profile of 60% Caucasians, 30% African Americans, and 10% Hispanic. What is the overall weighted average frequency for this genotype?

Solution 13.10

Each genotype frequency will be multiplied by the fraction of the population made up by each ethnic group. These values are then added together to give the weighted average. The calculation is as follows:

$$\begin{aligned}\text{weighted average} &= (0.60 \times 0.003) + (0.30 \times 0.002) + (0.10 \times 0.014)\\ &= 0.0018 + 0.0006 + 0.0014 = 0.0038\end{aligned}$$

Therefore, the weighted average genotype frequency for a 5, 6 STR type for the city's population is 0.0038. This can be converted to a '1 in' number by dividing 1 by 0.0038:

$$\frac{1}{0.0038} = 263$$

Therefore, 1 in approximately 263 randomly chosen individuals in the city would be expected to have the 5, 6 STR genotype.

13.7 THE MULTIPLICATION RULE

The more loci used for DNA typing, the better the typing is able to distinguish between individuals. Most DNAs are typed for multiple markers and then the genotype frequencies are multiplied together to give the frequency of the entire profile. This is called the **Multiplication Rule**. For its use to be valid, the loci must be in linkage equilibrium. That is, the markers are on different chromosomes or far enough apart on the same chromosome that they segregate independently; they are not linked.

Problem 13.11 A DNA is typed for five different, unlinked loci. The following genotype frequencies for each locus are determined. What is the genotype frequency for the entire profile?

Locus	Genotype frequency
1	0.021
2	0.003
3	0.014
4	0.001
5	0.052

Solution 13.11

To determine the genotype frequency for the complete profile, the genotype frequencies are multiplied together:

$$\textit{Overall frequency} = 0.021 \times 0.003 \times 0.014 \times 0.001 \times 0.052 = 4.6 \times 10^{-11}$$

Therefore, the genotype frequency for the entire profile is 4.6×10^{-11}. This can be converted to a '1 in' number by dividing 1 by 4.6×10^{-11}:

$$\frac{1}{4.6 \times 10^{-11}} = 2.2 \times 10^{10}$$

Therefore, from a randomly sampled population, 1 in 2.2×10^{10} individuals might be expected to have this particular genotype combination.

13.8 THE PATERNITY INDEX (PI)

The **Paternity Index** (PI) describes the probability of paternity. It is calculated as

$$\mathrm{PI} = \frac{\mathrm{X}}{\mathrm{Y}}$$

where X is the probability that the alleged father is, in fact, the father and Y is the probability that any randomly selected man of the same race is the father.

The PI calculation depends on the pattern of inheritance – which alleles were contributed by the mother and which by the alleged father (Table 13.2).

The PI can be interpreted to mean that there is a one chance in the PI value that a random, unrelated man of the same race is the biological father.

Table 13.2 The genotypes of the mother, child, and alleged father and the numerator (X) and denominator (Y) values for the calculation of the paternity index (PI) (PI = X/Y). *A*, *B*, *C*, and *D* are the possible alleles and *p* denotes probability (allele frequency). For example, p_A denotes the frequency for the *A* allele within the population data for the race of the alleged father.

Genotypes			Numerator	Denominator
Mother	**Child**	**Alleged father**		
AA	*AA*	*AA*	1	p_A
AA	*AA*	*AB*	1/2	p_A
AA	*AA*	*BC*	0	p_A
AB	*AA*	*AA*	1/2	$p_A/2$
AB	*AA*	*AB*	1/4	$p_A/2$
AB	*AA*	*AC*	1/4	$p_A/2$
AB	*AA*	*BC*	0	$p_A/2$
AA	*AB*	*AB*	1/4	$p_B/2$
AA	*AB*	*BB*	1	p_B
AA	*AB*	*BC*	1/2	p_B
AA	*AB*	*CD*	0	P_A
AB	*AB*	*AA*	1/2	$(p_A + p_B)/2$
AB	*AB*	*AB*	1/2	$(p_A + p_B)/2$
AB	*AB*	*BC*	1/4	$(p_A + p_B)/2$
AB	*AB*	*AC*	1/4	$(p_A + p_B)/2$
AB	*AC*	*AC*	1/2	p_C
AB	*AC*	*CD*	1/4	$p_C/2$
AB	*AC*	*BC*	1/4	$p_C/2$
AB	*BC*	*CC*	1/2	$p_C/2$
AB	*BB*	*AB*	1/4	$p_B/2$
AB	*BC*	*BC*	1/2	p_C
AB	*BC*	*CD*	1/4	$p_C/2$
AB	*AB*	*CD*	0	$(p_A + p_B)/2$
AC	*AB*	*BB*	1/2	$p_B/2$
AC	*AB*	*BD*	1/4	$p_B/2$
AC	*AB*	*BC*	1/4	$p_B/2$
AC	*AB*	*CD*	0	$p_B/2$

Problem 13.12 The locus D1S80 is being used to do a quick test for paternity. D1S80 is a VNTR that, within the human population, can carry anywhere from 14 to 41 copies of a 16 bp DNA sequence. The mother carries 18 and 21 repeats, the child carries 18 and 31 repeats, and the alleged father, an African American, carries the 24 and 31 repeat alleles. Within the African American population, 24 repeats has an allele frequency of 0.234 and 31 repeats has an allele frequency of 0.054 (Budowle et al., 1995). What is the PI?

Solution 13.12

We first identify the pattern of allele distribution:

Individual	Mother	Child	Alleged father
D1S80 Alleles	18,21	18,31	24,31
Allele Pattern	*AB*	*AC*	*CD*

Table 13.2 shows that, for this pattern of inheritance, ¼ should be used as the numerator and $p_C/2$ as the denominator. The frequency for the 31-repeat allele is 0.054. The PI is calculated as

$$\mathrm{PI} = \frac{1/4}{p_C/2}$$

$$= \frac{0.25}{0.054/2} = \frac{0.25}{0.027} = 9.259$$

Therefore, the PI for this example is 9.259 and we can say that there is a one chance in 9.259 that a random, unrelated man of the same race is the biological father.

13.8.1 Calculating the paternity index (PI) when the mother's genotype is not available

A PI can still be calculated if the mother's genotype is not known. The pattern of inheritance and the corresponding PI calculation are shown in Table 13.3.

Table 13.3 Calculation of the paternity index (PI) when the mother's genotype is unavailable.

Child's genotype	Alleged father's genotype	Paternity index calculation
AA	*AA*	$1/p_A$
AA	*AB*	$1/2p_A$
AB	*AB*	$(p_A + p_B)/4(p_A \times p_B)$
AB	*BB*	$1/2p_B$
AB	*BC*	$1/4p_B$

Problem 13.13 A Caucasian woman has been raised as an orphan. Her mother died giving birth to her and she never knew her father. Through some detective work, she has tracked down and made contact with a Caucasian man she suspects of being her biological father. He agrees to a quick genetic test that would exclude him as her father. D1S80 alleles are amplified. She has a genotype of 22,32 (22 repeats and 32 repeats). The alleged father has a genotype of 20,32. These alleles have the following frequency in the Caucasian population (Budowle et al., 1995). What is the PI?

Allele	Allele frequency
20	0.018
22	0.038
32	0.006

Solution 13.13

Since the woman and the alleged father share an allele, he cannot be excluded as the biological father. The allele pattern is as follows (**Note:** *B* is the 32 repeat allele and *C* is the 20 repeat allele):

Individual	Child's genotype	Alleged father's genotype
Genotype	22,32	20,32
Allele Pattern	*AB*	*BC*

For this pattern of allele inheritance and from Table 13.3, we should use the following PI calculation:

$$\text{PI} = 1/4p_B$$

Placing the frequency of the 32-repeat allele into the calculation gives us

$$\text{PI} = 1/4p_B = 1/4 \times 0.006 = 1/0.024 = 41.67$$

Therefore, the PI is 41.67 and there is a one chance in 41.67 that a Caucasian man chosen at random is the actual biological father.

13.8.2 The combined paternity index (CPI)

If multiple loci are used to determine paternity (and this is almost always the case because it increases the discriminating power), then the product of all the individual PI values for each locus is the **combined paternity index (CPI)**. It can be interpreted to mean that there is a one chance in the CPI value that a random, unrelated man of the same race is the biological father.

With the CPI, we can also determine a **probability of paternity** – the probability that the alleged father is, in fact, the biological father. It is calculated as

$$\text{Probability of Paternity} = (\text{CPI}/\text{CPI} + 1) \times 100$$

The probability of paternity assumes that the **prior probability** in a paternity test is, without any testing, 0.50. In other words, there is a 50% chance that any untested man is the father and a 50% chance that he is not the father.

Problem 13.14 Short tandem repeats are evaluated in a paternity case. A total of nine loci give the following PI values. What is the CPI and what is the probability of paternity?

STR Locus	Paternity index
D3S1358	2.76
vWA	4.15
FGA	1.37
D8S1179	3.52
D21S11	1.94
D18S51	2.89
D5S818	3.68
D13S317	2.12
D7S820	5.38

Solution 13.14

The CPI is obtained by multiplying all the PI values together:

$$\begin{aligned}\text{CPI} &= 2.76 \times 4.15 \times 1.37 \times 3.52 \times 1.94 \times 2.89 \times 3.68 \times 2.12 \times 5.38 \\ &= 12998\end{aligned}$$

Therefore, the CPI is 12 998. The odds in favor of paternity are 12 998 to 1. Or, there is a one chance in 12 998 that a randomly selected Caucasian is the actual biological father.

The probability of paternity is

$$\text{Probability of Paternity} = (12998/12999)100 = 99.9923\%$$

Therefore, all else being equal, the probability of paternity is 99.9923%.

CHAPTER SUMMARY

The number of possible genotypes from n alleles is calculated as

$$\text{number of genotypes} = \frac{n(n+1)}{2}$$

Genotype frequency is calculated as the number of individuals with a particular genotype divided by the number of individuals sampled. Allele frequency is calculated as the number of times an allele appears within sampled homozygous and heterozygous individuals divided by twice the number of individuals sampled.

The Hardy–Weinberg equation allows the conversion of allele frequencies into genotype frequencies. A homozygous genotype frequency is calculated as the square of the allele frequency. A heterozygous genotype frequency is calculated as two times the products of the two different alleles.

The chi-square test can be used to compare two datasets having observed and expected values using the equation

$$\chi^2 = \sum \frac{(\text{observed} - \text{expected})^2}{\text{expected}}$$

The sample variance, the degree to which a dataset is spread out, is calculated using the equation

$$s^2 = \frac{\sum (O - E)^2}{n - 1}$$

The sample standard deviation, the square root of the sample variance and another way to look at data spread from expected values, is calculated as the square root of the sample variance:

$$s = \sqrt{\frac{\sum(O - E)^2}{n - 1}}$$

The power of inclusion (P_i) gives the probability of two people chosen at random having the same genotype. It is calculated as the sum of the squares of the expected genotype frequencies. The power of discrimination (P_d) gives a measure of how likely it is that two individuals will have different genotypes. It is calculated as 1 minus the P_i.

A weighted average for genotype frequencies is calculated to get an idea of how frequent a particular genotype is given the race makeup of the location in which the DNA evidence was collected. It is calculated as the product of the genotype frequency (from the database) and the percent of a particular race within the population under study.

The Multiplication Rule allows for the determination of an overall genotype frequency by multiplying all the individual genotype frequencies together.

Paternity testing is used to determine whether an alleged man is the biological father of a child. A PI, calculated in a way that accounts for the possible contribution of alleles from the mother and alleged father, gives an idea of the chance that the alleged father is, in fact, the biological father. The combined PI is the product of all the individual PI values for each locus tested. The probability of paternity is calculated as

$$\text{Probability of Paternity} = (\text{CPI/CPI} + 1) \times 100$$

It is the probability that the alleged man is the biological father.

REFERENCES

Budowle, B., F.S. Baechtel, J.B. Smerick, K.W. Presley, A.M. Giusti, G. Parsons, M.C. Alevy, and R. Chakraborty (1995). D1S80 population data in African Americans, Causcasians, Southeastern Hispanics, Southwestern Hispanics, and Orientals. *J. Forensic Sci.* 40:38–44.

FURTHER READING

Buckleton, J., and J. Curran (2008). A discussion of the merits of random man not excluded and likelihood ratios. *Forensic Sci. International: Genetics* 2:343–348.

Buckleton, J., J.-A. Bright, and S.J. Walsh (2009). Database crime to crime match rate calculation. *Forensic Sci. International: Genetics* 3:200–201.

Budowle, B., T.R. Moretti, A.L. Baumstark, D.A. Defenbaugh, and K.M. Keys (1999). Population data on the thirteen CODIS core short tandem repeat loci in African Americans, U.S. Caucasians, Hispanics, Bahamians, Jamaicans, and Trinidadians. *J. Forensic Sci.* 44:1277–1286.

Drabek, J. (2009). Validation of software for calculating the likelihood ratio for parentage and kinship. *Forensic Sci. International: Genetics* 3:112–118.

Fung, W.K., and Y.-Q. Hu (2008). *Statistical DNA Forensics: Theory, Methods and Computation.* John Wiley & Sons, Ltd., West Sussex, England.

Kaiser, L., G.A.F. Seber, and J.M. Opitz (1983). Paternity testing: I. Calculation of paternity indexes. *Amer. J. Med Genet.* 15:323–329.

Li, C.C., and A. Chakravarti (1985). Basic fallacies in the formulation of the paternity index. *Am. J. Hum. Genet.* 37:809–818.

Lucy, D. (2005). *Introduction to Statistics for Forensic Scientists.* John Wiley & Sons, Ltd., West Sussex, England.

Reisner, E.G., and P. Reading (1983). Application of probability of paternity calculations to an alleged incestuous relationship. *J. Forensic Sci.* 28:1030–1034.

Song, Y.S., A. Patil, E.E. Murphy, and M. Slatkin (2009). Average probability that a "cold hit" in a DNA database search results in an erroneous attribution. *J. Forensic Sci.* 54:22–27.

Thompson, W.C., F. Taroni, and C.G.G. Aitken (2003). How the probability of a false positive affects the value of DNA evidence. *J. Forensic Sci.* 48:47–54.

Turchi, C., M. Pesaresi, F. Alessandrini, V. Onofri, A. Arseni, and A. Tagliabracci (2004). Unusual association of three rare alleles and a mismatch in a case of paternity testing. *J. Forensic Sci.* 49:260–262.

Tvedebrink, T., P.S. Eriksen, H.S. Mogensen, and N. Morling (2009). Estimating the probability of allelic drop-out of STR alleles in forensic genetics. *Forensic Sci. International: Genetics* 3:222–226.

Appendix A

Using Microsoft Excel's graphing utility

Excel, from Microsoft, is a software program that allows the user to create and format spreadsheets as typically employed in business accounting. The program also includes, however, an excellent chart and graphing utility that can be easily applied to data collected from any number of different types of experiments performed in the biotechnology laboratory.

Provided here is an example of how Excel can be used to plot a standard curve for a 100 bp ladder of DNA fragments the sizes of which are converted to log values vs. the distance those fragments migrated during agarose gel electrophoresis. Linear regression analysis is also used to determine a line of best fit along with the equation describing that line from which the sizes of DNA fragments of unknown length can be determined.

1. On a computer equipped with Microsoft Office, open the Excel program by clicking on its icon. On machines with Microsoft Windows, you may have to navigate to the program through **Start** > **Programs** > **Microsoft Excel**.

Calculations for Molecular Biology and Biotechnology. DOI: 10.1016/B978-0-12-375690-9.00020-6
© 2010 Elsevier Inc. All rights reserved.

2. Open the Excel Workbook.

3. In column A, row 1, enter 'bp.' In column B, row 1, enter 'Distance (x).' In column C, row 1, enter 'log bp (y).'
4. In column A, beneath the 'bp' heading, enter the size of each fragment of the 100 bp ladder, starting with the 1500 bp fragment (entered into box A2). In the rows in column B beneath the 'Distance (x)' heading, enter the distance, in cm, that each fragment band of the 100 bp ladder migrated from the well.

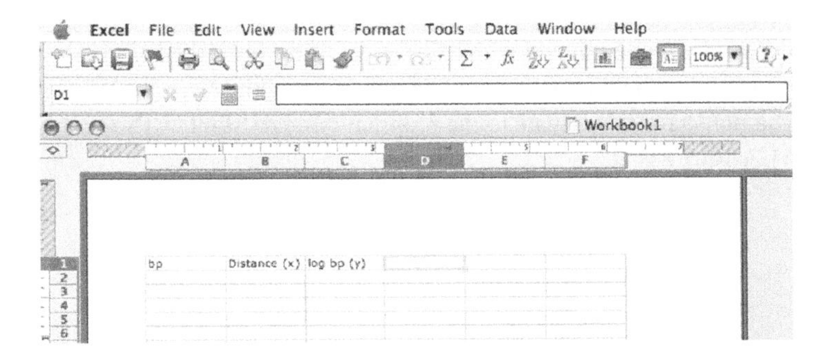

5. Click on the first box under the 'log bp (y)' heading to highlight it. Click the *fx* function icon in the toolbar at the top of the spreadsheet to bring up the **Paste Function** box. In the **Function category:** list, select 'Math & Trig.' In the **Function name:** list, select 'LOG 10.' Click **OK**. The **LOG 10** dialogue box will appear.

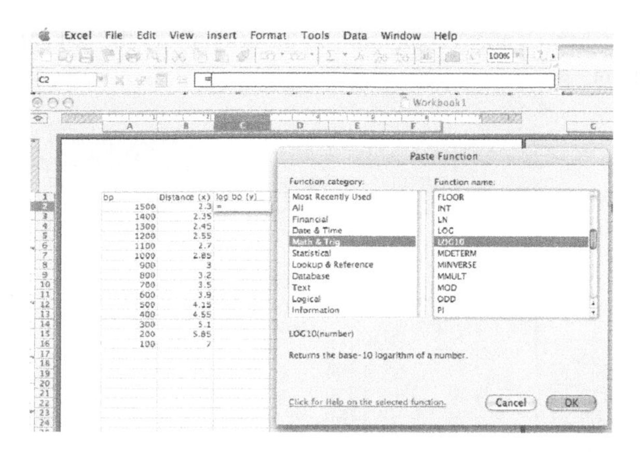

6. Click anywhere within the **LOG 10** dialogue box and drag it out onto the spreadsheet workspace so that the column entries are visible. Click on '1500' in box A2. 'A2' will then appear in the **Number** box. Click **OK**. The log value of 1500 will appear in the highlighted box in the 'log bp (y)' column.

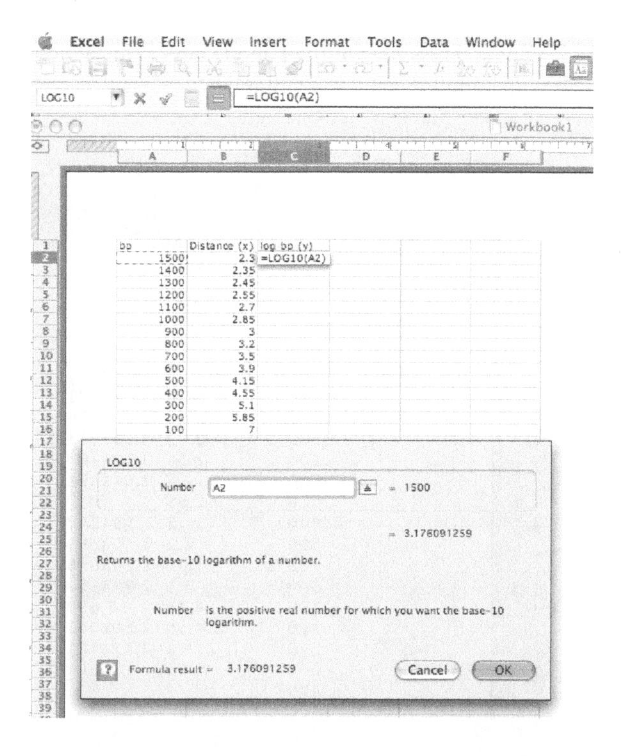

7. Highlight the entire C column from position C2 to the last entry in the column.

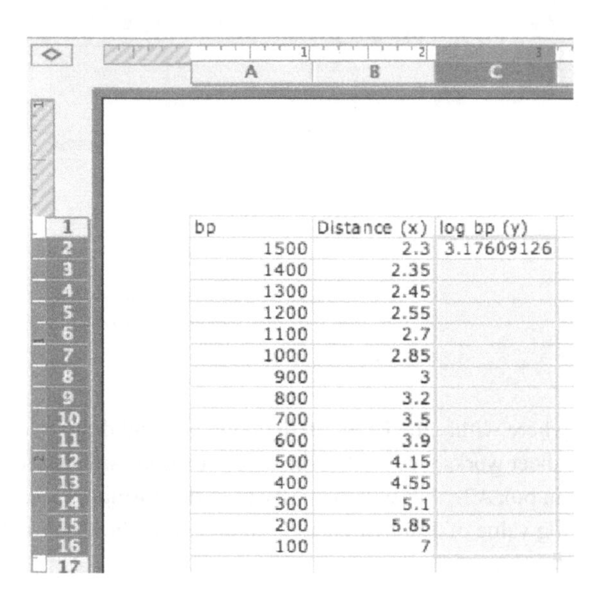

bp	Distance (x)	log bp (y)
1500	2.3	3.17609126
1400	2.35	
1300	2.45	
1200	2.55	
1100	2.7	
1000	2.85	
900	3	
800	3.2	
700	3.5	
600	3.9	
500	4.15	
400	4.55	
300	5.1	
200	5.85	
100	7	

8. On the keyboard, press the **Control** and **D** keys. This action will fill down all remaining log values in Column C.

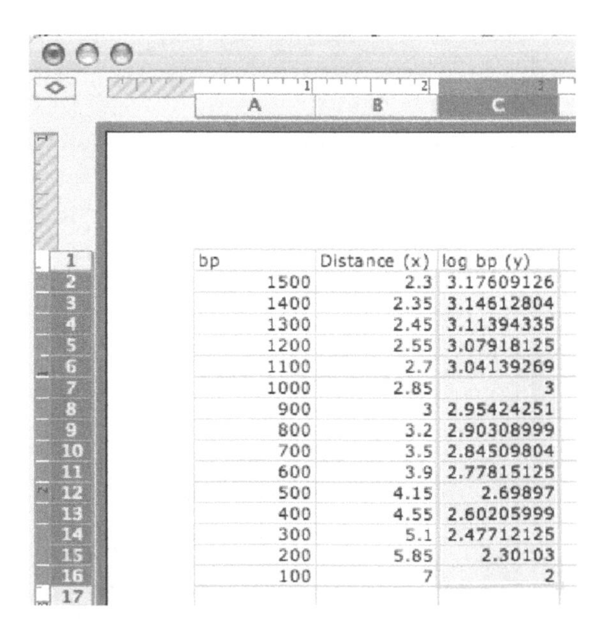

bp	Distance (x)	log bp (y)
1500	2.3	3.17609126
1400	2.35	3.14612804
1300	2.45	3.11394335
1200	2.55	3.07918125
1100	2.7	3.04139269
1000	2.85	3
900	3	2.95424251
800	3.2	2.90308999
700	3.5	2.84509804
600	3.9	2.77815125
500	4.15	2.69897
400	4.55	2.60205999
300	5.1	2.47712125
200	5.85	2.30103
100	7	2

9. Without selecting row 1 (the column heading row) or column 1 (the 'bp' column), highlight the values in the 'Distance (x)' and 'log bp (y)' columns.

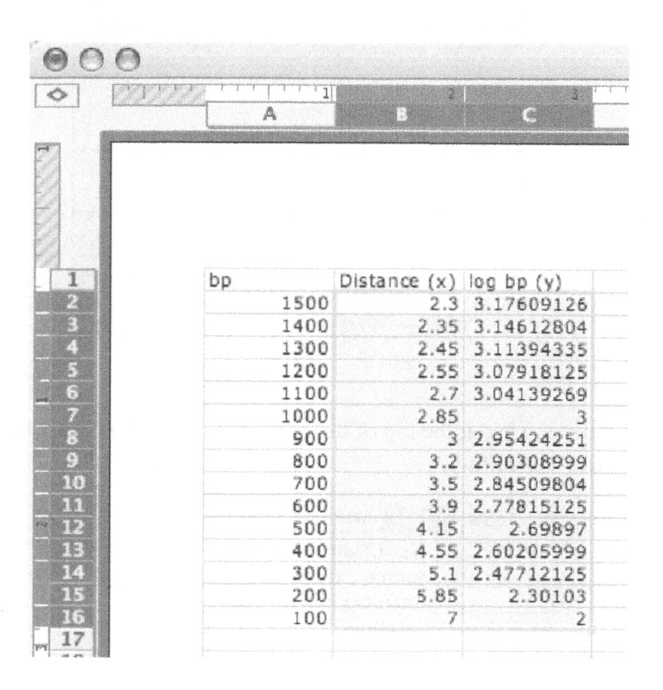

bp	Distance (x)	log bp (y)
1500	2.3	3.17609126
1400	2.35	3.14612804
1300	2.45	3.11394335
1200	2.55	3.07918125
1100	2.7	3.04139269
1000	2.85	3
900	3	2.95424251
800	3.2	2.90308999
700	3.5	2.84509804
600	3.9	2.77815125
500	4.15	2.69897
400	4.55	2.60205999
300	5.1	2.47712125
200	5.85	2.30103
100	7	2

10. In the menu bar, click on the **Chart Wizard** icon button (it has the appearance of a bar graph) to bring up the **Chart Wizard-Chart Type** window.

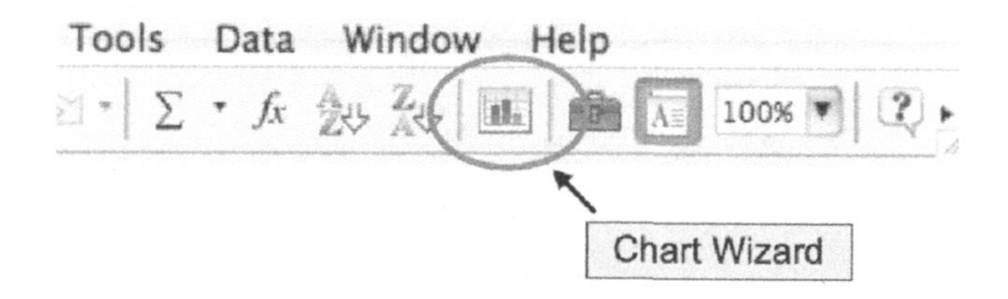

11. In the **Chart Wizard-Chart Type** window, select the 'XY (Scatter)' chart type and then click **Finish**. A chart will appear over the spreadsheet showing the data plotted on a graph.

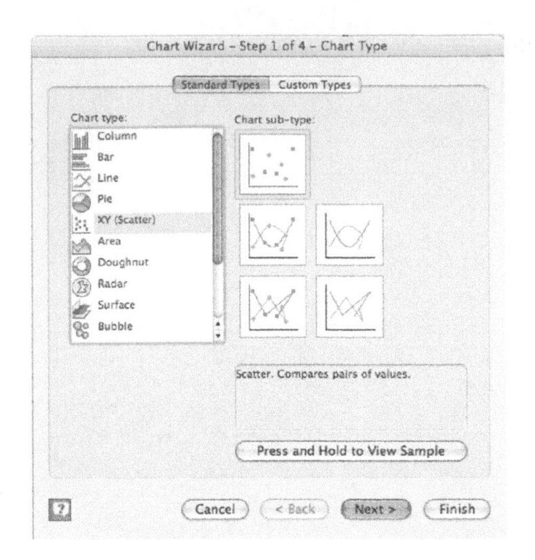

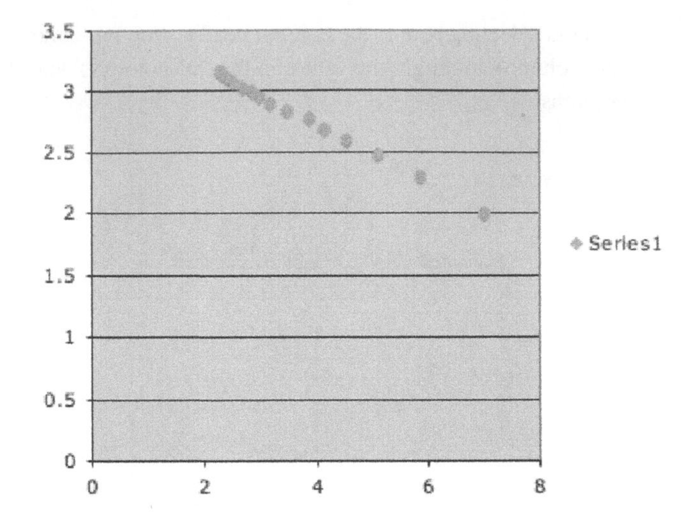

12. From the menu bar at the top of the page, pull down the **Chart** list and select **Add Trendline…**

13. In the **Add Trendline** popup window that appears, make sure the 'Linear' **Trend/Regression type** is selected.

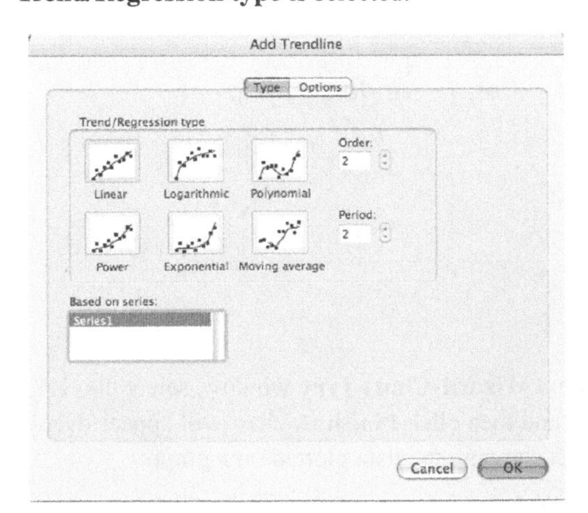

14. Click on the **Options** tab in the **Add Trendline** window. In the **Options** window that appears, check the boxes next to 'Display equation on chart' and 'Display r-squared value on chart.' Click **OK**. The line of best fit will be drawn on the graph. The regression line equation and the r-squared value will also appear.

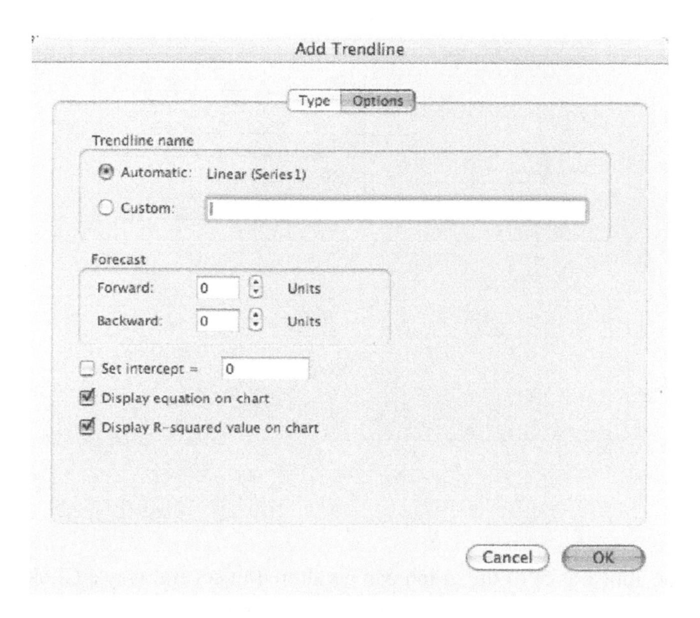

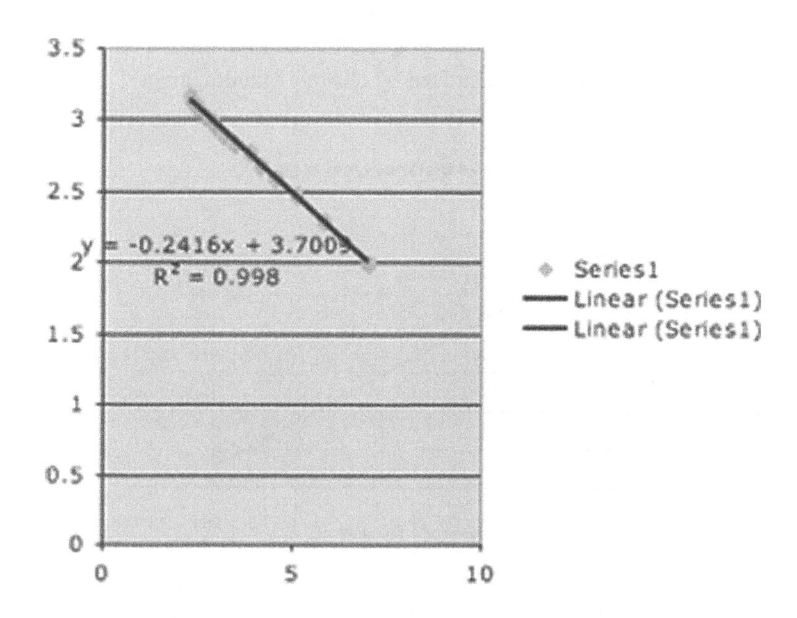

15. You can use **Chart Options** under the **Chart** pull-down menu to label the chart axes, remove or modify the legend, add or remove gridlines, and to attach a chart title. (**Note:** The chart must be selected on the spreadsheet to gain access to the **Chart** pull-down menu.)

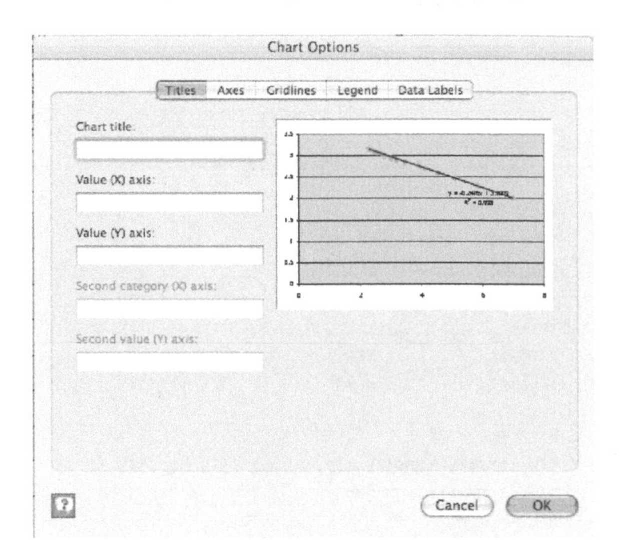

16. The appearance of the graph can be altered in several ways. Click along the x or y axes to change the numbering scheme and their minimum and maximum values. Clicking on the points or the line through those points will allow you to alter their qualities. The regression equation can be moved by clicking and dragging.

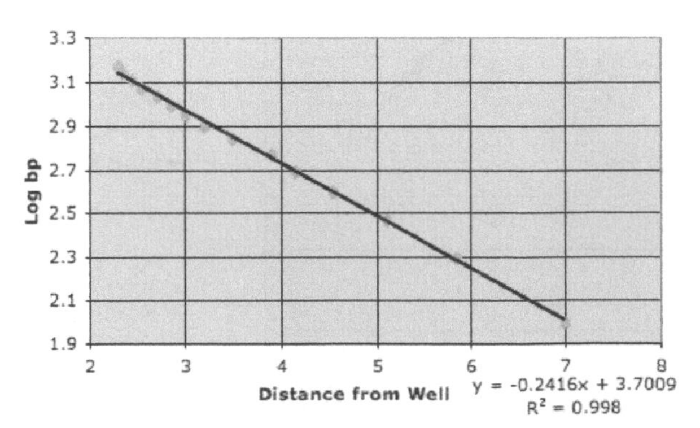

Index

S

T

V

Z

WITHDRAWN FROM LIBRARY

Lightning Source UK Ltd.
Milton Keynes UK
UKOW06f0203120814

236754UK00010B/222/P

9 780123 756909